Diagnosis, Management and Emerging Strategies for Chemotherapy-Induced Neuropathy

Maryam Lustberg • Charles Loprinzi
Editors

Diagnosis, Management and Emerging Strategies for Chemotherapy-Induced Neuropathy

A MASCC Book

 Springer

Editors
Maryam Lustberg
Yale University School of Medicine
New Haven, CT, USA

Charles Loprinzi
Mayo Clinic
Rochester, MN, USA

ISBN 978-3-030-78662-5   ISBN 978-3-030-78663-2   (eBook)
https://doi.org/10.1007/978-3-030-78663-2

This Springer imprint is published by the registered company Springer Nature Switzerland AG.
The registered company address is: Gewerbestrasse 11, 6330 Cham, Switzerland

Preface

When one of us was approached by Springer, to edit a book concerning chemotherapy-induced peripheral neuropathy (CIPN), the initial thought was to pass on this opportunity, given other competing priorities. However, understanding that this was an important topic and that no similar product was available, this proposal was further considered.

This further consideration led to a decision that the book might benefit from having two co-editors, as opposed to having only a single editor. When one of us approached the other, it was an easy decision for both of us, as we have worked, and are working, together on a number of other projects.

Both of us have been intimately involved with Multinational Association for Supportive Care in Cancer (MASCC) for a long period of time, including leading the Neurologic Complications Working Group for this association. This led to making this a MASCC-supported book.

The goal of the book was to provide a broad overview of CIPN, covering topics related to the natural history of CIPN, risk factors that predisposed patients to develop CIPN, means of diagnosing and evaluating clinical CIPN, basic science research regarding both the prevention of CIPN and the treatment of established CIPN, clinical research regarding both the prevention of CIPN and the treatment of established CIPN, physical and occupational therapeutic approaches to CIPN, understanding the patient perspective, and thoughts regarding future directions.

We hope that readers find this book to be helpful.

New Haven, CT   Maryam Lustberg
Rochester, MN   Charles Loprinzi

Contents

Natural History of Chemotherapy-Induced Peripheral Neuropathy

Andreas A. Argyriou, Aakash Desai, and Charles Loprinzi

Abstract

Chemotherapy-induced neuropathy, ranking among the most common toxic neuropathies, primarily affects the sensory nerve modalities. Taxanes and oxaliplatin commonly cause an acute pain problem, something that presents soon after each individual dose and then generally improves over a course of days. In addition, these drugs and several other neurotoxic chemotherapy drugs commonly cause a more gradually appearing and more chronic neuropathy that primarily involves distal extremities. There are a number of similarities regarding the peripheral neuropathy caused by many chemotherapy drugs, noting that there are some distinct differences in them, also. When neurotoxic chemotherapy is stopped, neuropathy problems often improve. However, neuropathy can become a prominent problem for years, in some patients, leading to marked disabilities.

Keywords

CIPN · Chemotherapy · Neuropathy · Taxanes · Platniums · Microtubule inhibitors

A. A. Argyriou
"Saint Andrew's" State General Hospital of Patras, Patras, Greece

A. Desai (✉)
Department of Medical Oncology, Mayo Clinic, Rochester, MN, USA
e-mail: Desai.Aakash@mayo.edu

C. Loprinzi
Mayo Clinic, Rochester, MN, USA
e-mail: cloprinzi@mayo.edu

M. Lustberg, C. Loprinzi (eds.), *Diagnosis, Management and Emerging Strategies for Chemotherapy-Induced Neuropathy*,
https://doi.org/10.1007/978-3-030-78663-2_1

1.1 Introduction

This chapter discusses the natural history of chemotherapy-induced neuropathy, caused by a number of chemotherapy agents. It starts with acute neuropathy troubles and then discusses the more problematic chronic neuropathy caused by these drugs.

1.2 Natural History of Acute Chemotherapy-Induced Neuropathy

1.2.1 Oxaliplatin-Induced Acute Chemotherapy-Induced Neuropathy

Together with the typical late, dose-dependent effects of platinum compounds, oxaliplatin is also able to induce acute neurotoxic effects in the majority of cancer patients during exposure to oxaliplatin-based chemotherapy at a dose ranging from 85 to 130 mg/m^2 [1, 2]. Acute oxaliplatin-induced peripheral neurotoxicity is typically characterized by the rapid onset of cold-induced paresthesias and/or dysesthesias in distal limbs, i.e., hands and feet, but also in the oropharynx, involving the perioral, pharyngeal and laryngeal regions. Oropharyngeal paresthesias can be triggered by consumption of cold beverages. According to data of large prospective studies, focusing on deciphering the incidence and clinical phenotype of oxaliplatin neuropathy, the symptoms of acute sensory oxaliplatin neuropathy in the distal hands and feet as well as in the oropharyngeal regions can be present in up to 95% of oxaliplatin-exposed patients at a dose of 85 mg/m^2 [3].

Apart from these cold-induced sensory symptoms, other less common symptoms of the acute, abnormal hyperexcitability state of peripheral sensory and motor nerve fibers, which remain unrelated to cold exposure, are encountered in a significant rate of oxaliplatin-exposed patients. According to the results of a prospective, multicenter study that sought to assess the incidence of uncommon acute oxaliplatin neurotoxicity symptoms in 100 colorectal cancer patients undergoing oxaliplatin-based chemotherapy [4], it was evident that among 84 patients experiencing acute oxaliplatin neuropathy, 45 (54.9%) also presented shortness of breath (32%), jaw spasm (26%), fasciculations (25%), cramps (20%), and difficulty in swallowing (18%). Voice (4%) and visual changes, ptosis, and pseudolaryngospasm (1%) have also rarely been reported. In particular, jaw spasms and cramps tended to have a paroxysmal character with attacks lasting from 1 to 5 min. Nonetheless, the intensity of symptoms was not strong enough to significantly interfere with function and/or to require chemotherapy dose modifications, thoroughly highlighting the relatively benign nature of acute oxaliplatin neuropathy in the majority of patients.

To date, the recording of the incidence and intensity of acute oxaliplatin neuropathy remain challenging, as there is no uniform definition, with different studies using different assessment tools. An oxaliplatin neuropathy questionnaire is a descriptive questionnaire in a yes/no response format, investigating the frequency of the 11 most common hyperexcitability symptoms of acute oxaliplatin neuropathy

Table 1.1 Description of the oxaliplatin neuropathy questionnaire

Symptoms	Absent	Present
Cold-induced perioral paresthesias	0	1
Cold-induced pharyngolaryngeal dysesthesia	0	1
Shortness of breath	0	1
Difficulty swallowing	0	1
Laryngospasm	0	1
Muscle cramps	0	1
Jaw stiffness	0	1
Visible fasciculations	0	1
Voice changes	0	1
Ptosis	0	1
Ocular changes	0	1
Total (sum):		

Table 1.2 Description of the oxaliplatin Sanofi Specific Scale (OSSS) and oxaliplatin-specific Levi's scale

Grading	Oxaliplatin Sanofi Specific Scale	Oxaliplatin-specific Levi's scale
0	No symptoms	No symptoms
1	Paresthesias/dysesthesias of short duration that resolve and do not interfere with function	Paresthesias/dysesthesias (induced by cold) with complete regression within 1 week
2	Paresthesias/dysesthesias, interfering with function, but not activities of daily living	Paresthesias/dysesthesias (induced by cold) with complete regression within 21 days
3	Paresthesias/dysesthesias with pain or with functional impairment that also interferes with daily living	Paresthesias/dysesthesias with incomplete regression at day 21
4	Persistent paresthesias/dysesthesias that are disabling or life-threatening	Paresthesias/dysesthesias with functional consequence

(Table 1.1). The severity of acute oxaliplatin neuropathy is scored based on the number of symptoms reported by the patients at each clinical assessment; the increased number of acute symptoms is considered an expression of an increased severity of acute oxaliplatin neuropathy. According to this tool, the severity of acute oxaliplatin neuropathy is classified as grade I (1–2 symptoms), grade II (3–4 symptoms), grade III (5–8 symptoms), and grade IV (9–11 symptoms). This assessment tool has been successfully used in various prospective studies or clinical trials enrolling oxaliplatin-treated patients [5, 6]. The oxaliplatin Sanofi Specific Scale (OSSS) is another outcome measure that has been used in the research setting of clinical trials [7]. Briefly, this tool measures the intensity of oxaliplatin-related paresthesias/dysesthesias in 0–4 score grading (Table 1.2); notably without taking into account the remaining nine hyperexcitability symptoms of acute oxaliplatin neuropathy. A modification of OSSS, coined as the oxaliplatin-specific Levi's scale

(Table 1.2), has also been used in neuroprotective trials assessing the severity of acute oxaliplatin neuropathy and its impact on functional ability [8, 9].

In any case, clinical practice shows that the vast majority of affected patients are usually able to complete the treatment plan without oxaliplatin dose modification. As such, acute oxaliplatin neuropathy attracts less attention and is considered of inferior clinical importance than the chronic form as it is usually transient and reversible within 48–72 h in most patients, although there is evidence of its attenuation in both duration and severity with repeated exposure to oxaliplatin and high cumulative oxaliplatin doses [10]. The latter view is supported by the results of a study on 346 patients treated with FOLFOX, in which 308 (89%) experienced at least one symptom of acute oxaliplatin neuropathy within the first cycle with a peak at day 3, later improvement, and incomplete remission between subsequent treatment courses. Notably, the results of this study also showed that despite the typically transitory character of acute oxaliplatin neuropathy symptoms, patients with more severe acute neuropathy during the first cycle of therapy are also those who will develop more severe chronic neurotoxicity [11]. This relationship was subsequently replicated from the results of a large prospective study on 200 oxaliplatin-treated patients for colorectal cancer, which showed that those patients with more symptoms of acute OIPN are more liable to develop more severe chronic neurotoxicity [12]. Results, all together, point at the higher susceptibility of some individuals with more severe acute oxaliplatin neuropathy to chronic peripheral nervous system damage. However, despite the results of studies demonstrating that patients with alterations of axonal excitability in early oxaliplatin treatment are more prone to developing dose-limiting neurotoxicity, a direct causal relationship between the degree of acute nerve dysfunction and the development of chronic neurotoxicity [13] cannot be definitely stated at this time.

The relevance of documenting the latter association is deemed crucial in order to accurately define whether the incidence and intensity of acute oxaliplatin neuropathy at early stages might be used as a clinical predictor for selecting patients who may benefit from neuroprotective strategies against the chronic form of oxaliplatin neuropathy, as previously suggested by studies using clinical or neurophysiological examination with nerve excitability and quantitative sensory testing [13–15].

The above-described clinical phenotypic characteristics of acute oxaliplatin neuropathy provide significant clues for comprehending its pathogenetic hallmarks. Overall, the rapid onset and transient nature of acute oxaliplatin neuropathy, over a few days, point towards a functional source of peripheral nerve damage as a result of a reversible interplay between cellular targets, such as ion channels.

Specifically, abnormalities in ion channels can evoke spontaneous (ectopic) discharge, repetitive firing, and overall neuronal hyperexcitability at sensory Aβ fibers conducting light touch. This effect is attributable to a reduction of the action potential initiation threshold. As such, the shift of damaged afferent neurons into hyperexcitability states is not due to synaptic actions, but rather to an increase in the intrinsic electrogenic properties of the neuronal membrane [16]. It is widely acknowledged that sodium channels play a major role in the generation of acute painful oxaliplatin neuropathy effects by determining neuronal excitability, rather than by affecting synaptic action [17, 18]. The latter theory is supported by evidence showing that oxaliplatin neuropathy-associated altered axonal refractoriness has been linked to axonal nodal voltage-gated sodium channel dysfunction and slowing in the kinetics of sodium channel inactivation; this effect may be exacerbated by exposure to cold [13, 19, 20].

Conversely, the cold-unrelated acute syndrome of jaw tightness, cramps, and spasms after oxaliplatin exposure, clinically resembling neuromyotonia or Isaac's syndrome, occurs as a hallmark of motor nerve hyperexcitability [21]. Neuromyotonia is clinically manifested with muscle cramps, spasms, and fasciculations as a result of muscular hyperactivity due to impairment of voltage-gated potassium channels [22]. In view of these clinical similarities between acute oxaliplatin neuropathy and neuromyotonia, it is strongly suggested that oxaliplatin interacts with neuronal or muscular ion channels located in the cellular membrane, thus generating neurotoxicity [21, 23]. Nonetheless, the abnormalities in sodium rather than in potassium channel kinetics play a pivotal role in modulating the severity of acute oxaliplatin neuropathy [24].

Studies applying neurophysiological methods, such as nerve excitability tests, are in keeping with the above-described pathogenetic mechanism of acute oxaliplatin neuropathy genesis, based on ion channel-interference [17, 25]. However, although the major clinical hallmark of acute oxaliplatin neuropathy consists of cold-induced sensory symptoms attributable to abnormalities in neuronal excitability, there is evidence from axonal excitability techniques that motor nerves demonstrate a much more increased refractoriness and reduced super excitability, associated with slowing or inactivation of the nodal voltage-gated sodium channel, compared to sensory axons [26]. Finally, both motor and sensory nerve acute excitability changes have been shown to be associated with alterations in sodium channel function [26, 27], thereby bolstering the view that sodium channel abnormalities are of paramount importance in mediating acute oxaliplatin neuropathy.

1.2.2 Taxane-Induced Acute Neuropathy

An acute pain syndrome after taxane exposure (TAPS) is a distinct form of nerve pathology and has a completely different clinical phenotype than that of classical, chronic, and dose-dependent taxane-induced peripheral neurotoxicity (TIPN). For years, the TAPS was described as being diffuse myalgias/arthralgias, as the symptoms were predominantly manifested in shoulder, hip, and paraspinal regions, noting that less prominent symptoms were also observed in more distal muscles. The symptoms occur in the week following the first taxane administration, often starting 2–3 days after chemotherapy onset and lasting for 5–7 days before complete resolution [28]. Patients commonly describe TAPS as having an "aching," "shooting," "stabbing," or "pulsating" character [29] while generally the burning, neuropathic component of pain is lacking during the acute constellation of TAPS symptoms [30]. It was not until 2007, 14 years after paclitaxel had been commonly used in

Table 1.3 Grading of myalgias and arthralgias using the National Cancer Institute Common Terminology Criteria for Adverse Events (NCICTCAE)

CTCAE term	Grade 1	Grade 2	Grade 3
Arthralgias (discomfort in a joint)	Mild pain	Moderate pain; limiting instrumental activities of daily living (ADL)	Severe pain; limiting self-care ADL
Myalgias (discomfort originating for a muscle or a group of muscles)	Mild pain	Moderate pain; limiting instrumental ADL	Severe pain; limiting self-care ADL

clinical practice, when this pain syndrome was claimed to be a form of acute neuropathy, as opposed to being a manifestation of muscle or joint pathology [30].

The incidence of TAPS greatly varies, mainly in relation to the specific taxane compound [31]. The grading of TAPS severity has not been commonly assessed well as it has not been well described by commonly used National Cancer Institute Common Terminology Criteria for Adverse Events (NCI CTCAE), where such may be assessed as myalgias and arthralgias (Table 1.3), because this is how it initially was described. The same situation applies to other various patient reported quality of life tools, including the Brief Pain Inventory or the cancer-specific quality of life questionnaire, QLQ-C30, developed by the European Organization for Research and Treatment of Cancer (EORTC), which includes five functions, nine symptoms, and a global health status. The use of the above-mentioned outcome measures is generally problematic as it is associated with significant inter-observer disagreement and with underestimation of TAPS incidence and severity. Having said this, there is an available tool that has been used to describe and measure this neuropathy syndrome, by patient reported outcome (PRO) means [11, 32–34]. This PRO consists of patients completing questionnaires daily for 5 days after each paclitaxel dose.

Nonetheless, depending on various risk factors, there is evidence that up to 90% of taxane-treated patients may develop TAPS [35]. TAPS generally occurs more frequently with paclitaxel than with docetaxel or nab-paclitaxel treatment and is more frequent with higher dosages and shorter drug infusion duration [36, 37]. The incidence of TAPS with paclitaxel depends on several clinical oncological parameters, related to both schedule and disease stage [38]. Specifically, weekly paclitaxel schemes seem to evoke less frequent and/or intense TAPS than the three-weekly regimens [33], most certainly because a much higher paclitaxel dose is given with the 3-weekly regimen. Moreover, the infusion rate appears to also play a role, with longer paclitaxel (140 mg/m^2) infusion, over 96 h, being associated with TAPS to a lesser extent compared to paclitaxel (250 mg/m^2) given over 3 h. Finally, paclitaxel-treated patients in the metastatic setting appear to be more liable to manifest TAPS compared to their counterparts who are treated in the adjuvant setting [31].

In comparison to paclitaxel, TAPS after docetaxel exposure is estimated to occur in up to 70% of patients, with higher dosing of 100 mg/m^2 being more harmful than

75 mg/m^2 every 3 weeks and use of the EC-D regimen (epirubicin + cyclophospha-mide followed by docetaxel) being safer than FEC-D (5-fluorouracil-epirubicin-cyclophosphamide followed by docetaxel) in terms of TAPS incidence and severity [39]. However, contrary to the paclitaxel-associated syndrome, docetaxel-treated patients in the adjuvant setting are at higher risk of developing more severe TAPS than patients with metastatic disease [40, 41]. Finally, nab-paclitaxel evokes TAPS in up to 45% of exposed patients and this effect appears to remain unrelated to the treatment setting (neoadjuvant vs adjuvant vs metastatic) or the administered scheme of weekly vs every 2 weeks [31]. Other risk factors, including the cancer type and concurrent agents such as corticosteroids and G-CSF use, may also affect the development of TAPS. Specifically, there is evidence of an increased TAPS inci-dence in castrate-resistant patients with prostate cancer when corticosteroids were not concurrently used with taxane-based chemotherapeutic regimens [31].

Generally, the incidence of TAPS might reach the "clinically significant" level but rarely causes treatment cessation. This view is advocated by the results of a multicenter, prospective, non-randomized study assessing the incidence and characteristics of TAPS in taxane-treated patients with breast ($n = 66$) or prostate ($n = 9$) cancer [31]. A total of 33/75 (44%) experienced TAPS either after the first cycle of taxane or after infusion of a subsequent chemotherapy treatment. However, TAPS was not severe enough to necessitate change in the treatment plan. It is notable that the vast majority of patients in this trial received docetaxel, known to cause less TAPS than paclitaxel.

As mentioned earlier, TAPS is clinically quite distinct from TIPN and has different temporal profiles. However, comparison of data from 176 paclitaxel-treated patients showed that a more pronounced TAPS can predispose to subsequent chronic TIPN [11]. The assumption of a causal relationship between TAPS and chronic TIPN was further suggested by the results of a study challenging the association between the severity of TAPS and eventual peripheral neuropathy symptoms in 81 cancer patients who were scheduled to receive paclitaxel and carboplatin every 3 weeks. The results showed that worse TAPS severities predispose to a more severe chronic TIPN, thoroughly supporting the view that TAPS is a form of nerve pathology [32], possibly as a result of sensitization of nociceptors, their fibers, or the spinothalamic system [30].

Tellingly, from the pathogenetic point of view, TAPS is likewise quite distinct from the dose-dependent chronic TIPN. Neuroinflammation via rapid infiltration of macrophages within DRG and peripheral nerves seems to be an important aspect of TAPS genesis [42]. In line with the latter view, there is evidence demonstrating that activation of the inflammatory pathways may be responsible for the genesis of neuropathic pain induced by paclitaxel and can also mediate structural axonal damage. Furthermore, paclitaxel seems to be able to increase sphingosine 1-phosphate receptor (S1P) and ceramide levels in astrocytes of the dorsal horn spinal cord [43]. It is anticipated that greater elucidation and a more in-depth understanding of the pathological mechanisms underlying TAPS would allow the identification of a mechanistic basis of symptomatic TAPS improvement.

1.3   Natural History of Chronic Neuropathy

Chronic CIPN can be defined as a clinical syndrome characterized by a dose-related, persistent (at least two subsequent cycles without a "symptoms free" interval), syndrome with symmetrical distal painful and/or non-painful paresthesia and dysesthesia. With respect to chronic neuropathic pain related to chemotherapy, the syndrome is often termed chemotherapy-induced peripheral neurotoxicity (CIPN). It is labeled as "peripheral" since it is a peripheral nervous system problem, as opposed to being a central nervous system problem [10].

CIPN may initially manifest itself after the early doses of neurotoxic chemotherapy in some patients while it may not become apparent until a patient has received multiple doses of neurotoxic chemotherapy in most patients. When it occurs, it generally does not wax and wane, as occurs with acute neuropathy (described above). Rather, it tends to persist between doses of chemotherapy and generally worsens with the continuation of chemotherapy. When the neurotoxic chemotherapy is stopped, the chronic neuropathy symptoms generally persist for some time. With prolonged treatment the neuronopathy can eventually evolve in a non-length-dependent pattern with evidence of sensory ataxia, severe gait deficit, and increased liability to falls [44, 45]. While it may improve in the months following neurotoxic chemotherapy completion, it can persist for years in many patients.

Patients with chronic CIPN usually complain about distally attenuated painful or painless paresthesias and dysesthesias in hands and feet in a stocking-and-glove distribution. Proportional to sensory loss, there is clinical evidence of altered proprioception and suppression and/or abolishment of deep tendon reflexes. Motor and autonomic modalities are rarely affected. Data are available from a trial which involved patients with substantial peripheral neuropathy who had previously received taxanes (49%), oxaliplatin (44%), carboplatin/cisplatin (20%), vinca alkaloids (8%), thalidomide (3%), or a combination of these drugs [45]. Virtually all patients with pain had substantial numbness and tingling. Symptoms in these patients were more severe in lower extremities, then upper extremities. Among all the patients studied in the trial designed to treat this chronic symptom, it was found that numbness and tingling were more problematic issues than was pain (Fig. 1.1). This may be why duloxetine has limited effectiveness for treating established CIPN, as duloxetine appears to mainly provide analgesia, as opposed to impacting upon numbness/tingling symptoms [46].

Although multiple publications have addressed the natural history of CIPN development related to a number of neurotoxic chemotherapy agents, it has been difficult to compare and contrast the natural history of CIPN for these different drugs, given the variety of outcome measures that neuropathy has been evaluated in these trials. However, over the last few years, an ongoing effort has been underway to evaluate the natural history of CIPN for a variety of drugs, utilizing the same instrument in each situation: the EORTC CIPN 20. This tool was used at baseline and serially after therapy initiation for the following drug regimens: weekly paclitaxel, paclitaxel/carboplatin, oxaliplatin, and cisplatin. In addition, the same instrument was utilized to evaluate patients receiving doxorubicin/cyclophosphamide as a

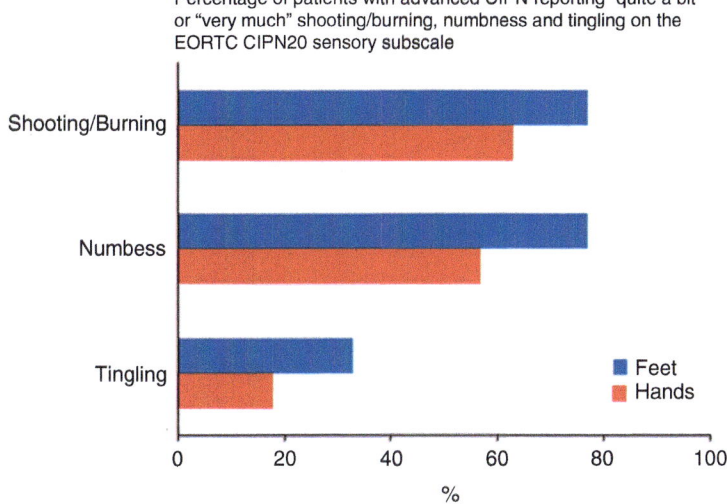

Fig. 1.1 Illustration of significant numbness, tingling, and shooting/burning pain in patients with substantial CIPN [45]

curative-intent treatment for breast cancer [47]. The reason for evaluating the use of this tool in this latter situation is because this regimen does not cause neurotoxicity and it was desired to understand what this tool would read in patients receiving non-neurotoxic chemotherapy. The results from this evaluation demonstrated that there were not any substantial changes in CIPN 20 scores in patients receiving chemotherapy that did not cause neurotoxicity, illustrating its ability to measure neuropathy problems, as opposed to generalized toxicity from chemotherapy [47].

The first publication in this series dealt with patients receiving weekly paclitaxel at a dose around 80 mg/m^2, a standard adjuvant therapy for patients with breast cancer [28]. This work illustrated that the neuropathology related to this therapy was mostly related to sensory deficits (having a worsening of 23 points) more than motor neuropathy (12 point worsening) or autonomic neuropathy (6 point worsening; $p < 0.03$). Similarly, to what was seen in patients with established CIPN, discussed above, numbness and tingling were closely related to each other and both were much more prominent than was pain. Data from another trial reported similar findings [48]. Both during the time while chemotherapy was administered and for 6 months thereafter, numbness, tingling, and pain symptoms were more prominent in lower extremities, than upper extremities, in keeping with length-dependent peripheral nerve damage (Fig. 1.2).

The next publication in this series reported on patients receiving paclitaxel and carboplatin, at 3 week intervals [32]. Noting that paclitaxel is thought to be the most neurotoxic component of this regimen, as opposed to carboplatin, similar results were seen in this trial, compared to the previously described study. Sensory neuropathy was more problematic than was motor or autonomic neuropathy; additionally,

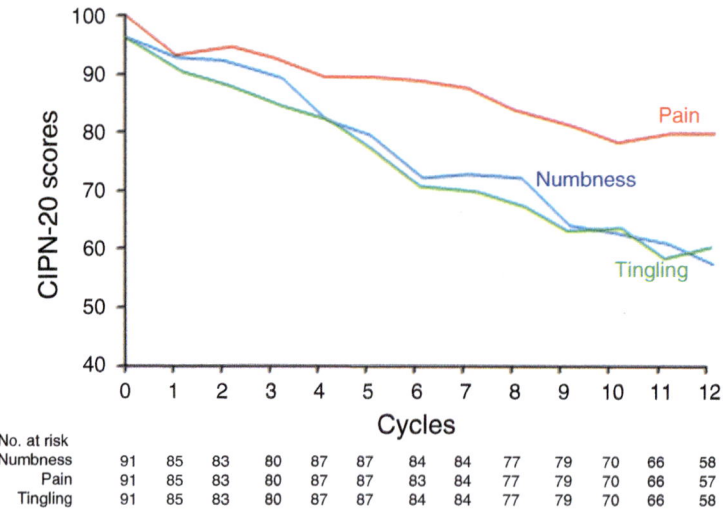

Fig. 1.2 Numbness, tingling, and pain neuropathy scores in lower extremities of patients receiving weekly paclitaxel. Lower scores illustrate more neuropathy

numbness and tingling, not pain, were the most common clinical components comprising the clinical phenotype of the sensory peripheral neurotoxicity.

Next, the natural history of oxaliplatin-based peripheral neurotoxicity was evaluated in a comparable manner [11, 49]. Findings were similar to what was seen with the prior two agents, with regard to oxaliplatin causing a predominant sensory chronic neuropathy with numbness and tingling being much more common, than pain; although there were some interesting differences between oxaliplatin and the paclitaxel-based regimens. One of these differences was that, while the patient was actively receiving oxaliplatin, their symptoms were more prominent in the upper extremities, as opposed to the lower extremities. After completion of active oxaliplatin therapy, the upper extremity symptoms improved more quickly, so that, in the months after finishing oxaliplatin therapy, sensory symptoms were more prominent in lower extremities. Another interesting contrast related to neuropathy symptoms in the first 3 months following cessation of neurotoxic chemotherapy. While symptom changes varied from person to person in both groups, on average, paclitaxel-based neuropathy improved in the 1st month following therapy cessation.

In contrast, oxaliplatin-treated patients, on average, had a worsening of their sensory neuropathy for about 3 months after oxaliplatin was stopped. This phenomenon has been called a coasting phenomenon. Previous reports had suggested that some patients did not get neuropathy until they stopped their oxaliplatin, with the implication that stopping the oxaliplatin actually caused the neuropathy. However, in this study, the slope of worsening neuropathy was similar in the 3 months prior to oxaliplatin cessation, when compared to the 3 months following oxaliplatin cessation. This is consistent with the notion that it takes 3 months, after each dose, for oxaliplatin-induced neuropathy to be fully manifested.

In 2020, a manuscript added cisplatin to the list of similarly evaluated agents, with regard to CIPN [47]. This revealed that cisplatin-induced neuropathy was more similar to oxaliplatin-induced neuropathy than it was to the neuropathy seen in patients who received paclitaxel. This makes sense, as they are both platinum compounds. Cisplatin, interestingly, does not evoke acute neurotoxicity resembling the typical oxaliplatin-induced acute effects. There were similarities between the two drugs, including the coasting phenomenon and that upper extremity symptoms are more prominent than lower extremities during the weeks of ongoing chemotherapy. In this study, cisplatin-induced neuropathy symptoms were less severe than were seen with paclitaxel or oxaliplatin. This might have been because the patients receiving cisplatin, in this trial, were younger males being treated for testicular cancer. After cisplatin was stopped, motor and autonomic neuropathy symptoms improved almost back to baseline. For sensory neuropathy symptoms, lower extremity numbness and tingling were the most problematic symptoms, 1 year following cisplatin cessation.

Vincristine, bortezomib, and thalidomide are other neurotoxic chemotherapy agents, commonly used for hematologic malignancies. Vinca alkaloids such as vincristine and vinorelbine, like other anticancer drugs, evoke a sensory length-dependent polyneuropathy [50]. Despite that, a distal, motor neuropathy, clinically manifested with foot drop, is more common than other drug classes [51]. Rare cranial neuropathy [52] (bilateral ptosis or abducens palsy [53]), autonomic involvement (neurogenic bladder, reversible esophageal dysphagia [54], and paralytic ileus [55]), and vocal cord paralysis [56] have also been reported. Vincristine is the most neurotoxic type of the clinically available vinca alkaloids. It leads to the development of a peripheral neuropathy, generally in patients who receive a cumulative dose of >4 mg/m^2 [57]. Usually, mild neurotoxicity subsides within 3 weeks of treatment discontinuation, while the more severe forms resolve much more slowly and can even persist for years [58]. However, off-therapy worsening of both neurotoxic symptoms and signs, resembling the coasting phenomenon, might unexpectedly occur in the first month after finishing vincristine therapy in up to 30% of patients, being more prevalent in high intensity cumulative dose of 12 mg vincristine [59]. Data from the HOVON-65/GMMG-HD4 trial, which evaluated mechanisms of peripheral neuropathy-associated with bortezomib and vincristine in patients with newly diagnosed multiple myeloma, showed that median time to development of vincristine-induced peripheral neuropathy was 37 days (range 0–171) [60]. Twenty-four percent of patients developed vincristine-induced peripheral neuropathy, with 7% developing grade 1 peripheral neuropathy before progressing to a higher grade. When patients developed vincristine-induced peripheral neuropathy, vincristine was generally discontinued and supportive treatments such as pregabalin were used in the trial. Another vinca alkaloid, Vinorelbine has been found to be more tumor specific and less toxic in comparison to vincristine [61]. Younger children tolerate relatively higher dosage of vincristine and develop less severe VIPN compared to adolescents and adults [62]. Also, Caucasian patients have greater incidence and severity of

VIPN than African-American patients [63]. Lastly, patients with hereditary neuropathies especially Charcot-Marie-Tooth (CMT) disease are also exceptionally sensitive to VIPN.

Bortezomib (BTZ), commonly used for treating multiple myeloma, commonly causes neuropathy [47]. This neuropathy usually becomes evident after 3–4 cycles of treatment, peaking at cycle 5 (cumulated dose 45 mg/m^2) followed by a plateau after which it does not evolve thereafter [64–66]. A sudden or quick progressive rather than gradual onset of severe neuropathy restricted to the first cycles has been commonly reported [66–69]. The mechanism of bortezomib-induced peripheral neuropathy is multifactorial with mitochondrial damage of dorsal root ganglia [70], dysregulation of mitochondrial calcium homeostasis [71], autoimmune inflammation [72], blockade of nerve growth factor mediated neuronal survival, and the myeloma itself all playing a role [70–73].

As compared to other chemotherapy-induced neuropathies, bortezomib-induced peripheral neuropathy (BIPN) portends a favorable outcome. The median time to recovery is approximately 3 months [65]. In a study of 256 patients with relapsed/refractory multiple myeloma treated with bortezomib, 35% of patients were found to have treatment emergent neuropathy with 13% and 0.4% having grade 3 and grade 4 neuropathy, respectively [74]. Although severe neuropathy was more frequent in the presence of baseline neuropathy, the overall occurrence was independent of baseline neuropathy or type of prior therapy. Even though almost 70–80% of patients recover, chronic painful peripheral neuropathy from bortezomib can be a major issue negatively impacting the quality of life in some multiple myeloma survivors.

A meta-analysis of four trials, involving a total of 911 patients receiving bortezomib, demonstrated that subcutaneous treatment administration was associated with a lower incidence of drug-induced neuropathy, compared to intravenous administration (41.4% vs. 16%), without interfering with anti-neoplastic activity [75]. Another systematic review found that, compared to IV administration, SC bortezomib had a significantly lower incidence of some all-grade or grade 3–4 AE, such as peripheral sensory neuropathy ($p < 0.05$) with no statistical difference in 1-year OS, 1-year progression free survival (PFS), or overall response rate (ORR) between SC and IV bortezomib [76].

Thalidomide-induced CIPN can cause both sensory and motor axonal polyneuropathies [77, 78]. Evidence exists that thalidomide causes mostly a sensory, axonal, length-dependent polyneuropathy that presents as painful paresthesias or numbness [79]. Observations of longitudinal cohorts of pediatric population showed that motor involvement is more pronounced than in adults [77]. The onset of thalidomide-induced peripheral neuropathy usually occurs at 12–24 weeks, with a range of 2–60 weeks; the incidence increases from 38% at 6 months to 73% at 12 months [80]. Furthermore, the risk of neuropathy has been found to be related to the daily dose, regardless of the treatment duration in some non-cancer settings [81]. The neuropathy is usually reversible after discontinuation of treatment. Generally, grade 1 or 2 events are expected to subside after a median duration of 3 weeks of treatment discontinuation [82].

A multicenter trial of dose escalating thalidomide with or without interferon on 75 patients with relapsed/refractory multiple myeloma showed that for those 31 patients who developed neuropathy, the median time to peripheral neurotoxicity onset was 24 weeks with incidence of 38% at 6 months and 73% at 12 months [80]. In general, in children/adolescents undergoing prolonged therapy, longitudinal clinical/neurophysiological monitoring is suggested [77]. Lenalidomide (at a dose of 30 mg orally or 15 mg twice daily [days 1–21 every 28 days]) has also been associated with a similar profile but neurotoxicity is less frequent and less severe than is seen with thalidomide [83–85].

Lastly, some other drugs, including ixabepilone, eribulin, and trastuzumab emtansine, cause some peripheral neuropathy, although not to the degree of the drugs noted above. Eribulin has been associated with a sensorimotor, mainly sensory polyneuropathy. A post-marketing observational study reported that approximately a quarter of patients receiving eribulin developed CIPN, being grade 1–2 in most cases; most patients did not need to stop eribulin because of CIPN [86]. A meta-analysis of large subset of patients revealed that the majority of patients receiving eribulin only developed a low-grade/moderate neuropathy [87]. Eribulin-induced neuropathy largely resolves 48 weeks off-therapy after eribulin cessation [82].

One clinical trial compared neuropathy associated with eribulin versus ixabepilone in patients with metastatic breast cancer [88]. While the total incidence of neuropathy was relatively similar with the two drugs (33% with eribulin and 48% with ixabepilone), the median time until the onset of neuropathy was about 12 weeks for ixabepilone versus 36 weeks for eribulin; fewer patients receiving eribulin discontinued treatment due to neuropathy, compared to patients receiving ixabepilone (3.9% vs. 18%). Many patients may stop these drugs due to disease progression and therefore true long term toxicity profile is not known.

In conclusion, chemotherapy-induced neuropathy is mostly a sensory neuropathic problem that usually presents in distal extremities. Some drugs, such as taxanes and oxaliplatin, can additionally cause an acute pain problem that presents soon after each individual dose and then tends to improve over a period of days. While there are many similarities regarding the peripheral neuropathy caused by many chemotherapy drugs, there are distinct differences in them, also. While neuropathy problems tend to improve following chemotherapy cessation, such can be a prominent problem for years, in some patients.

References

1. Cavaletti G, Marmiroli P (2020) Management of oxaliplatin-induced peripheral sensory neuropathy. Cancer 12(6):1370
2. Argyriou AA, Polychronopoulos P, Iconomou G, Chroni E, Kalofonos HP (2008) A review on oxaliplatin-induced peripheral nerve damage. Cancer Treat Rev 34(4):368–377
3. Staff NP, Cavaletti G, Islam B, Lustberg M, Psimaras D, Tamburin S (2019) Platinum-induced peripheral neurotoxicity: from pathogenesis to treatment. J Peripher Nerv Syst 24:S26–S39

4. Lucchetta M, Lonardi S, Bergamo F et al (2012) Incidence of atypical acute nerve hyperexcitability symptoms in oxaliplatin-treated patients with colorectal cancer. Cancer Chemother Pharmacol 70(6):899–902

5. Velasco R, Bruna J, Briani C et al (2014) Early predictors of oxaliplatin-induced cumulative neuropathy in colorectal cancer patients. J Neurol Neurosurg Psychiatry 85(4):392–398

6. Bruna J, Videla S, Argyriou AA et al (2018) Efficacy of a novel sigma-1 receptor antagonist for oxaliplatin-induced neuropathy: a randomized, double-blind, placebo-controlled phase IIa clinical trial. Neurotherapeutics 15(1):178–189

7. Glimelius B, Manojlovic N, Pfeiffer P et al (2018) Persistent prevention of oxaliplatin-induced peripheral neuropathy using calmangafodipir (PledOx®): a placebo-controlled randomised phase II study (PLIANT). Acta Oncol 57(3):393–402

8. Lévi FA, Zidani R, Vannetzel J-M et al (1994) Chronomodulated versus fixed-infusion—rate delivery of ambulatory chemotherapy with oxaliplatin, fluorouracil, and folinic acid (Leucovorin) in patients with colorectal cancer metastases: a randomized multi-institutional trial. JNCI J Natl Cancer Inst 86(21):1608–1617

9. Durand J, Deplanque G, Montheil V et al (2012) Efficacy of venlafaxine for the prevention and relief of oxaliplatin-induced acute neurotoxicity: results of EFFOX, a randomized, double-blind, placebo-controlled phase III trial. Ann Oncol 23(1):200–205

10. Argyriou AA, Bruna J, Marmiroli P, Cavaletti G (2012) Chemotherapy-induced peripheral neurotoxicity (CIPN): an update. Crit Rev Oncol Hematol 82(1):51–77

11. Pachman DR, Qin R, Seisler D et al (2016) Comparison of oxaliplatin and paclitaxel-induced neuropathy (Alliance A151505). Support Care Cancer 24(12):5059–5068

12. Argyriou AA, Cavaletti G, Briani C et al (2013) Clinical pattern and associations of oxaliplatin acute neurotoxicity: a prospective study in 170 patients with colorectal cancer. Cancer 119 (2):438–444

13. Park SB, Goldstein D, Lin CS-Y, Krishnan AV, Friedlander ML, Kiernan MC (2009) Acute abnormalities of sensory nerve function associated with oxaliplatin-induced neurotoxicity. J Clin Oncol 27(8):1243–1249

14. Gebremedhn EG, Shortland PJ, Mahns DA (2018) The incidence of acute oxaliplatin-induced neuropathy and its impact on treatment in the first cycle: a systematic review. BMC Cancer 18 (1):410

15. Tanishima H, Tominaga T, Kimura M, Maeda T, Shirai Y, Horiuchi T (2017) Hyperacute peripheral neuropathy is a predictor of oxaliplatin-induced persistent peripheral neuropathy. Support Care Cancer 25(5):1383–1389

16. North RY, Lazaro TT, Dougherty PM (2018) Ectopic spontaneous afferent activity and neuropathic pain. Neurosurgery; 65(CN_suppl_1):49–54

17. Argyriou AA, Park SB, Bruna J, Cavaletti G (2019) Voltage-gated sodium channel dysfunction and the search for other satellite channels in relation to acute oxaliplatin-induced peripheral neurotoxicity. J Peripher Nerv Syst 24(4):360–361

18. Argyriou AA, Antonacopoulou AG, Alberti P et al (2019) Liability of the voltage-gated potassium channel KCNN3 repeat polymorphism to acute oxaliplatin-induced peripheral neurotoxicity. J Peripher Nerv Syst 24(4):298–303

19. Park SB, Lin CS, Kiernan MC (2012) Nerve excitability assessment in chemotherapy-induced neurotoxicity. JoVE (J Visual Exp) (62):e3439

20. Krishnan AV, Goldstein D, Friedlander M, Kiernan MC (2006) Oxaliplatin and axonal Na+ channel function in vivo. Clin Cancer Res 12(15):4481–4484

21. Webster RG, Brain KL, Wilson RH, Grem JL, Vincent A (2005) Oxaliplatin induces hyperexcitability at motor and autonomic neuromuscular junctions through effects on voltage-gated sodium channels. Br J Pharmacol 146(7):1027–1039

22. Katirji B (2019) Peripheral nerve hyperexcitability. Handb Clin Neurol 161:281–290. Elsevier

23. Wilson RH, Lehky T, Thomas RR, Quinn MG, Floeter MK, Grem JL (2002) Acute oxaliplatin-induced peripheral nerve hyperexcitability. J Clin Oncol 20(7):1767–1774

24. Jacobson D, Herson PS, Neelands TR, Maylie J, Adelman JP (2002) SK channels are necessary but not sufficient for denervation-induced hyperexcitability. Muscle Nerve 26(6):817–822
25. Bennedsgaard K, Ventzel L, Grafe P et al (2020) Cold aggravates abnormal excitability of motor axons in oxaliplatin-treated patients. Muscle Nerve 61(6):796–800
26. Park SB, Lin CS-Y, Krishnan AV, Goldstein D, Friedlander ML, Kiernan MC (2009) Oxaliplatin-induced neurotoxicity: changes in axonal excitability precede development of neuropathy. Brain 132(10):2712–2723
27. Sittl R, Lampert A, Huth T et al (2012) Anticancer drug oxaliplatin induces acute cooling-aggravated neuropathy via sodium channel subtype NaV1. 6-resurgent and persistent current. Proc Natl Acad Sci USA 109(17):6704–6709
28. Loprinzi CL, Reeves BN, Dakhil SR et al (2011) Natural history of paclitaxel-associated acute pain syndrome: prospective cohort study NCCTG N08C1. J Clin Oncol 29(11):1472
29. Asthana R, Zhang L, Wan BA et al (2020) Pain descriptors of taxane acute pain syndrome (TAPS) in breast cancer patients—a prospective clinical study. Support Care Cancer 28 (2):589–598
30. Loprinzi CL, Maddocks-Christianson K, Wolf SL et al (2007) The paclitaxel acute pain syndrome: sensitization of nociceptors as the putative mechanism. Cancer J 13(6):399–403
31. Fernandes R, Mazzarello S, Hutton B et al (2016) Taxane acute pain syndrome (TAPS) in patients receiving taxane-based chemotherapy for breast cancer—a systematic review. Support Care Cancer 24(8):3633–3650
32. Reeves BN, Dakhil SR, Sloan JA et al (2012) Further data supporting that paclitaxel-associated acute pain syndrome is associated with development of peripheral neuropathy: North Central Cancer Treatment Group trial N08C1. Cancer 118(20):5171–5178
33. Pachman DR, Dockter T, Zekan PJ et al (2017) A pilot study of minocycline for the prevention of paclitaxel-associated neuropathy: ACCRU study RU221408I. Support Care Cancer 25 (11):3407–3416
34. Shinde SS, Seisler D, Soori G et al (2016) Can pregabalin prevent paclitaxel-associated neuropathy?— An ACCRU pilot trial. Support Care Cancer 24(2):547–553
35. Tamburin S, Park SB, Alberti P, Demichelis C, Schenone A, Argyriou AA (2019) Taxane and epothilone-induced peripheral neurotoxicity: from pathogenesis to treatment. J Peripher Nerv Syst 24:S40–S51
36. Chiu N, Chiu L, Chow R et al (2017) Taxane-induced arthralgia and myalgia: a literature review. J Oncol Pharm Pract 23(1):56–67
37. Moulder SL, Holmes FA, Tolcher AW et al (2010) A randomized phase 2 trial comparing 3-hour versus 96-hour infusion schedules of paclitaxel for the treatment of metastatic breast cancer. Cancer 116(4):814–821
38. Velasco R, Bruna J (2015) Taxane-induced peripheral neurotoxicity. Toxics 3(2):152–169
39. Schönherr A, Aivazova-Fuchs V, Annecke K et al (2012) Toxicity analysis in the ADEBAR trial: sequential anthracycline-taxane therapy compared with FEC120 for the adjuvant treatment of high-risk breast cancer. Breast Care 7(4):289–295
40. Martin M, Lluch A, Segui M et al (2006) Toxicity and health-related quality of life in breast cancer patients receiving adjuvant docetaxel, doxorubicin, cyclophosphamide (TAC) or 5-fluorouracil, doxorubicin and cyclophosphamide (FAC): impact of adding primary prophylactic granulocyte-colony stimulating factor to the TAC regimen. Ann Oncol 17(8):1205–1212
41. Lee KS, Ro J, Nam B-H et al (2008) A randomized phase-III trial of docetaxel/capecitabine versus doxorubicin/cyclophosphamide as primary chemotherapy for patients with stage II/III breast cancer. Breast Cancer Res Treat 109(3):481–489
42. Peters CM, Jimenez-Andrade JM, Jonas BM et al (2007) Intravenous paclitaxel administration in the rat induces a peripheral sensory neuropathy characterized by macrophage infiltration and injury to sensory neurons and their supporting cells. Exp Neurol 203(1):42–54
43. Janes K, Little JW, Li C et al (2014) The development and maintenance of paclitaxel-induced neuropathic pain require activation of the sphingosine 1-phosphate receptor subtype 1. J Biol Chem 289(30):21082–21097

44. Argyriou AA, Bruna J, Anastopoulou GG, Velasco R, Litsardopoulos P, Kalofonos HP (2020) Assessing risk factors of falls in cancer patients with chemotherapy-induced peripheral neurotoxicity. Support Care Cancer 28(4):1991–1995

45. Wolf SL, Barton DL, Qin R et al (2012) The relationship between numbness, tingling, and shooting/burning pain in patients with chemotherapy-induced peripheral neuropathy (CIPN) as measured by the EORTC QLQ-CIPN20 instrument, N06CA. Support Care Cancer 20(3):625–632

46. Loprinzi CL, Lacchetti C, Bleeker J et al (2020) Prevention and management of chemotherapy-induced peripheral neuropathy in survivors of adult cancers: ASCO guideline update. J Clin Oncol 38(28):3325–3348

47. Albany C, Dockter T, Wolfe E et al (2021) Cisplatin-associated neuropathy characteristics compared with those associated with other neurotoxic chemotherapy agents (Alliance A151724). Support Care Cancer 29(2):833–840

48. Dougherty PM, Cata JP, Cordella JV, Burton A, Weng H-R (2004) Taxol-induced sensory disturbance is characterized by preferential impairment of myelinated fiber function in cancer patients. Pain 109(1-2):132–142

49. Pachman DR, Qin R, Seisler DK et al (2015) Clinical course of oxaliplatin-induced neuropathy: results from the randomized phase III trial N08CB (Alliance). J Clin Oncol 33(30):3416

50. Argyriou AA, Kyritsis AP, Makatsoris T, Kalofonos HP (2014) Chemotherapy-induced peripheral neuropathy in adults: a comprehensive update of the literature. Cancer Manag Res 6:135

51. Bokemeyer C, Berger CC, Kuczyk MA, Schmoll H-J (1996) Evaluation of long-term toxicity after chemotherapy for testicular cancer. J Clin Oncol 14(11):2923–2932

52. Talebian A, Goudarzi RM, Mohammadzadeh M, Mirzadeh AS (2014) Vincristine-induced cranial neuropathy. Iran J Child Neurol 8(1):66

53. Toker E, Yenice O, Oğüt MS (2004) Isolated abducens nerve palsy induced by vincristine therapy. J AAPOS 8(1):69–71

54. Wang WS (2000) Vincristine-induced dysphagia suggesting esophageal motor dysfunction: a case report. Jpn J Clin Oncol 30(11):515–518

55. Leker RR, Peretz T, Hubert A, Lossos A (1997) Vincristine-induced paralytic ileus in Parkinson's disease. Parkinsonism Relat Disord 3(2):109–110

56. Naithani R, Dolai TK, Kumar R (2009) Bilateral vocal cord paralysis following treatment with vincristine. Indian Pediatr 46(1)

57. Tay CG, Lee VWM, Ong LC, Goh KJ, Ariffin H, Fong CY (2017) Vincristine-induced peripheral neuropathy in survivors of childhood acute lymphoblastic leukaemia. Pediatr Blood Cancer 64(8):e26471

58. Okada N, Hanafusa T, Sakurada T et al (2014) Risk factors for early-onset peripheral neuropathy caused by vincristine in patients with a first administration of R-CHOP or R-CHOP-like chemotherapy. J Clin Med Res 6(4):252

59. Verstappen C, Koeppen S, Heimans J et al (2005) Dose-related vincristine-induced peripheral neuropathy with unexpected off-therapy worsening. Neurology 64(6):1076–1077

60. Broyl A, Corthals SL, Jongen JL et al (2010) Mechanisms of peripheral neuropathy associated with bortezomib and vincristine in patients with newly diagnosed multiple myeloma: a prospective analysis of data from the HOVON-65/GMMG-HD4 trial. Lancet Oncol 11(11):1057–1065

61. Gregory R, Smith I (2000) Vinorelbine—a clinical review. Br J Cancer 82(12):1907–1913

62. Frost BM, Lönnerholm G, Koopmans P et al (2003) Vincristine in childhood leukaemia: no pharmacokinetic rationale for dose reduction in adolescents. Acta Paediatr 92(5):551–557

63. Diouf B, Crews KR, Lew G et al (2015) Association of an inherited genetic variant with vincristine-related peripheral neuropathy in children with acute lymphoblastic leukemia. JAMA 313(8):815–823

64. Cho J, Kang D, Lee JY, Kim K, Kim SJ (2014) Impact of dose modification on intravenous bortezomib-induced peripheral neuropathy in multiple myeloma patients. Support Care Cancer 22(10):2669–2675

65. Corso A, Mangiacavalli S, Varettoni M, Pascutto C, Zappasodi P, Lazzarino M (2010) Bortezomib-induced peripheral neuropathy in multiple myeloma: a comparison between previously treated and untreated patients. Leuk Res 34(4):471–474
66. Rampen A, Jongen J, Van Heuvel I, Scheltens-de Boer M, Sonneveld P, van den Bent M (2013) Bortezomib-induced polyneuropathy. Age 57:1.29
67. Badros A, Goloubeva O, Dalal JS et al (2007) Neurotoxicity of bortezomib therapy in multiple myeloma: a single-center experience and review of the literature. Cancer 110(5):1042–1049
68. Lanzani F, Mattavelli L, Frigeni B et al (2008) Role of a pre-existing neuropathy on the course of bortezomib-induced peripheral neurotoxicity. J Peripher Nerv Syst 13(4):267–274
69. Popat R, Oakervee H, Williams C et al (2009) Bortezomib, low-dose intravenous melphalan, and dexamethasone for patients with relapsed multiple myeloma. Br J Haematol 144 (6):887–894
70. Cavaletti G, Gilardini A, Canta A et al (2007) Bortezomib-induced peripheral neurotoxicity: a neurophysiological and pathological study in the rat. Exp Neurol 204(1):317–325
71. Landowski TH, Megli CJ, Nullmeyer KD, Lynch RM, Dorr RT (2005) Mitochondrial-mediated disregulation of Ca2+ is a critical determinant of Velcade (PS-341/bortezomib) cytotoxicity in myeloma cell lines. Cancer Res 65(9):3828–3836
72. Ravaglia S, Corso A, Piccolo G et al (2008) Immune-mediated neuropathies in myeloma patients treated with bortezomib. Clin Neurophysiol 119(11):2507–2512
73. Ropper AH, Gorson KC (1998) Neuropathies associated with paraproteinemia. N Engl J Med 338(22):1601–1607
74. Richardson PG, Briemberg H, Jagannath S et al (2006) Frequency, characteristics, and reversibility of peripheral neuropathy during treatment of advanced multiple myeloma with bortezomib. J Clin Oncol 24(19):3113–3120
75. Liu Z, Xia H, Li C, Xia L (2019) Incidence and Risk of Peripheral Neuropathy Caused by Intravenous and Subcutaneous Injection of Bortezomib. Zhongguo Shi Yan Xue Ye Xue Za Zhi 27(5):1654–1663
76. Ye Z, Chen J, Xuan Z, Yang W, Chen J (2019) Subcutaneous bortezomib might be standard of care for patients with multiple myeloma: a systematic review and meta-analysis. Drug Des Dev Ther 13:1707
77. Liew WK, Pacak CA, Visyak N, Darras BT, Bousvaros A, Kang PB (2016) Longitudinal patterns of thalidomide neuropathy in children and adolescents. J Pediatr 178:227–232
78. Chaudhry V, Cornblath DR, Corse A, Freimer M, Simmons-O'Brien E, Vogelsang G (2002) Thalidomide-induced neuropathy. Neurology 59(12):1872–1875
79. Attal N, Bouhassira D, Gautron M et al (2009) Thermal hyperalgesia as a marker of oxaliplatin neurotoxicity: a prospective quantified sensory assessment study. PAIN® 144(3):245–252
80. Mileshkin L, Stark R, Day B, Seymour JF, Zeldis JB, Prince HM (2006) Development of neuropathy in patients with myeloma treated with thalidomide: patterns of occurrence and the role of electrophysiologic monitoring. J Clin Oncol 24(27):4507–4514
81. Bastuji-Garin S, Ochonisky S, Bouche P et al (2002) Incidence and risk factors for thalidomide neuropathy: a prospective study of 135 dermatologic patients. J Invest Dermatol 119 (5):1020–1026
82. Palumbo A, Facon T, Sonneveld P et al (2008) Thalidomide for treatment of multiple myeloma: 10 years later. Blood 111(8):3968–3977
83. Briani C, Torre CD, Campagnolo M et al (2013) Lenalidomide in patients with chemotherapy-induced polyneuropathy and relapsed or refractory multiple myeloma: results from a single-centre prospective study. J Peripher Nerv Syst 18(1):19–24
84. Nozza A, Terenghi F, Gallia F et al (2017) Lenalidomide and dexamethasone in patients with POEMS syndrome: results of a prospective, open-label trial. Br J Haematol 179(5):748–755

85. Aguiar PM, de Mendonça LT, Colleoni GWB, Storpirtis S (2017) Efficacy and safety of bortezomib, thalidomide, and lenalidomide in multiple myeloma: an overview of systematic reviews with meta-analyses. Crit Rev Oncol Hematol 113:195–212

86. Sakata Y, Matsuoka T, Ohashi S, Koga T, Toyoda T, Ishii M (2019) Use of a healthcare claims database for post-marketing safety assessments of Eribulin in Japan: a comparative assessment with a prospective post-marketing surveillance study. Drugs Real World Outcomes 6(1):27–35

87. Zhao B, Zhao H, Zhao J (2018) Incidence and clinical parameters associated with eribulin mesylate-induced peripheral neuropathy. Crit Rev Oncol Hematol 128:110–117

88. Vahdat LT, Garcia AA, Vogel C et al (2013) Eribulin mesylate versus ixabepilone in patients with metastatic breast cancer: a randomized Phase II study comparing the incidence of peripheral neuropathy. Breast Cancer Res Treat 140(2):341–351

Predisposing Factors for the Development of Chemotherapy-Induced Peripheral Neuropathy (CIPN)

Daniel L. Hertz, Cindy Tofthagen, and Sara Faithfull

Abstract

This chapter summarizes the current knowledge of predisposing factors of CIPN development. These predisposing factors can be classified as intrinsic (i.e., demographics, genetics) or extrinsic (i.e., lifestyle, neurotoxic treatment) to the patient. Intrinsic factors that increase a patient's CIPN risk include older age, African American race, and diabetes. Other factors such as vitamin D deficiency and genetics may also increase risk but have not been validated. Objective and subjective indicators of CIPN prior to, or early in, treatment predict CIPN severity at the end of treatment but this information is not consistently used to inform patient management. Extrinsic factors including lifestyle and neurotoxic regimen affect CIPN risk. Healthy lifestyle choices including physical activity and better nutrition may protect against CIPN. The predominant predictor of CIPN is cumulative treatment with a neurotoxic chemotherapeutic agent. Different regimens have different CIPN risk, and in the case of paclitaxel there is strong evidence that systemic drug exposure is a major contributor to CIPN. Further research is needed to validate these predisposing factors and determine their effect on CIPN onset, severity, and duration. Prospective studies are also needed to test strategies to use these predictive factors to inform personalized treatment

D. L. Hertz (✉)
Department of Clinical Pharmacy, University of Michigan College of Pharmacy, Ann Arbor, MI, USA
e-mail: dlhertz@med.umich.edu

C. Tofthagen
Department of Nursing, Division of Nursing Research, Mayo Clinic, Jacksonville, FL, USA
e-mail: Tofthagen.Cindy@mayo.edu

S. Faithfull
School of Health Sciences, University of Surrey, Guildford, Surrey, UK
e-mail: s.faithfull@surrey.ac.uk

decisions to prevent severe, life altering CIPN and optimize long-term outcomes in patients with cancer.

Keywords

Demographics · Lifestyle · Nutrient deficiency · Cumulative dosing · Pharmacogenetics · Pharmacokinetics · Biomarker

2.1 Introduction

CIPN risk factors could be used to identify patients who could consider less neurotoxic treatment regimens or may benefit from enhanced CIPN monitoring. Predisposing factors can generally be classified as intrinsic (i.e., demographics, genetics) or extrinsic (i.e., lifestyle, neurotoxic treatment) to the patient (Fig. 2.1). This chapter will summarize the current knowledge of predisposing factors. As with most fields, this field is rapidly changing as more data are collected; future work will advance our understanding of how these factors differ across patients or interact in ways that are not currently understood. There is immense heterogeneity across studies, including defining CIPN as a single phenotype or by its characteristic subtypes of sensory, motor, autonomic, and painful neuropathy. CIPN can be also be defined by different aspects of its timecourse including its onset trajectory, maximal severity, or post-treatment duration, and biomarkers could be assessed at various time points to predict this timecourse (Fig. 2.2). It is likely that different biomarkers at different timepoints have different predictive effects on CIPN

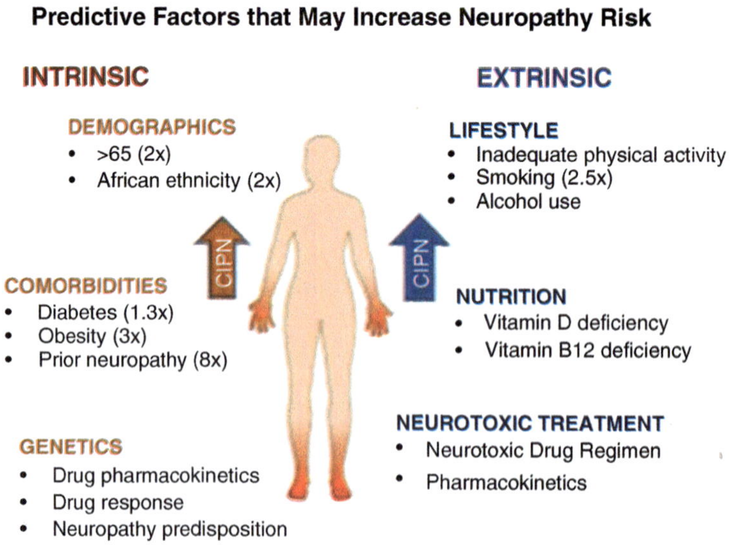

Predictive Factors that May Increase Neuropathy Risk

INTRINSIC

DEMOGRAPHICS
- >65 (2x)
- African ethnicity (2x)

COMORBIDITIES
- Diabetes (1.3x)
- Obesity (3x)
- Prior neuropathy (8x)

GENETICS
- Drug pharmacokinetics
- Drug response
- Neuropathy predisposition

EXTRINSIC

LIFESTYLE
- Inadequate physical activity
- Smoking (2.5x)
- Alcohol use

NUTRITION
- Vitamin D deficiency
- Vitamin B12 deficiency

NEUROTOXIC TREATMENT
- Neurotoxic Drug Regimen
- Pharmacokinetics

Fig. 2.1 Predictive factors that may increase chemotherapy-induced peripheral neuropathy risk

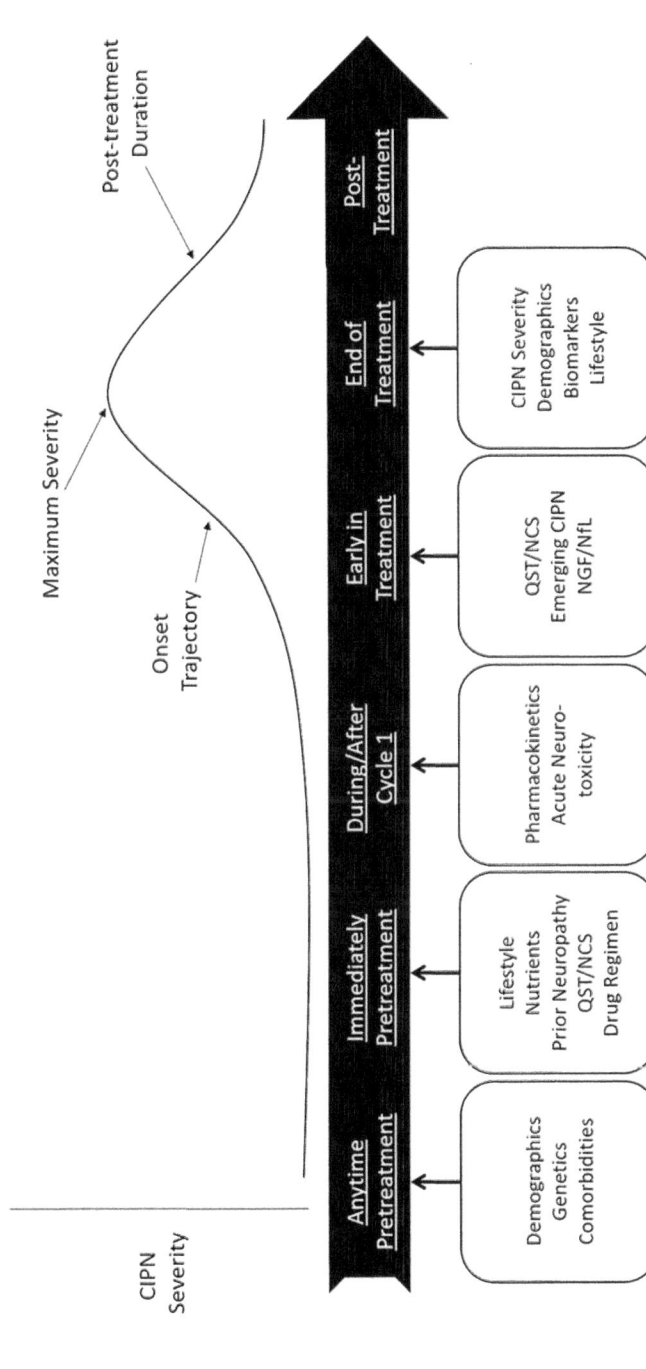

Fig. 2.2 The timecourse of chemotherapy-induced peripheral neuropathy and predictive biomarkers that could be collected prior to, during, and at the end of treatment. The trajectory, severity, and onset of CIPN differ between regimens and may be differentially affected by predictive biomarkers. Pre-treatment biomarkers would likely be most predictive of trajectory and severity. Biomarkers collected at the end of treatment would only be useful to predict the duration of CIPN. Any of these predictive biomarkers might eventually be shown to be useful to inform treatment decisions and CIPN monitoring for individual patients. Abbreviations: CIPN, chemotherapy-induced peripheral neuropathy; NCS, nerve conduction studies; NfL, neurofilament light chain; NGF, nerve growth factor; QST, quantitative sensory testing

subtypes and timecourse, but there are insufficient data at this time since most studies have used a general CIPN severity endpoint.

2.2   Intrinsic Factors

Intrinsic factors such as demographics, comorbidities, and genetics may be important predictors of CIPN. In this section we will review the available data to identify which of these factors could be used to identify patients with elevated CIPN risk.

2.2.1   Demographics

2.2.1.1 Age and Race

Patients who are over 60 years old when they receive neurotoxic chemotherapy are more likely to experience CIPN [1–3], though differences in how older age (i.e., >60 > 70) was defined make it challenging to estimate the magnitude of effect. There is evidence that age effects duration more than severity [4]. The association with age may be related to neurological aging and pre-existing conditions that impact an individual's recovery. Large studies of cancer survivors have confirmed the association of age and CIPN [5, 6] and prospective longitudinal studies indicate that older age increases post-treatment duration [1, 7]. The effect of age on CIPN may be particularly pronounced within CIPN subtypes, as older patients have reported a greater loss of cold sensation and sensory perception in their hands and feet [3]. Age should be considered when making treatment decisions based on CIPN risk since even mild-to-moderate CIPN can have substantive impact on quality of life in older adults [8].

Race has been identified as a potential contributor to severity of CIPN. Analyses of patient registries and prospective clinical studies have consistently reported that patients who identify as African American have a higher CIPN risk [9–12].

There is also evidence that non-Chinese Asians, primarily from Malay and Indian origin, may have a higher risk of CIPN than Chinese or Caucasian patients [7], but more studies are needed to confirm this relationship.

2.2.1.2 Diabetes and Other Comorbidities

Many CIPN studies exclude patients with diabetes mellitus because diabetes causes neurological damage, limiting our understanding of the impact of diabetes on CIPN risk. A retrospective population database review found that patients with diabetes had twice the odds of experiencing CIPN and this effect was concentrated in patients with diabetic complications [5]. Obese patients also seem to have elevated CIPN risk [13–18], as do patients with a higher overall comorbidity burden [19]. Thyroid dysfunction, metabolic and infectious diseases (hepatitis B or C and poliomyelitis, HIV) have been implicated in increasing CIPN [7, 20] whereas a large analysis indicates that patients with autoimmune disease are about half as likely to experience CIPN (OR = 0.49, 95% CI: 0.24–1.02, $p = 0.06$) [5].

2.2.1.3 Neuropathy Prior to Treatment

Some patients with cancer have subclinical neuropathy prior to treatment [21, 22]; there is evidence that these patients have substantially higher CIPN risk (OR = 8.36, 95% CI: 1.74–40.13, $p < 0.001$) [7]. CIPN risk is also higher in patients with worse pre-treatment neurological function [23] and touch sensation [24–26]. The mechanism for this may be that patients with subclinical neuropathy have fewer Meissner corpuscles in their tissues, which may reduce their ability to recover sensory levels and cause more clinically overt CIPN [27].

2.2.2 Physiological Biomarkers

In addition to clinical variables, it may be possible to predict CIPN risk based on physiological biomarkers including nutrients (Table 2.1) and genetics.

2.2.2.1 Nutrients

25-hydroxy vitamin D (vitamin D) and its metabolites have neuroprotective properties [34] and vitamin D deficiency is involved with several etiologies of neuropathy [35]. Low vitamin D levels prior to paclitaxel chemotherapy have been associated with increased CIPN [29, 30]. This has also been reported in patients

Table 2.1 Summary of studies supporting nutritional deficiencies as predictors of CIPN

Nutritional marker	Cancer	Neurotoxic agent(s)	Study design	n	Major findings	Ref
Vitamin D	Multiple myeloma	Bortezomib and/or thalidomide	Multi-center, cross-sectional	109	Vitamin D deficient patients more likely to have motor and sensory CIPN	[28]
Vitamin D	Breast	Paclitaxel	Case–control	70	Vitamin D levels were significantly lower in patients with CIPN	[29]
Vitamin D	Breast	Paclitaxel	Observational	38	Vitamin D deficient patients had a greater increase in patient-reported CIPN	[30]
Anemia, Magnesium	Colorectal	Oxaliplatin	Retrospective	169	Incidence of CIPN higher in patients with pre-treatment anemia, hypoalbuminemia, or hypomagnesemia	[31]
Anemia, Magnesium	Colorectal	Oxaliplatin	Descriptive	130	Anemia and hypomagnesemia were associated with greater CIPN	[32]
Anemia	Lymphoma	Vincristine	Retrospective cohort	40	Anemia at baseline was predictive of severe CIPN	[33]

treated with bortezomib or thalidomide [28], which is particularly interesting since bortezomib decreases serum vitamin D [36]. Vitamin B12 deficiency, which can be either a true deficiency or a functional deficiency [37], is another cause of polyneuropathy that may play a role in CIPN. There are case series of patients with functional vitamin B12 deficiency, whose CIPN improved from supplementation [37]; however, further studies are needed to determine the effects of B vitamins on CIPN [29, 38]. Deficiencies in iron and folic acid contribute to the development of anemia [39], which has been reported to be a risk factor for CIPN [31–33]. Intake of magnesium, which supports neuromuscular function by reducing neuronal excitability [40], has been associated with less CIPN in patients who received oxaliplatin [31, 41] and capecitabine [41]. Low vitamin E levels before and during cisplatin-based chemotherapy have also been associated with CIPN risk [42]. These data have been used to justify prospective clinical trials, trials that have been unsuccessful in demonstrating CIPN prevention from supplementation with calcium/magnesium [43–45], vitamin E [46–51], omega-3 fatty acids [29, 52], acetyl-l-carnitine [53–57], alpha-lipoic acid [58, 59], or glutamine [60–63]. Current guidelines do not recommend any nutritional supplements or dietary interventions for CIPN prevention or treatment [64].

2.2.2.2 Metabolomics and Proteomics

Metabolomics and proteomics are novel approaches for measuring an array of compounds in a biofluid that may reflect nutritional and health status, and these techniques could be used to discover CIPN biomarkers [65]. A metabolomics study reported that patients with CIPN had low pre-treatment levels of three essential amino acids; histidine, phenylalanine, and threonine [66]. Other small studies have reported possible metabolomics signatures of vincristine-induced neuropathy [67] and a proteomics signature of paclitaxel-induced neuropathy [68]. These omics-based approaches may provide clues about nutritional interventions to reduce CIPN risk, particularly in patients with pre-treatment deficiencies. However, routine use of this approach has not been validated for routine use.

2.2.2.3 Genetics

Pharmacogenetics is the study of whether inherited variants, or polymorphisms, in the germline genome affect response to medication. Pharmacogenetics studies often investigate candidate polymorphisms in enzymes or transporters that may affect drug concentrations or in genes involved in drug response (Table 2.2, Fig. 2.3). Alternatively, pharmacogenetic analyses can use an omics approach to simultaneously test many polymorphisms distributed throughout the entire genome in a genome-wide association study (GWAS). Discovery-phase pharmacogenetic studies are typically conducted with liberal statistical methodology and reported associations require robust validation in multiple independent studies, prior to clinical translation. Unless stated otherwise, the associations described in this section should be considered as being in a discovery phase and should not be used to inform patient care.

Readers interested in a comprehensive review of CIPN pharmacogenetics studies and their limitations are directed to this systematic review and meta-analysis [121].

Table 2.2 Candidate genes investigated for associations with CIPN

Drug	Gene	Mechanism	References
Paclitaxel	CYP2C8	Pharmacokinetics	[9, 69–73]
	CYP3A4/5	Pharmacokinetics	[71, 74]
	ABCB1	Pharmacokinetics	[72, 75–77]
	TUBB2A	Drug mechanism	[77, 78]
	EPHA	Hereditary neuropathy	[72, 77, 79–83]
Docetaxel	GSTP1	Reactive oxygen species	[84–86]
	VAC14	Hereditary neuropathy	[87]
Platinums	GSTP1	Pharmacokinetics	[88–96]
	ABCs	Pharmacokinetics	[97–99]
	ERCC1/2	Drug mechanism	[91, 100–105]
	DCLRE1A	Drug mechanism	[106]
Vincas	CYP3A5	Pharmacokinetics	[107–111]
	ABCB1	Pharmacokinetics	[108–111]
	CEP72	Drug mechanism	[112–117]
Thalidomide	CYP2C19	Pharmacokinetics	[118–120]

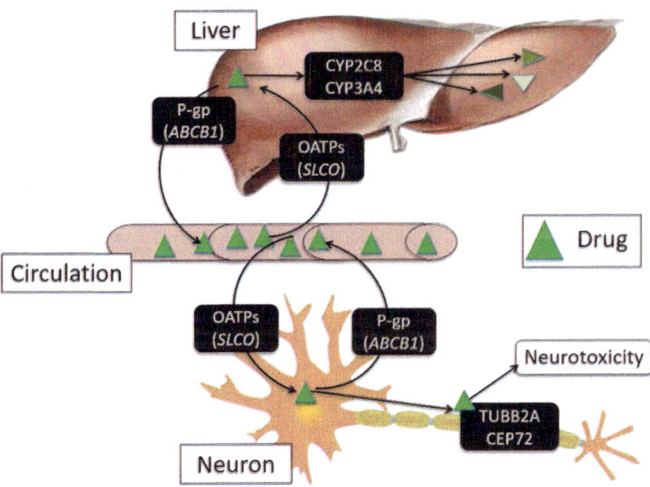

Fig. 2.3 Candidate pharmacogenetics often focus on genes involved in the distribution (ABCB1, SLCO), metabolism (CYP2C8, CYP3A4), or mechanism (TUBB2A, CEP72) of the drug of interest. Although many associations have been reported, none have been definitively validated and translated into clinical practice

Pharmacogenetic predictors of paclitaxel-induced neuropathy have been extensively studied [122, 123]. Early studies primarily focused on enzymes responsible for paclitaxel metabolism, including CYP2C8. *CYP2C8*3* may diminish paclitaxel metabolic activity [124] and several studies suggest that patients carrying

CYP2C8*3 have increased CIPN risk [9, 69–73], consistent with the increased CIPN risk in patients with higher systemic paclitaxel concentrations, covered later in this chapter. However, recent evidence suggests that *CYP2C8*3* carriers may have lower systemic paclitaxel exposure [125], similar to its effect on other drugs [126]; at this time CYP2C8 genotype should not be used to predict paclitaxel pharmacokinetics or CIPN risk. Paclitaxel is also metabolized by CYP3A4 and CYP3A5, which have diminished-activity variants (i.e., *CYP3A4*22* and *CYP3A5*3*) that have been reported to increase neuropathy risk [71, 74]. In addition to metabolic enzymes, paclitaxel is a substrate for several transporters including the efflux transporter P-glycoprotein (P-gp). P-gp is encoded by the *ABCB1* gene, which has several polymorphisms that have been reported to affect CIPN risk, including the common *ABCB1*2* one [72, 75–77]. Gene candidates have also been selected based on paclitaxel pharmacology or neuropathy pathophysiology, including studies indicating effects of polymorphisms in paclitaxel's molecular target, B-tubulin class IIa (*TUBB2A*) [77, 78]. In addition to these candidate-gene studies, several GWAS of paclitaxel-induced neuropathy have reported associations for polymorphisms in genes that had not been investigated in candidates gene studies [127–132]. Interestingly, many studies have reported associations for polymorphisms in genes related to inherited neuropathy conditions [73, 133–135], particularly polymorphisms in the EphrinA (*EPHA*) gene family [72, 77, 79–83].

While there has been much less research on the genetic predictors of docetaxel-induced neuropathy, similar to paclitaxel, most studies have focused on the genes involved in docetaxel pharmacokinetics including *CYP3A4/5*, *ABCB1*, and the uptake *SLCO1B3* transporter [136–139]. Unlike paclitaxel, though, the relationship between docetaxel pharmacokinetics and neuropathy has not been well established. Several studies have reported associations for polymorphisms in *GSTP1* [84–86], which could be due to the role of this enzyme in managing reactive oxygen species. The only GWAS of docetaxel-induced neuropathy identified a variant in *VAC14* [87], another gene related to hereditary neuropathy.

Candidate genes in platinum pharmacokinetics have included the enzymes responsible for secondary metabolism via glutathione conjugation (*GSTP1*, *GSTT1*, *GSTM1*) and uptake transporters (*ABCC1/2*, *ABCG2*). Polymorphisms in *GSTP1*, particularly the non-synonymous I105V variant, have been reported to be associated with platin-induced neuropathy [88–95] but a meta-analysis did not confirm the association [96]. Similarly, studies have reported that variants in the ABC transporters affect peripheral neuropathy risk [97–99] but this has also not been validated. Besides pharmacokinetics candidates, many studies have investigated polymorphisms in the genes responsible for DNA repair including *ERCC1/ERCC2* and *XRCC1/XRCC3*. *ERCC1* rs11615 (Asn118Asn) may be associated with CIPN risk per some studies [91, 100–104], but a meta-analysis did not confirm the association [105]. Another large ($n = 2183$) study using a panel of candidate genes reported a potential association for a non-synonymous (Asp317His) polymorphism in *DCLRE1A* [106], that has not been verified. A GWAS study reported potential associations for variants in several genes without strong biological rationale [140] that have failed attempted replication [141–143]. Finally, a GWAS of long-term neuropathy in cisplatin treated cancer survivors identified an association for *RPRD1B* [144] that requires validation in independent cohorts.

Studies have found that patients with an inactive variant in CYP3A5 (*CYP3A5*3*), the enzyme responsible for vincristine metabolism, have higher neuropathy risk [107]. However, replication of this association has been unsuccessful [108–110] and it is not clear that patient's carrying *CYP3A5*3* have higher systemic vincristine exposure [111] or that exposure is associated with neuropathy risk. Similar to the results for other agents, some studies report that variants in *ABCB1* affect vincristine pharmacokinetics [145] or neuropathy [109] but other studies have failed to replicate these findings [108–111]. A GWAS reported that pediatric patients who are homozygous carriers of the *CEP72* promoter variant rs904627 have increased risk of vincristine-induced neuropathy [112]. This finding was replicated in an analysis of young adults [113] and another pediatric cohort, followed by a successful meta-analysis [114]. Other studies have not replicated the association [115–117]. Attempted validation of this association is ongoing in a pharmacogenetics substudy embedded within the ongoing Total Therapy XVII trial (NCT03117751), in which patients are randomized to standard or rs904627-guided vincristine regimens.

There have been several discovery-phase pharmacogenetics studies of bortezomib-induced neuropathy that used large panels of candidate genes [146–149] or GWAS [150–152]. Thalidomide is used in combination with bortezomib in some regimens. There is evidence that *CYP2C19* pharmacogenetics affects thalidomide pharmacokinetics [118, 119] but not neuropathy [120]. Other studies using candidate genetic panels have reported associations with thalidomide-induced neuropathy [153–155]. No genetic predictors of peripheral neuropathy from bortezomib and/or thalidomide have been validated for clinical translation.

2.2.3 Early Indicators of Emerging CIPN

CIPN onset after a single cycle or early in treatment may be indicative of a trajectory toward severe CIPN by the end of treatment. Paclitaxel and oxaliplatin cause acute neurotoxicity symptoms that can present as early as the first cycle. Paclitaxel produces an acute pain syndrome that mimics arthralgia and myalgia, whereas oxaliplatin causes pain and dysesthesias in the hands and oropharynx upon exposure to cold temperatures [156]. Emergence of these acute toxicities early in treatment is indicative of eventual CIPN severity [157–159], more so with oxaliplatin than paclitaxel. Objective measures of neuronal function including quantitative sensory testing (QST) and nerve conduction studies (NCS) may also indicate concerning CIPN trajectories.

Oxaliplatin reduces sensory nerve action potential amplitudes [160] and the decrease at the midpoint of treatment can predict CIPN severity at the end of treatment [161]. Other objective measures that can be collected early in treatment that seem to predict eventual CIPN severity include spleen enlargement during oxaliplatin treatment [162], diminished vibration and deep tendon reflexes [163], depletion of nerve growth factor [163, 164], or increases in the neuron-specific protein neurofilament light chain [165–167]. Finally, CIPN severity midway through treatment with paclitaxel and oxaliplatin predicts severity at the end of treatment [161, 168]. Prospective studies are needed to determine whether, when, and how to intervene based on these early indicators to prevent CIPN. One study investigating QST to guide CIPN management found that QST changes may occur too late in treatment to be clinically useful [169].

2.2.4 Summary of Intrinsic Factors

Current data indicates higher CIPN risk in patients who are older, African American, diabetic, or have subclinical PN prior to treatment. Objective and subjective indicators of CIPN after a single or several cycles also predict CIPN severity but has not been successfully used to inform patient management. Associations for other potential biomarkers including metabolomics and genetics are in discovery phase and have not been validated for clinical translation. Future work in this area is needed to replicate previously reported associations in larger and more diverse samples, confirm effects with other neurotoxic regimens, and prospectively test interventions in high-risk patients to demonstrate how to use these predictive biomarkers to prevent CIPN and improve treatment outcomes.

2.3 Extrinsic Factors

This section discusses the factors that are extrinsic to the patient that may affect CIPN risk including patients' lifestyle choices and their neurotoxic chemotherapeutic regimen.

2.3.1 Lifestyle

A retrospective analysis indicates that women who have lower levels of moderate to vigorous physical activity (MVPA) prior to treatment are more likely to have long-term CIPN [13].

Another large study found that patients who spent more than 5 h/week on MVPA were 60% less likely to experience CIPN [14]. Prospective clinical trials have also indicated a protective effect of exercise on CIPN, further supporting this association and its potential clinical usefulness [170, 171].

Measuring alcohol and smoking intake is challenging, limiting reliability of study results. Some studies indicate alcohol use is a CIPN risk factor [31, 172, 173], while others suggest the opposite [7]. Smoking has also been reported to increase CIPN [7]. Additional research is needed to confirm which of these potentially modifiable behaviors predict CIPN risk.

2.3.2 Nutrition

Although nutritional status is a likely contributor to CIPN, the relationships between specific nutritional factors and CIPN are not fully understood [5, 174]. Micronutrients have a potential role in CIPN through anti-inflammatory, antioxidant, and neuroprotective mechanisms [34, 175]. Increased CIPN risk among diabetics, obese individuals, and regular alcohol users, all of whom tend to have worse nutrition, suggests that nutrition may be a contributing factor to CIPN [176]. Multivitamin use before and during taxane chemotherapy has been associated with CIPN protection [5, 174] but initiating an antioxidant (beta-carotene, selenium, vitamin C, vitamin E, and zinc) during treatment has been reported to increase CIPN risk [14] and decrease survival [177]. Patients with higher consumption of grains and citrus have also been reported to have increased risk of neuropathy from paclitaxel treatment [178]. Further work is needed to verify and mechanistically explain these findings, as the current evidence is insufficient to justify measuring nutrient levels and correcting nutrient insufficiencies for CIPN risk reduction.

2.3.3 Neurotoxic Treatment

2.3.3.1 Chemotherapy Regimen

Prospective randomized clinical trials comparing different doses with the same schedule [179, 180] or similar doses with differing numbers of cycles [181] consistently show that CIPN increase with cumulative treatment [182]. Although there are no established maximum cumulative dosing limits for neurotoxic chemotherapy, the cumulative dose threshold above which CIPN occurs has been estimated for several neurotoxic drugs [183–185] (Table 2.3). Chemotherapy dose reductions or delays are common in patients experiencing moderate CIPN to prevent further symptom progression [12, 186–188], meaning that some retrospective analyses find that patients with severe CIPN receive lower cumulative doses [187, 189, 190].

Table 2.3 Cumulative threshold dose for development of CIPN symptoms for neurotoxic chemotherapy drugs [183–185]

Chemotherapy class	Drug	Threshold dose
Taxane	Paclitaxel	>300 mg/m^2
	Docetaxel	>100 mg/m^2
Platinum	Oxaliplatin	>540 mg/m^2
	Cisplatin	>350 mg/m^2
Vinca alkaloid	Vincristine	>4 mg/m^2
Immunomodulatory/antiangiogenic agent	Thalidomide	>20 g
Proteasome inhibitor	Bortezomib	>16 mg/m^2

Note: These thresholds provide a general estimate of the cumulative dose at which CIPN occurs for different neurotoxic agents. They are not maximum cumulative dosing guidelines and do not reflect the substantial inter-patient variability in CIPN onset due to the other factors described in this chapter

There are also differences in CIPN rates between agents in the same class and between dosing regimens of the same drug, though direct comparison is difficult as regimens have different doses, frequencies, and durations of treatment. Smaller, weekly paclitaxel doses seem to be somewhat less neurotoxic, even though the weekly regimens have a slightly greater intensity (mg/m^2/day) and higher total dose administered [191–195]. Similar comparisons have been made for docetaxel but the differences in CIPN between the regimens are less distinct [196–198]. CIPN severity is dependent on platinum dose, frequency, and duration of administration [199–201], though oxaliplatin-induced cold sensitivity seems to be dose independent [202]. Among the vinca-alkaloids, vincristine has been identified as the most neurotoxic and vinorelbine the least neurotoxic [203]. Single vincristine doses above 2.0 mg [204] and cumulative doses exceeding 12 mg [205] are associated with greater CIPN. Cumulative doses of thalidomide up to 20 mg are associated with progressively increasing risk of CIPN [206].

2.3.3.2 Pharmacokinetics

Pharmacokinetics describes systemic drug concentrations within the body, including drug absorption, distribution, and elimination. The amount of drug in the body at a given time or the duration the drug remains within the body can be related to treatment outcomes (Fig. 2.4). This section summarizes the evidence supporting an association between systemic concentrations of neurotoxic drugs and CIPN, including prospective trials testing whether adjusting dosing to achieve therapeutic exposure improves treatment outcomes.

At least eight studies have reported that paclitaxel pharmacokinetics is associated with CIPN (Table 2.4). Larger systemic area under the curve (AUC) is associated with greater CIPN [74, 210, 212, 213] but AUC is unlikely to be clinically useful due to the need for repeated sampling. The amount of time the patient's systemic concentration remains above a threshold of 0.05μM ($T_{c > 0.05}$) can be estimated from a single sample collected the day after infusion [214] and has been associated with neuropathy in multiple studies [74, 209, 211]. Two prospective randomized clinical trials have demonstrated that exposure-guided paclitaxel dosing significantly reduces CIPN without diminishing efficacy in patients with non-small lung cancer [215, 216]. Exposure-guided paclitaxel dosing may improve outcomes but has not been widely adopted in clinical practice, potentially due to the inconvenience of collecting a next-day sample. A sample collected during or at the end of paclitaxel infusion would be much less inconvenient for patients. The maximum concentration (C_{max}) collected right at the end of infusion is associated with CIPN [74, 172, 208] but no prospective studies have individualized dosing based on this measure.

There is no evidence that docetaxel pharmacokinetics is associated with CIPN. Prospective exposure-guided docetaxel dosing studies reduce myelosuppression [217, 218] and may reduce overall toxicity [219], but reductions in CIPN have not been reported. Accumulation of platinum compounds in neural tissue is presumed to cause CIPN [201]. Residual systemic cisplatin concentrations have been associated with CIPN in cancer survivors [220, 221] but this effect is likely due to confounding by age [222]. Several studies have not identified a relationship between systemic

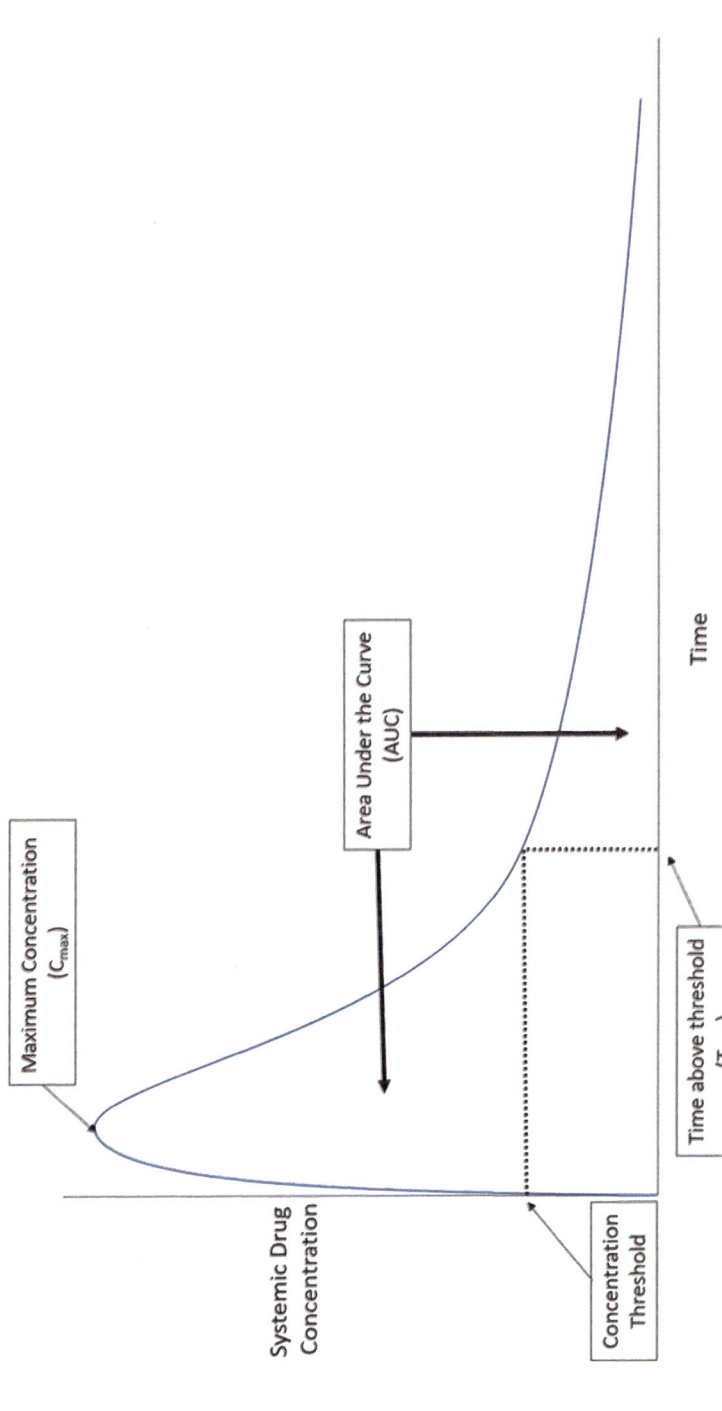

Fig. 2.4 The curve indicates the systemic paclitaxel concentration (*Y*-axis) over time (*X*-axis), beginning with a rapid increase during infusion, a rapid decrease post-infusion, and a long terminal elimination phase. Maximum concentration (C_{max}) is collected right at the end of infusion. Area under the curve (AUC) is estimated by the cumulative amount of drug in the body from the start of infusion until it has all been removed. Time above threshold ($T_c > xx$) describes the amount of time (hours) that the patient's systemic concentration remains above a threshold concentration (indicated by the hatched line). All three of these paclitaxel pharmacokinetic parameters have been associated with peripheral neuropathy severity

Table 2.4 Studies identifying association of paclitaxel pharmacokinetics with peripheral neuropathy

n	Tumor	Association	Ref
32	Mixed	C_{ss} correlated with PN	[207]
96	Ovarian	$T_{c > 0.05}$ higher in patients with PN	[208]
295	Mixed	$T_{c > 0.05}$ higher in patients with PN	[209]
261	Mixed	AUC, C_{max}, and $T_{c > 0.05}$ all associated with PN	[74]
9	Breast	AUC higher in patients with PN	[210]
24	Mixed	$T_{c > 0.05}$ higher in patients with PN	[211]
38	Ovarian	AUC correlated with PN	[212]
38	Ovarian	AUC correlated with PN inconvenience	[212]
60	Breast	C_{max} and $T_{c > 0.05}$ higher in patients with PN-induced treatment alteration	[172]

Acronyms: AUC, area under the curve; C_{max}, maximum concentration at end of infusion; C_{ss}, steady-state concentration during 24-h infusion; hr, hour; PN, peripheral neuropathy; $T_{c > 0.05}$, time systemic concentration remains above 0.05μM

platinum concentrations during treatment and CIPN [223–225]. Studies have reported greater CIPN in patients with higher systemic concentrations of vincristine [226, 227] or its major metabolite [107], however, this has not been consistently demonstrated [108, 145, 228, 229]. There has been little work investigating the association of CIPN with the pharmacokinetics of other neurotoxic agents, though published reports do not suggest an association for vinblastine [230, 231] or bortezomib [232].

2.3.4   Summary of Extrinsic Factors

Extrinsic factors including physical activity and nutrition may be modifiable risk factors for CIPN. The predominant predictor of CIPN is cumulative treatment with a neurotoxic chemotherapeutic agent.

There are differences in CIPN risk between agents within the same class and even between different regimens of the same agent. Finally, paclitaxel pharmacokinetics is strongly predictive of CIPN but the association is less clear for other neurotoxic agents. Additional research is needed to validate these findings and prospectively test strategies to optimize treatment to reduce CIPN while maintaining or enhancing treatment efficacy.

2.4   Conclusions

There are numerous intrinsic and extrinsic factors that influence risk of developing CIPN; however, these factors are not yet well understood. Risk factors may vary with use of different neurotoxic agents.

Certainly, the drugs, doses, and frequency of chemotherapy administration contribute to risk. However, there are several patient related factors that are likely to influence individual risk, including non-modifiable factors like age, race, and genetics, and modifiable factors such as unhealthy weight and lack of exercise.

References

1. Bandos H, Melnikow J, Rivera DR, Swain SM, Sturtz K, Fehrenbacher L, Wade JL 3rd, Brufsky AM, Julian TB, Margolese RG, McCarron EC, Ganz PA (2018) Long-term peripheral neuropathy in breast cancer patients treated with adjuvant chemotherapy: NRG oncology/ NSABP B-30. J Natl Cancer Inst 110(2). https://doi.org/10.1093/jnci/djx162
2. Kober K, Mastick J, Paul S, Topp K, Smoot B, Abrams G, Chen L, Conley Y, Chesney M, Bolla K, Mausisa G, Azor M, Wong M, Schumacher M, Levine J, Miaskowski C (2017) (431) Characteristics of chemotherapy induced neuropathy (CIN) in cancer survivors who received taxol. J Pain 18(4):S82. https://doi.org/10.1016/j.jpain.2017.02.281
3. Wong ML, Cooper BA, Paul SM, Abrams G, Topp K, Kober KM, Chesney MA, Mazor M, Schumacher MA, Conley YP, Levine JD, Miaskowski C (2019) Age-related differences in patient-reported and objective measures of chemotherapy-induced peripheral neuropathy among cancer survivors. Support Care Cancer 27(10):3905–3912. https://doi.org/10.1007/s00520-019-04695-3. Epub 2019 Feb 15
4. Bulls HW, Hoogland AI, Kennedy B, James BW, Arboleda BL, Apte S, Chon HS, Small BJ, Gonzalez BD, Jim HSL (2019) A longitudinal examination of associations between age and chemotherapy-induced peripheral neuropathy in patients with gynecologic cancer. Gynecol Oncol 152(2):310–315. https://doi.org/10.1016/j.ygyno.2018.12.002. Epub 2018 Dec 14
5. Hershman DL, Till C, Wright JD, Awad D, Ramsey SD, Barlow WE, Minasian LM, Unger J (2016) Comorbidities and risk of chemotherapy-induced peripheral neuropathy among participants 65 years or older in southwest oncology group clinical trials. J Clin Oncol 34(25):3014–3022. https://doi.org/10.1200/jco.2015.66.2346
6. Raphael MJ, Fischer HD, Fung K, Austin PC, Anderson GM, Booth CM, Singh S (2017) Neurotoxicity outcomes in a population-based cohort of elderly patients treated with adjuvant oxaliplatin for colorectal cancer. Clin Colorectal Cancer 16(4):397–404.e1. https://doi.org/10.1016/j.clcc.2017.03.013. Epub 2017 Mar 24
7. Molassiotis A, Cheng HL, Leung KT, Li YC, Wong KH, Au JSK, Sundar R, Chan A, Ng TR, Suen LKP, Chan CW, Yorke J, Lopez V (2019) Risk factors for chemotherapy-induced peripheral neuropathy in patients receiving taxane- and platinum-based chemotherapy. Brain Behav 9(6):e01312. https://doi.org/10.1002/brb3.1312. Epub 2019 May 7
8. Kalsi T, Babic-Illman G, Fields P, Hughes S, Maisey N, Ross P, Wang Y, Harari D (2014) The impact of low-grade toxicity in older people with cancer undergoing chemotherapy. Br J Cancer 111(12):2224–2228. https://doi.org/10.1038/bjc.2014.496. Epub 2014 Sep 30
9. Hertz DL, Roy S, Motsinger-Reif AA, Drobish A, Clark LS, McLeod HL, Carey LA, Dees EC (2013) CYP2C8*3 increases risk of neuropathy in breast cancer patients treated with paclitaxel. Ann Oncol 24(6):1472–1478. https://doi.org/10.1093/annonc/mdt018
10. Lewis MA, Zhao F, Jones D, Loprinzi CL, Brell J, Weiss M, Fisch MJ (2015) Neuropathic symptoms and their risk factors in medical oncology outpatients with colorectal vs. breast, lung, or prostate cancer: results from a prospective multicenter study. J Pain Symptom Manage 49(6):1016–1024. http://www.ncbi.nlm.nih.gov/entrez/query.fcgi?cmd=Retrieve&db=PubMed&dopt=Citation&list_uids=25596011
11. Schneider BP, Shen F, Jiang G, O'Neill A, Radovich M, Li L, Gardner L, Lai D, Foroud T, Sparano JA, Sledge GW Jr, Miller KD (2017) Impact of genetic ancestry on outcomes in ECOG-ACRIN-E5103. JCO Precis Oncol. https://doi.org/10.1200/PO.17.00059. Epub 2017 Aug 21

12. Speck RM, Sammel MD, Farrar JT, Hennessy S, Mao JJ, Stineman MG, DeMichele A (2013) Impact of chemotherapy-induced peripheral neuropathy on treatment delivery in nonmetastatic breast cancer. J Oncol Pract 9(5):e234–e240. https://doi.org/10.1200/jop. 2012.000863

13. Cox-Martin E, Trahan LH, Cox MG, Dougherty PM, Lai EA, Novy DM (2017) Disease burden and pain in obese cancer patients with chemotherapy-induced peripheral neuropathy. Support Care Cancer 25(6):1873–1879. https://doi.org/10.1007/s00520-017-3571-5. Epub 2017 Jan 26

14. Greenlee H, Hershman DL, Shi Z, Kwan ML, Ergas IJ, Roh JM, Kushi LH (2017) BMI, lifestyle factors and taxane-induced neuropathy in breast cancer patients: the pathways study. J Natl Cancer Inst 109(2). https://doi.org/10.1093/jnci/djw206

15. Moore DC, Ringley JT, Nix D, Muslimani A (2020) Impact of body mass index on the incidence of bortezomib-induced peripheral neuropathy in patients with newly diagnosed multiple myeloma. Clin Lymphoma Myeloma Leuk 20(3):168–173. https://doi.org/10.1016/j.clml.2019.08.012. Epub 2019 Sep 18

16. Petrovchich I, Kober KM, Wagner L, Paul SM, Abrams G, Chesney MA, Topp K, Smoot B, Schumacher M, Conley YP, Hammer M, Levine JD, Miaskowski C (2019) Deleterious effects of higher body mass index on subjective and objective measures of chemotherapy-induced peripheral neuropathy in cancer survivors. J Pain Symptom Manage 58(2):252–263. https://doi.org/10.1016/j.jpainsymman.2019.04.029. Epub 2019 Apr 30

17. Sajdyk TJ, Boyle FA, Foran KS, Tong Y, Pandya P, Smith EML, Ho RH, Wells E, Renbarger JL (2019) Obesity as a potential risk factor for vincristine-induced peripheral neuropathy. J Pediatr Hematol Oncol 18(10)

18. Winters-Stone KM, Horak F, Jacobs PG, Trubowitz P, Dieckmann NF, Stoyles S, Faithfull S (2017) Falls, functioning, and disability among women with persistent symptoms of chemotherapy-induced peripheral neuropathy. J Clin Oncol 35(23):2604–2612. https://doi.org/10.1200/jco.2016.71.3552

19. Bano N, Ikram R (2019) Effect of diabetes on neurological adverse effects and chemotherapy induced peripheral neuropathy in advanced colorectal cancer patients treated with different FOLFOX regimens. Pak J Pharm Sci 32(1):125–130

20. Nyrop KA, Deal AM, Reeder-Hayes KE, Shachar SS, Reeve BB, Basch E, Choi SK, Lee JT, Wood WA, Anders CK, Carey LA, Dees EC, Jolly TA, Kimmick GG, Karuturi MS, Reinbolt RE, Speca JC, Muss HB (2019) Patient-reported and clinician-reported chemotherapy-induced peripheral neuropathy in patients with early breast cancer: current clinical practice. Cancer 125 (17):2945–2954. https://doi.org/10.1002/cncr.32175. Epub 2019 May 15

21. Boyette-Davis JA, Eng C, Wang XS, Cleeland CS, Wendelschafer-Crabb G, Kennedy WR, Simone DA, Zhang H, Dougherty PM (2012) Subclinical peripheral neuropathy is a common finding in colorectal cancer patients prior to chemotherapy. Clin Cancer Res 18 (11):3180–3187. https://doi.org/10.1158/1078-0432.CCR-12-0205. Epub 2012 Apr 10

22. de Carvalho Barbosa M, Kosturakis AK, Eng C, Wendelschafer-Crabb G, Kennedy WR, Simone DA, Wang XS, Cleeland CS, Dougherty PM (2014) A quantitative sensory analysis of peripheral neuropathy in colorectal cancer and its exacerbation by oxaliplatin chemotherapy. Cancer Res 74(21):5955–5962. https://doi.org/10.1158/0008-5472.CAN-14-2060. Epub 2014 Sep 2

23. Griffith KA, Zhu S, Johantgen M, Kessler MD, Renn C, Beutler AS, Kanwar R, Ambulos N, Cavaletti G, Bruna J, Briani C, Argyriou AA, Kalofonos HP, Yerges-Armstrong LM, Dorsey SG (2017) Oxaliplatin-induced peripheral neuropathy and identification of unique severity groups in colorectal cancer. J Pain Symptom Manage 54(5):701–706.e1. https://doi.org/10.1016/j.jpainsymman.2017.07.033. Epub 2017 Jul 23

24. Reddy SM, Vergo MT, Paice JA, Kwon N, Helenowski IB, Benson AB, Mulcahy MF, Nimeiri HS, Harden RN (2016) Quantitative sensory testing at baseline and during cycle 1 oxaliplatin infusion detects subclinical peripheral neuropathy and predicts clinically overt chronic

neuropathy in gastrointestinal malignancies. Clin Colorectal Cancer 15(1):37–46. https://doi.org/10.1016/j.clcc.2015.07.001. Epub 2015 Jul 26

25. Vichaya EG, Wang XS, Boyette-Davis JA, Mendoza TR, He Z, Thomas SK, Shah N, Williams LA, Cleeland CS, Dougherty PM (2013) Subclinical pretreatment sensory deficits appear to predict the development of pain and numbness in patients with multiple myeloma undergoing chemotherapy. Cancer Chemother Pharmacol 71(6):1531–1540. https://doi.org/10.1007/s00280-013-2152-7

26. Wang XS, Shi Q, Dougherty PM, Eng C, Mendoza TR, Williams LA, Fogelman DR, Cleeland CS (2016) Prechemotherapy touch sensation deficits predict oxaliplatin-induced neuropathy in patients with colorectal cancer. Oncology 90(3):127–135. https://doi.org/10.1159/000443377. Epub 2016 Feb 17

27. Kennedy WR, Selim MM, Brink TS, Hodges JS, Wendelschafer-Crabb G, Foster SX, Nolano M, Provitera V, Simone DA (2011) A new device to quantify tactile sensation in neuropathy. Neurology 76(19):1642–1649. https://doi.org/10.1212/WNL.0b013e318219fadd

28. Wang J, Udd KA, Vidisheva A, Swift RA, Spektor TM, Bravin E, Ibrahim E, Treisman J, Masri M, Berenson JR (2016) Low serum vitamin D occurs commonly among multiple myeloma patients treated with bortezomib and/or thalidomide and is associated with severe neuropathy. Support Care Cancer 24(7):3105–3110. https://doi.org/10.1007/s00520-016-3126-1

29. Grim J, Ticha A, Hyspler R, Valis M, Zadak Z (2017) Selected risk nutritional factors for chemotherapy-induced polyneuropathy. Nutrients 9(6). https://doi.org/10.3390/nu9060535

30. Jennaro TS, Fang F, Kidwell KM, Smith EML, Vangipuram K, Burness ML, Griggs JJ, Van Poznak C, Hayes DF, Henry NL, Hertz DL (2020) Vitamin D deficiency increases severity of paclitaxel-induced peripheral neuropathy. Breast Cancer Res Treat 180(3):707–714. https://doi.org/10.1007/s10549-020-05584-8. Epub 2020 Mar 12

31. Vincenzi B, Frezza AM, Schiavon G, Spoto C, Silvestris N, Addeo R, Catalano V, Graziano F, Santini D, Tonini G (2013) Identification of clinical predictive factors of oxaliplatin-induced chronic peripheral neuropathy in colorectal cancer patients treated with adjuvant Folfox IV. Support Care Cancer 21(5):1313–1319. https://doi.org/10.1007/s00520-012-1667-5. Epub 2012 Nov 30

32. Shahriari-Ahmadi A, Fahimi A, Payandeh M, Sadeghi M (2015) Prevalence of oxaliplatin-induced chronic neuropathy and influencing factors in patients with colorectal cancer in Iran. Asian Pac J Cancer Prev 16(17):7603–7606. https://doi.org/10.7314/apjcp.2015.16.17.7603

33. Saito T, Okamura A, Inoue J, Makiura D, Doi H, Yakushijin K, Matsuoka H, Sakai Y, Ono R (2019) Anemia is a novel predictive factor for the onset of severe chemotherapy-induced peripheral neuropathy in lymphoma patients receiving rituximab plus cyclophosphamide, doxorubicin, vincristine, and prednisolone therapy. Oncol Res 27(4):469–474. https://doi.org/10.3727/096504018X15267574931782. Epub 2018 May 19

34. Gil Á, Plaza-Diaz J, Mesa MD (2018) Vitamin D: classic and novel actions. Ann Nutr Metab 72(2):87–95. https://doi.org/10.1159/000486536. Epub 2018 Jan 18

35. Lv WS, Zhao WJ, Gong SL, Fang DD, Wang B, Fu ZJ, Yan SL, Wang YG (2015) Serum 25-hydroxyvitamin D levels and peripheral neuropathy in patients with type 2 diabetes: a systematic review and meta-analysis. J Endocrinol Invest 38(5):513–518. https://doi.org/10.1007/s40618-014-0210-6

36. Kaiser MF, Heider U, Mieth M, Zang C, von Metzler I, Sezer O (2013) The proteasome inhibitor bortezomib stimulates osteoblastic differentiation of human osteoblast precursors via upregulation of vitamin D receptor signalling. Eur J Haematol 90(4):263–272. https://doi.org/10.1111/ejh.12069. Epub 2013 Feb 15

37. Solomon LR (2016) Functional vitamin B12 deficiency in advanced malignancy: implications for the management of neuropathy and neuropathic pain. Support Care Cancer 24(8):3489–3494. https://doi.org/10.1007/s00520-016-3175-5. Epub 2016 Mar 22

38. Dudeja S, Gupta S, Sharma S, Jain A, Jain P, Aneja S, Chandra J (2019) Incidence of vincristine induced neurotoxicity in children with acute lymphoblastic leukemia and its

correlation with nutritional deficiencies. Pediatr Hematol Oncol 36(6):344–351. https://doi. org/10.1080/08880018.2019.1637981. Epub 2019 Sep 13

39. Emiroglu C, Görpelioglu S, Aypak C (2019) The relationship between nutritional status, anemia and other vitamin deficiencies in the elderly receiving home care. J Nutr Health Aging 23(7):677–682. https://doi.org/10.1007/s12603-019-1215-9

40. Kirkland AE, Sarlo GL, Holton KF (2018) The role of magnesium in neurological disorders. Nutrients 10(6):730. https://doi.org/10.3390/nu10060730

41. Wesselink E, Winkels RM, van Baar H, Geijsen A, van Zutphen M, van Halteren HK, Hansson BME, Radema SA, de Wilt JHW, Kampman E, Kok DEG (2018) Dietary intake of magnesium or calcium and chemotherapy-induced peripheral neuropathy in colorectal cancer patients. Nutrients 10(4):398. https://doi.org/10.3390/nu10040398

42. Bove L, Picardo M, Maresca V, Jandolo B, Pace A (2001) A pilot study on the relation between cisplatin neuropathy and vitamin E. J Exp Clin Cancer Res 20(2):277–280

43. Gamelin L, Boisdron-Celle M, Delva R, Guérin-Meyer V, Ifrah N, Morel A, Gamelin E (2004) Prevention of oxaliplatin-related neurotoxicity by calcium and magnesium infusions: a retrospective study of 161 patients receiving oxaliplatin combined with 5-Fluorouracil and leucovorin for advanced colorectal cancer. Clin Cancer Res 10(12 Pt 1):4055–4061. https://doi.org/10.1158/1078-0432.CCR-03-0666

44. Grothey A, Nikcevich DA, Sloan JA, Kugler JW, Silberstein PT, Dentchev T, Wender DB, Novotny PJ, Chitaley U, Alberts SR, Loprinzi CL (2011) Intravenous calcium and magnesium for oxaliplatin-induced sensory neurotoxicity in adjuvant colon cancer: NCCTG N04C7. J Clin Oncol 29(4):421–427. https://doi.org/10.1200/JCO.2010.31.5911. Epub 2010 Dec 28

45. Loprinzi CL, Qin R, Dakhil SR, Fehrenbacher L, Flynn KA, Atherton P, Seisler D, Qamar R, Lewis GC, Grothey A (2014) Phase III randomized, placebo-controlled, double-blind study of intravenous calcium and magnesium to prevent oxaliplatin-induced sensory neurotoxicity (N08CB/Alliance). J Clin Oncol 32(10):997–1005. https://doi.org/10.1200/jco.2013.52.0536

46. Afonseca SO, Cruz FM, Cubero Dde I, Lera AT, Schindler F, Okawara M, Souza LF, Rodrigues NP, Giglio A (2013) Vitamin E for prevention of oxaliplatin-induced peripheral neuropathy: a pilot randomized clinical trial. Sao Paulo Med J 131(1):35–38. https://doi.org/10.1590/s1516-31802013000100006

47. Argyriou AA, Chroni E, Koutras A, Iconomou G, Papapetropoulos S, Polychronopoulos P, Kalofonos HP (2006) A randomized controlled trial evaluating the efficacy and safety of vitamin E supplementation for protection against cisplatin-induced peripheral neuropathy: final results. Support Care Cancer 14(11):1134–1140. https://doi.org/10.1007/s00520-006-0072-3. Epub 2006 Apr 19

48. Huang H, He M, Liu L, Huang L (2016) Vitamin E does not decrease the incidence of chemotherapy-induced peripheral neuropathy: a meta-analysis. Contemp Oncol (Pozn) 20 (3):237–241. https://doi.org/10.5114/wo.2016.61567. Epub 2016 Aug 4

49. Kottschade LA, Sloan JA, Mazurczak MA, Johnson DB, Murphy BP, Rowland KM, Smith DA, Berg AR, Stella PJ, Loprinzi CL (2011) The use of vitamin E for the prevention of chemotherapy-induced peripheral neuropathy: results of a randomized phase III clinical trial. Support Care Cancer 19(11):1769–1777. https://doi.org/10.1007/s00520-010-1018-3. Epub 2010 Oct 9

50. Pace A, Giannarelli D, Galiè E, Savarese A, Carpano S, Della Giulia M, Pozzi A, Silvani A, Gaviani P, Scaioli V, Jandolo B, Bove L, Cognetti F (2010) Vitamin E neuroprotection for cisplatin neuropathy: a randomized, placebo-controlled trial. Neurology 74(9):762–766. https://doi.org/10.1212/WNL.0b013e3181d5279e

51. Salehi Z, Roayaei M (2015) Effect of vitamin E on oxaliplatin-induced peripheral neuropathy prevention: a randomized controlled trial. Int J Prev Med 6:104. https://doi.org/10.4103/2008-7802.169021. eCollection 2015

52. Ghoreishi Z, Esfahani A, Djazayeri A, Djalali M, Golestan B, Ayromlou H, Hashemzade S, Asghari Jafarabadi M, Montazeri V, Keshavarz SA, Darabi M (2012) Omega-3 fatty acids are

protective against paclitaxel-induced peripheral neuropathy: a randomized double-blind placebo controlled trial. BMC Cancer 12:355. https://doi.org/10.1186/1471-2407-12-355

53. Bianchi G, Vitali G, Caraceni A, Ravaglia S, Capri G, Cundari S, Zanna C, Gianni L (2005) Symptomatic and neurophysiological responses of paclitaxel- or cisplatin-induced neuropathy to oral acetyl-L-carnitine. Eur J Cancer 41(12):1746–1750. https://doi.org/10.1016/j.ejca.2005.04.028

54. Callander N, Markovina S, Eickhoff J, Hutson P, Campbell T, Hematti P, Go R, Hegeman R, Longo W, Williams E, Asimakopoulos F, Miyamoto S (2014) Acetyl-L-carnitine (ALCAR) for the prevention of chemotherapy-induced peripheral neuropathy in patients with relapsed or refractory multiple myeloma treated with bortezomib, doxorubicin and low-dose dexamethasone: a study from the Wisconsin Oncology Network. Cancer Chemother Pharmacol 74 (4):875–882. https://doi.org/10.1007/s00280-014-2550-5. Epub 2014 Aug 29

55. Campone M, Berton-Rigaud D, Joly-Lobbedez F, Baurain JF, Rolland F, Stenzl A, Fabbro M, van Dijk M, Pinkert J, Schmelter T, de Bont N, Pautier P (2013) A double-blind, randomized phase II study to evaluate the safety and efficacy of acetyl-L-carnitine in the prevention of sagopilone-induced peripheral neuropathy. Oncologist 18(11):1190–1191. https://doi.org/10.1634/theoncologist.2013-0061. Epub 2013 Oct 8

56. Hershman DL, Unger JM, Crew KD, Minasian LM, Awad D, Moinpour CM, Hansen L, Lew DL, Greenlee H, Fehrenbacher L, Wade JL 3rd, Wong SF, Hortobagyi GN, Meyskens FL, Albain KS (2013) Randomized double-blind placebo-controlled trial of acetyl-L-carnitine for the prevention of taxane-induced neuropathy in women undergoing adjuvant breast cancer therapy. J Clin Oncol 31(20):2627–2633. https://doi.org/10.1200/jco.2012.44.8738

57. Hershman DL, Unger JM, Crew KD, Till C, Greenlee H, Minasian LM, Moinpour CM, Lew DL, Fehrenbacher L, Wade JL 3rd, Wong SF, Fisch MJ, Lynn Henry N, Albain KS (2018) Two-year trends of taxane-induced neuropathy in women enrolled in a randomized trial of acetyl-L-carnitine (SWOG S0715). J Natl Cancer Inst. https://doi.org/10.1093/jnci/djx259

58. Desideri I, Francolini G, Becherini C, Terziani F, Delli Paoli C, Olmetto E, Loi M, Perna M, Meattini I, Scotti V, Greto D, Bonomo P, Sulprizio S, Livi L (2017) Use of an alpha lipoic, methylsulfonylmethane and bromelain dietary supplement (Opera((R))) for chemotherapy-induced peripheral neuropathy management, a prospective study. Med Oncol 34(3):46. https://doi.org/10.1007/s12032-017-0907-4

59. Guo Y, Jones D, Palmer JL, Forman A, Dakhil SR, Velasco MR, Weiss M, Gilman P, Mills GM, Noga SJ, Eng C, Overman MJ, Fisch MJ (2014) Oral alpha-lipoic acid to prevent chemotherapy-induced peripheral neuropathy: a randomized, double-blind, placebo-controlled trial. Support Care Cancer 22(5):1223–1231. https://doi.org/10.1007/s00520-013-2075-1. Epub 2013 Dec 22

60. Loven D, Levavi H, Sabach G, Zart R, Andras M, Fishman A, Karmon Y, Levi T, Dabby R, Gadoth N (2009) Long-term glutamate supplementation failed to protect against peripheral neurotoxicity of paclitaxel. Eur J Cancer Care 18(1):78–83. https://doi.org/10.1111/j.1365-2354.2008.00996.x

61. Stubblefield MD, Vahdat LT, Balmaceda CM, Troxel AB, Hesdorffer CS, Gooch CL (2005) Glutamine as a neuroprotective agent in high-dose paclitaxel induced peripheral neuropathy: a clinical and electrophysiologic study. Clin Oncol (R Coll Radiol) 17(4):271–276

62. Vahdat L, Papadopoulos K, Lange D, Leuin S, Kaufman E, Donovan D, Frederick D, Bagiella E, Tiersten A, Nichols G, Garrett T, Savage D, Antman K, Hesdorffer CS, Balmaceda C (2001) Reduction of paclitaxel-induced peripheral neuropathy with glutamine. Clin Cancer Res 7(5):1192–1197. http://clincancerres.aacrjournals.org/content/7/5/1192.abstract

63. Wang WS, Lin JK, Lin TC, Chen WS, Jiang JK, Wang HS, Chiou TJ, Liu JH, Yen CC, Chen PM (2007) Oral glutamine is effective for preventing oxaliplatin-induced neuropathy in colorectal cancer patients. Oncologist 12(3):312–319. https://doi.org/10.1634/theoncologist.12-3-312

64. Loprinzi CL, Lacchetti C, Bleeker J, Cavaletti G, Chauhan C, Hertz DL, Kelley MR, Lavino A, Lustberg MB, Paice JA, Schneider BP, Lavoie Smith EM, Smith ML, Smith TJ,

Wagner-Johnston N, Hershman DL (2020) Prevention and management of chemotherapy-induced peripheral neuropathy in survivors of adult cancers: ASCO guideline update. J Clin Oncol 14(10):01399

65. O'Gorman A, Brennan L (2017) The role of metabolomics in determination of new dietary biomarkers. Proc Nutr Soc 76(3):295–302. https://doi.org/10.1017/S0029665116002974. Epub 2017 Jan 16

66. Sun Y, Kim JH, Vangipuram K, Hayes DF, Smith EML, Yeomans L, Henry NL, Stringer KA, Hertz DL (2018) Pharmacometabolomics reveals a role for histidine, phenylalanine, and threonine in the development of paclitaxel-induced peripheral neuropathy. Breast Cancer Res Treat. https://doi.org/10.1007/s10549-018-4862-3

67. Verma P, Devaraj J, Skiles JL, Sajdyk T, Ho RH, Hutchinson R, Wells E, Li L, Renbarger J, Cooper B, Ramkrishna D (2020) A metabolomics approach for early prediction of vincristine-induced peripheral neuropathy. Sci Rep 10(1):9659. https://doi.org/10.1038/s41598-020-66815-y

68. Chen EI, Crew KD, Trivedi M, Awad D, Maurer M, Kalinsky K, Koller A, Patel P, Kim Kim J, Hershman DL (2015) Identifying predictors of taxane-induced peripheral neuropathy using mass spectrometry-based proteomics technology. PLoS One 10(12):e0145816. https://doi.org/10.1371/journal.pone.0145816

69. Green H, Soderkvist P, Rosenberg P, Mirghani RA, Rymark P, Lundqvist EA, Peterson C (2009) Pharmacogenetic studies of Paclitaxel in the treatment of ovarian cancer. Basic Clin Pharmacol Toxicol 104(2):130–137. https://doi.org/10.1111/j.1742-7843.2008.00351.x

70. Hertz DL, Motsinger-Reif AA, Drobish A, Winham SJ, McLeod HL, Carey LA, Dees EC (2012) CYP2C8*3 predicts benefit/risk profile in breast cancer patients receiving neoadjuvant paclitaxel. Breast Cancer Res Treat 134(1):401–410. https://doi.org/10.1007/s10549-012-2054-0

71. Leskela S, Jara C, Leandro-Garcia L, Martinez A, Garcia-Donas J, Hernando S, Hurtado A, Vicario JCC, Montero-Conde C, Landa I, Lopez-Jimenez E, Cascon A, Milne RL, Robledo M, Rodriguez-Antona C (2011) Polymorphisms in cytochromes P450 2C8 and 3A5 are associated with paclitaxel neurotoxicity. Pharmacogenomics J 11(2):121–129. http://www.nature.com/tpj/journal/v11/n2/suppinfo/tpj201013s1.html. https://doi.org/10.1038/tpj.2010.13

72. Boora GK, Kanwar R, Kulkarni AA, Abyzov A, Sloan J, Ruddy KJ, Banck MS, Loprinzi CL, Beutler AS (2016) Testing of candidate single nucleotide variants associated with paclitaxel neuropathy in the trial NCCTG N08C1 (Alliance). Cancer Med 5(4):631–639. https://doi.org/10.1002/cam4.625

73. Lam SW, Frederiks CN, van der Straaten T, Honkoop AH, Guchelaar HJ, Boven E (2016) Genotypes of CYP2C8 and FGD4 and their association with peripheral neuropathy or early dose reduction in paclitaxel-treated breast cancer patients. Br J Cancer 115(11):1335–1342. https://doi.org/10.1038/bjc.2016.326

74. de Graan A-JM, Elens L, Sprowl JA, Sparreboom A, Friberg LE, van der Holt B, de Raaf PJ, de Bruijn P, Engels FK, Eskens FALM, Wiemer EAC, Verweij J, Mathijssen RHJ, van Schaik RHN (2013) CYP3A4*22 genotype and systemic exposure affect paclitaxel-induced neuro-toxicity. Clin Cancer Res 19(12):3316–3324. http://clincancerres.aacrjournals.org/content/19/12/3316.abstract. https://doi.org/10.1158/1078-0432.CCR-12-3786

75. Sissung TM, Mross K, Steinberg SM, Behringer D, Figg WD, Sparreboom A, Mielke S (2006) Association of ABCB1 genotypes with paclitaxel-mediated peripheral neuropathy and neutro-penia. Eur J Cancer 42(17):2893–2896. http://www.sciencedirect.com/science/article/B6T68-4KTVNX8-3/2/4073a0f32f3a1e153aa1ce4b00b3abbf

76. Tanabe Y, Shimizu C, Hamada A, Hashimoto K, Ikeda K, Nishizawa D, Hasegawa J, Shimomura A, Ozaki Y, Tamura N, Yamamoto H, Yunokawa M, Yonemori K, Takano T, Kawabata H, Tamura K, Fujiwara Y (2017) Paclitaxel-induced sensory peripheral neuropathy is associated with an ABCB1 single nucleotide polymorphism and older age in Japanese. Cancer Chemother Pharmacol 79(6):1179–1186. https://doi.org/10.1007/s00280-017-3314-9

77. Abraham JE, Guo Q, Dorling L, Tyrer J, Ingle S, Hardy R, Vallier AL, Hiller L, Burns R, Jones L, Bowden S, Dunn J, Poole C, Caldas C, Pharoah PDP, Earl HM (2014) Replication of genetic polymorphisms reported to be associated with taxane-related sensory neuropathy in patients with early breast cancer treated with paclitaxel. Clin Cancer Res 20(9):2466–2475

78. Leandro-García LJ, Leskelä S, Inglada-Pérez L, Landa I, de Cubas AA, Maliszewska A, Comino-Méndez I, Letón R, Gómez-Graña Á, Torres R, Ramírez JC, Álvarez S, Rivera J, Martínez C, Lozano ML, Cascón A, Robledo M, Rodríguez-Antona C (2012) Hematologic β-tubulin VI isoform exhibits genetic variability that influences paclitaxel toxicity. Cancer Res 72(18):4744–4752. http://cancerres.aacrjournals.org/content/72/18/4744.abstract. https://doi.org/10.1158/0008-5472.CAN-11-2861

79. Wilkinson DG (2001) Multiple roles of EPH receptors and ephrins in neural development. Nat Rev Neurosci 2(3):155–164. https://doi.org/10.1038/35058515

80. Leandro-Garcia LJ, Inglada-Perez L, Pita G, Hjerpe E, Leskela S, Jara C, Mielgo X, Gonzalez-Neira A, Robledo M, Avall-Lundqvist E, Green H, Rodriguez-Antona C (2013) Genome-wide association study identifies ephrin type A receptors implicated in paclitaxel induced peripheral sensory neuropathy. J Med Genet 50(9):599–605. https://doi.org/10.1136/jmedgenet-2012-101466

81. Kroetz DL, Baldwin RM, Owzar K, Jiang C, Zembutsu H, Kubo M, Nakamura Y, Shulman LN, Ratain MJ, Cancer and Leukemia Group B (2010) Inherited genetic variation in EPHA5, FGD4, and NRDG1 and paclitaxel (P)-induced peripheral neuropathy (PN): results from a genome-wide association study (GWAS) in CALGB 40101. ASCO Meeting Abstr 28 (15_suppl):3021. http://meeting.ascopubs.org/cgi/content/abstract/28/15_suppl/3021

82. Marcath LA, Kidwell KM, Vangipuram K, Gersch CL, Rae JM, Burness ML, Griggs JJ, Van Poznak C, Hayes DF, Smith EML, Henry NL, Beutler AS, Hertz DL (2020) Genetic variation in EPHA contributes to sensitivity to paclitaxel-induced peripheral neuropathy. Br J Clin Pharmacol 86(5):880–890. https://doi.org/10.1111/bcp.14192. Epub 2020 Feb 4

83. Apellaniz-Ruiz M, Tejero H, Inglada-Perez L, Sanchez-Barroso L, Gutierrez-Gutierrez G, Calvo I, Castelo B, Redondo A, Garcia-Donas J, Romero-Laorden N, Sereno M, Merino M, Curras-Freixes M, Montero-Conde C, Mancikova V, Avall-Lundqvist E, Green H, Al-Shahrour F, Cascon A, Robledo M, Rodriguez-Antona C (2017) Targeted sequencing reveals low-frequency variants in EPHA genes as markers of paclitaxel-induced peripheral neuropathy. Clin Cancer Res 23(5):1227–1235. https://doi.org/10.1158/1078-0432.ccr-16-0694

84. Mir O, Alexandre J, Tran A, Durand JP, Pons G, Treluyer JM, Goldwasser F (2009) Relationship between GSTP1 Ile105Val polymorphism and docetaxel-induced peripheral neuropathy: clinical evidence of a role of oxidative stress in taxane toxicity. Ann Oncol 20 (4):736–740. http://annonc.oxfordjournals.org/cgi/content/abstract/20/4/736. https://doi.org/10.1093/annonc/mdn698

85. Eckhoff L, Feddersen S, Knoop AS, Ewertz M, Bergmann TK (2015) Docetaxel-induced neuropathy: a pharmacogenetic case-control study of 150 women with early-stage breast cancer. Acta Oncologica (Stockholm, Sweden) 54(4):530–537. https://doi.org/10.3109/0284186X.2014.969846

86. van Rossum AGJ, Kok M, McCool D, Opdam M, Miltenburg NC, Mandjes IAM, van Leeuwen-Stok E, Imholz ALT, Portielje JEA, Bos M, van Bochove A, van Werkhoven E, Schmidt MK, Oosterkamp HM, Linn SC (2017) Independent replication of polymorphisms predicting toxicity in breast cancer patients randomized between dose-dense and docetaxel-containing adjuvant chemotherapy. Oncotarget 8(69):113531–113542. https://doi.org/10.18632/oncotarget.22697. eCollection 2017 Dec 26

87. Hertz DL, Owzar K, Lessans S, Wing C, Jiang C, Kelly WK, Patel J, Halabi S, Furukawa Y, Wheeler HE, Sibley AB, Lassiter C, Weisman L, Watson D, Krens SD, Mulkey F, Renn CL, Small EJ, Febbo PG, Shterev I, Kroetz DL, Friedman PN, Mahoney JF, Carducci MA, Kelley MJ, Nakamura Y, Kubo M, Dorsey SG, Dolan ME, Morris MJ, Ratain MJ, McLeod HL (2016) Pharmacogenetic discovery in CALGB (Alliance) 90401 and mechanistic validation of

a VAC14 polymorphism that increases risk of docetaxel-induced neuropathy. Clin Cancer Res 22(19):4890–4900. https://doi.org/10.1158/1078-0432.ccr-15-2823

88. Li QF, Yao RY, Liu KW, Lv HY, Jiang T, Liang J (2010) Genetic polymorphism of GSTP1: prediction of clinical outcome to oxaliplatin/5-FU-based chemotherapy in advanced gastric cancer. J Korean Med Sci 25(6):846–852. https://doi.org/10.3346/jkms.2010.25.6.846. Epub 2010 May 24

89. Goekkurt E, Al-Batran SE, Hartmann JT, Mogck U, Schuch G, Kramer M, Jaeger E, Bokemeyer C, Ehninger G, Stoehlmacher J (2009) Pharmacogenetic analyses of a phase III trial in metastatic gastroesophageal adenocarcinoma with fluorouracil and leucovorin plus either oxaliplatin or cisplatin: a study of the arbeitsgemeinschaft internistische onkologie. J Clin Oncol 27(17):2863–2873. https://doi.org/10.1200/JCO.2008.19.1718. Epub 2009 Mar 30

90. McLeod HL, Sargent DJ, Marsh S, Green EM, King CR, Fuchs CS, Ramanathan RK, Williamson SK, Findlay BP, Thibodeau SN, Grothey A, Morton RF, Goldberg RM (2010) Pharmacogenetic predictors of adverse events and response to chemotherapy in metastatic colorectal cancer: results from North American Gastrointestinal Intergroup Trial N9741. J Clin Oncol 28(20):3227–3233. https://doi.org/10.1200/JCO.2009.21.7943. Epub 2010 Jun 7

91. Inada M, Sato M, Morita S, Kitagawa K, Kawada K, Mitsuma A, Sawaki M, Fujita K, Ando Y (2010) Associations between oxaliplatin-induced peripheral neuropathy and polymorphisms of the ERCC1 and GSTP1 genes. Int J Clin Pharmacol Ther 48(11):729–734. https://doi.org/10.5414/cpp48729

92. Chen YC, Tzeng CH, Chen PM, Lin JK, Lin TC, Chen WS, Jiang JK, Wang HS, Wang WS (2010) Influence of GSTP1 I105V polymorphism on cumulative neuropathy and outcome of FOLFOX-4 treatment in Asian patients with colorectal carcinoma. Cancer Sci 101 (2):530–535. https://doi.org/10.1111/j.1349-7006.2009.01418.x. Epub 2009 Oct 28

93. Kumamoto K, Ishibashi K, Okada N, Tajima Y, Kuwabara K, Kumagai Y, Baba H, Haga N, Ishida H (2013) Polymorphisms of GSTP1, ERCC2 and TS-3'UTR are associated with the clinical outcome of mFOLFOX6 in colorectal cancer patients. Oncol Lett 6(3):648–654. https://doi.org/10.3892/ol.2013.1467. Epub 2013 July 15

94. Joerger M, Burgers SA, Baas P, Smit EF, Haitjema TJ, Bard MP, Doodeman VD, Smits PH, Vincent A, Huitema AD, Beijnen JH, Schellens JH (2012) Germline polymorphisms in patients with advanced nonsmall cell lung cancer receiving first-line platinum-gemcitabine chemotherapy: a prospective clinical study. Cancer 118(9):2466–2475. https://doi.org/10.1002/cncr.26562. Epub 2011 Sep 28

95. Oldenburg J, Kraggerud SM, Brydøy M, Cvancarova M, Lothe RA, Fossa SD (2007) Association between long-term neuro-toxicities in testicular cancer survivors and polymorphisms in glutathione-s-transferase-P1 and -M1, a retrospective cross sectional study. J Transl Med 5:70. https://doi.org/10.1186/1479-5876-5-70

96. Peng Z, Wang Q, Gao J, Ji Z, Yuan J, Tian Y, Shen L (2013) Association between GSTP1 Ile105Val polymorphism and oxaliplatin-induced neuropathy: a systematic review and meta-analysis. Cancer Chemother Pharmacol 72(2):305–314. https://doi.org/10.1007/s00280-013-2194-x. Epub 2013 May 22

97. Cecchin E, D'Andrea M, Lonardi S, Zanusso C, Pella N, Errante D, De Mattia E, Polesel J, Innocenti F, Toffoli G (2013) A prospective validation pharmacogenomic study in the adjuvant setting of colorectal cancer patients treated with the 5-fluorouracil/leucovorin/oxaliplatin (FOLFOX4) regimen. Pharmacogenomics J 13(5):403–409. https://doi.org/10.1038/tpj.2012.31. Epub 2012 Aug 7

98. Custodio A, Moreno-Rubio J, Aparicio J, Gallego-Plazas J, Yaya R, Maurel J, Higuera O, Burgos E, Ramos D, Calatrava A, Andrada E, López R, Moreno V, Madero R, Cejas P, Feliu J (2014) Pharmacogenetic predictors of severe peripheral neuropathy in colon cancer patients treated with oxaliplatin-based adjuvant chemotherapy: a GEMCAD group study. Ann Oncol 25(2):398–403. https://doi.org/10.1093/annonc/mdt546. Epub 2013 Dec 18

99. Johnson C, Pankratz VS, Velazquez AI, Aakre JA, Loprinzi CL, Staff NP, Windebank AJ, Yang P (2015) Candidate pathway-based genetic association study of platinum and platinum-

taxane related toxicity in a cohort of primary lung cancer patients. J Neurol Sci 349 (1–2):124–128. https://doi.org/10.1016/j.jns.2014.12.041. Epub 2015 Jan 5

100. Paré L, Marcuello E, Altés A, del Río E, Sedano L, Salazar J, Cortés A, Barnadas A, Baiget M (2008) Pharmacogenetic prediction of clinical outcome in advanced colorectal cancer patients receiving oxaliplatin/5-fluorouracil as first-line chemotherapy. Br J Cancer 99(7):1050–1055. https://doi.org/10.1038/sj.bjc.6604671. Epub 2008 Sep 16

101. Chua W, Goldstein D, Lee CK, Dhillon H, Michael M, Mitchell P, Clarke SJ, Iacopetta B (2009) Molecular markers of response and toxicity to FOLFOX chemotherapy in metastatic colorectal cancer. Br J Cancer 101(6):998–1004. https://doi.org/10.1038/sj.bjc.6605239. Epub 2009 Aug 11

102. Chang PM, Tzeng CH, Chen PM, Lin JK, Lin TC, Chen WS, Jiang JK, Wang HS, Wang WS (2009) ERCC1 codon 118 C→T polymorphism associated with ERCC1 expression and outcome of FOLFOX-4 treatment in Asian patients with metastatic colorectal carcinoma. Cancer Sci 100(2):278–283. https://doi.org/10.1111/j.1349-7006.2008.01031.x

103. Kjersem JB, Thomsen M, Guren T, Hamfjord J, Carlsson G, Gustavsson B, Ikdahl T, Indrebø G, Pfeiffer P, Lingjærde O, Tveit KM, Wettergren Y, Kure EH (2016) AGXT and ERCC2 polymorphisms are associated with clinical outcome in metastatic colorectal cancer patients treated with 5-FU/oxaliplatin. Pharmacogenomics J 16(3):272–279. https://doi.org/10.1038/tpj.2015.54. Epub 2015 Aug 11

104. Lee KH, Chang HJ, Han SW, Oh DY, Im SA, Bang YJ, Kim SY, Lee KW, Kim JH, Hong YS, Kim TW, Park YS, Kang WK, Shin SJ, Ahn JB, Kang GH, Jeong SY, Park KJ, Park JG, Kim TY (2013) Pharmacogenetic analysis of adjuvant FOLFOX for Korean patients with colon cancer. Cancer Chemother Pharmacol 71(4):843–851. https://doi.org/10.1007/s00280-013-2075-3. Epub 2013 Jan 13

105. Qian YY, Liu XY, Wu Q, Song X, Chen XF, Liu YQ, Pei D, Shen LZ, Shu YQ (2014) The ERCC1 C118T polymorphism predicts clinical outcomes of colorectal cancer patients receiving oxaliplatin-based chemotherapy: a meta-analysis based on 22 studies. Asian Pac J Cancer Prev 15(19):8383–8390. https://doi.org/10.7314/apjcp.2014.15.19.8383

106. Madi A, Fisher D, Maughan TS, Colley JP, Meade AM, Maynard J, Humphreys V, Wasan H, Adams RA, Idziaszczyk S, Harris R, Kaplan RS, Cheadle JP (2018) Pharmacogenetic analyses of 2183 patients with advanced colorectal cancer; potential role for common dihydropyrimidine dehydrogenase variants in toxicity to chemotherapy. Eur J Cancer 102:31–39. https://doi.org/10.1016/j.ejca.2018.07.009. Epub 2018 Aug 13

107. Egbelakin A, Ferguson MJ, MacGill EA, Lehmann AS, Topletz AR, Quinney SK, Li L, McCammack KC, Hall SD, Renbarger JL (2011) Increased risk of vincristine neurotoxicity associated with low CYP3A5 expression genotype in children with acute lymphoblastic leukemia. Pediatr Blood Cancer 56(3):361–367. https://doi.org/10.1002/pbc.22845

108. Guilhaumou R, Solas C, Bourgarel-Rey V, Quaranta S, Rome A, Simon N, Lacarelle B, Andre N (2011) Impact of plasma and intracellular exposure and CYP3A4, CYP3A5, and ABCB1 genetic polymorphisms on vincristine-induced neurotoxicity. Cancer Chemother Pharmacol 68(6):1633–1638. https://doi.org/10.1007/s00280-011-1745-2

109. Ceppi F, Langlois-Pelletier C, Gagne V, Rousseau J, Ciolino C, De Lorenzo S, Kevin KM, Cijov D, Sallan SE, Silverman LB, Neuberg D, Kutok JL, Sinnett D, Laverdiere C, Krajinovic M (2014) Polymorphisms of the vincristine pathway and response to treatment in children with childhood acute lymphoblastic leukemia. Pharmacogenomics 15(8):1105–1116. https://doi.org/10.2217/pgs.14.68

110. Hartman A, van Schaik RH, van der Heiden IP, Broekhuis MJ, Meier M, den Boer ML, Pieters R (2010) Polymorphisms in genes involved in vincristine pharmacokinetics or pharmacodynamics are not related to impaired motor performance in children with leukemia. Leuk Res 34(2):154–159. https://doi.org/10.1016/j.leukres.2009.04.027

111. Guilhaumou R, Simon N, Quaranta S, Verschuur A, Lacarelle B, Andre N, Solas C (2011) Population pharmacokinetics and pharmacogenetics of vincristine in paediatric patients treated for solid tumour diseases. Cancer Chemother Pharmacol 68(5):1191–1198. https://doi.org/10.1007/s00280-010-1541-4

112. Diouf B, Crews KR, Lew G, Pei D, Cheng C, Bao J, Zheng JJ, Yang W, Fan Y, Wheeler HE, Wing C, Delaney SM, Komatsu M, Paugh SW, McCorkle JR, Lu X, Winick NJ, Carroll WL, Loh ML, Hunger SP, Devidas M, Pui CH, Dolan ME, Relling MV, Evans WE (2015) Association of an inherited genetic variant with vincristine-related peripheral neuropathy in children with acute lymphoblastic leukemia. JAMA 313(8):815–823. https://doi.org/10.1001/jama.2015.0894

113. Stock W, Diouf B, Crews KR, Pei D, Cheng C, Laumann K, Mandrekar SJ, Luger S, Advani A, Stone RM, Larson RA, Evans WE (2017) An inherited genetic variant in CEP72 promoter predisposes to vincristine-induced peripheral neuropathy in adults with acute lymphoblastic leukemia. Clin Pharmacol Ther 101(3):391–395. https://doi.org/10.1002/cpt.506

114. Wright GEB, Amstutz U, Drogemoller BI, Shih J, Rassekh SR, Hayden MR, Carleton BC, Ross CJD (2018) Pharmacogenomics of vincristine-induced peripheral neuropathy implicates pharmacokinetic and inherited neuropathy genes. Clin Pharmacol Ther. https://doi.org/10.1002/cpt.1179

115. Gutierrez-Camino A, Martin-Guerrero I, Lopez-Lopez E, Echebarria-Barona A, Zabalza I, Ruiz I, Guerra-Merino I, Garcia-Orad A (2015) Lack of association of the CEP72 rs924607 TT genotype with vincristine-related peripheral neuropathy during the early phase of pediatric acute lymphoblastic leukemia treatment in a Spanish population. Pharmacogenet Genomics. https://doi.org/10.1097/FPC.0000000000000191

116. Zgheib NK, Ghanem KM, Tamim H, Aridi C, Shahine R, Tarek N, Saab R, Abboud MR, El-Solh H, Muwakkit SA (2018) Genetic polymorphisms in candidate genes are not associated with increased vincristine-related peripheral neuropathy in Arab children treated for acute childhood leukemia: a single institution study. Pharmacogenet Genomics 28(8):189–195. https://doi.org/10.1097/fpc.0000000000000345

117. Li L, Sajdyk T, Smith EML, Chang CW, Li C, Ho RH, Hutchinson R, Wells E, Skiles JL, Winick N, Martin PL, Renbarger JL (2019) Genetic variants associated with vincristine-induced peripheral neuropathy in two populations of children with acute lymphoblastic leukemia. Clin Pharmacol Ther 105(6):1421–1428. https://doi.org/10.1002/cpt.1324

118. Ando Y, Price DK, Dahut WL, Cox MC, Reed E, Figg WD (2002) Pharmacogenetic associations of CYP2C19 genotype with in vivo metabolisms and pharmacological effects of thalidomide. Cancer Biol Ther 1(6):669–673. https://doi.org/10.4161/cbt.318

119. Matsuzawa N, Nakamura K, Matsuda M, Ishida F, Ohmori S (2012) Influence of cytochrome P450 2C19 gene variations on pharmacokinetic parameters of thalidomide in Japanese patients. Biol Pharm Bull 35(3):317–320. https://doi.org/10.1248/bpb.35.317

120. Feng R, Xu PP, Chen BL, Mao R, Zhang SH, Qiu Y, Zeng ZR, Chen MH, He Y (2020) CYP2C19 polymorphism has no correlation with the efficacy and safety of thalidomide in the treatment of immune-related bowel disease. J Dig Dis 21(2):98–103. https://doi.org/10.1111/1751-2980.12842

121. Cliff J, Jorgensen AL, Lord R, Azam F, Cossar L, Carr DF, Pirmohamed M (2017) The molecular genetics of chemotherapy-induced peripheral neuropathy: a systematic review and meta-analysis. Crit Rev Oncol Hematol 120:127–140. https://doi.org/10.1016/j.critrevonc.2017.09.009. Epub 2017 Sep 25

122. Frederiks CN, Lam SW, Guchelaar HJ, Boven E (2015) Genetic polymorphisms and paclitaxel- or docetaxel-induced toxicities: a systematic review. Cancer Treat Rev 41(10):935–950. https://doi.org/10.1016/j.ctrv.2015.10.010

123. Hertz DL (2013) Germline pharmacogenetics of paclitaxel for cancer treatment. Pharmacogenomics 14(9):1065–1084. https://doi.org/10.2217/pgs.13.90

124. Bergmann TK, Brasch-Andersen C, Green H, Mirza M, Pedersen RS, Nielsen F, Skougaard K, Wihl J, Keldsen N, Damkier P, Friberg LE, Peterson C, Vach W, Karlsson MO, Brosen K (2011) Impact of CYP2C8*3 on paclitaxel clearance: a population pharmacokinetic and pharmacogenomic study in 93 patients with ovarian cancer. Pharmacogenomics J 11(2):113–120. https://doi.org/10.1038/tpj.2010.19

125. Marcath LA, Kidwell KM, Robinson AC, Vangipuram K, Burness ML, Griggs JJ, Poznak CV, Schott AF, Hayes DF, Henry NL, Hertz DL (2019) Patients carrying CYP2C8*3 have shorter systemic paclitaxel exposure. Pharmacogenomics 20(2):95–104. https://doi.org/10.2217/pgs-2018-0162

126. Marcath LA, Pasternak AL, Hertz DL (2019) Challenges to assess substrate-dependent allelic effects in CYP450 enzymes and the potential clinical implications. Pharmacogenomics J 19 (6):501–515. https://doi.org/10.1038/s41397-019-0105-1. Epub 2019 Oct 15

127. Schneider BP, Li L, Miller K, Flockhart D, Radovich M, Hancock BA, Kassem N, Foroud T, Koller DL, Badve SS, Li Z, Partridge AH, O'Neill AM, Sparano JA, Dang CT, Northfelt DW, Smith ML, Railey E, Sledge GW (2011) Genetic associations with taxane-induced neuropathy by a genome-wide association study (GWAS) in E5103. ASCO Meeting Abstr 29 (15_suppl):1000. http://meeting.ascopubs.org/cgi/content/abstract/29/15_suppl/1000

128. Bergmann TK, Vach W, Feddersen S, Eckhoff L, Gréen H, Herrstedt J, Brosen K (2012) GWAS-based association between RWDD3 and TECTA variants and paclitaxel induced neuropathy could not be confirmed in Scandinavian ovarian cancer patients. Acta Oncol 52 (4):871–874. https://doi.org/10.3109/0284186X.2012.707787

129. Chua KC, Xiong C, Ho C, Mushiroda T, Jiang C, Mulkey F, Lai D, Schneider BP, Rashkin SR, Witte JS, Friedman PN, Ratain MJ, McLeod HL, Rugo HS, Shulman LN, Kubo M, Owzar K, Kroetz DL (2020) Genome-wide meta-analysis validates a role for S1PR1 in microtubule targeting agent-induced sensory peripheral neuropathy. Clin Pharmacol Ther 20 (10)

130. Kulkarni AA, Boora G, Kanwar R, Ruddy KJ, Banck MS, Le-Lindqwister N, Therneau TM, Loprinzi CL, Beutler AS (2015) RWDD3 and TECTA variants not linked to paclitaxel induced peripheral neuropathy in North American trial Alliance N08C1. Acta Oncol 54 (8):1227–1229. https://doi.org/10.3109/0284186x.2014.985388

131. Schneider BP, Li L, Radovich M, Shen F, Miller KD, Flockhart DA, Jiang G, Vance G, Gardner L, Vatta M, Bai S, Lai D, Koller D, Zhao F, O'Neill A, Smith ML, Railey E, White C, Partridge A, Sparano J, Davidson NE, Foroud T, Sledge GW Jr (2015) Genome-wide association studies for taxane-induced peripheral neuropathy in ECOG-5103 and ECOG-1199. Clin Cancer Res 21(22):5082–5091. https://doi.org/10.1158/1078-0432.CCR-15-0586

132. Sucheston-Campbell LE, Clay-Gilmour AI, Barlow WE, Budd GT, Stram DO, Haiman CA, Sheng X, Yan L, Zirpoli G, Yao S, Jiang C, Owzar K, Hershman D, Albain KS, Hayes DF, Moore HC, Hobday TJ, Stewart JA, Rizvi A, Isaacs C, Salim M, Gralow JR, Hortobagyi GN, Livingston RB, Kroetz DL, Ambrosone CB (2018) Genome-wide meta-analyses identifies novel taxane-induced peripheral neuropathy-associated loci. Pharmacogenet Genomics 28 (2):49–55. https://doi.org/10.1097/fpc.0000000000000318

133. Baldwin RM, Owzar K, Zembutsu H, Chhibber A, Kubo M, Jiang C, Watson D, Eclov RJ, Mefford J, McLeod HL, Friedman PN, Hudis CA, Winer EP, Jorgenson EM, Witte JS, Shulman LN, Nakamura Y, Ratain MJ, Kroetz DL (2012) A genome-wide association study identifies novel loci for paclitaxel-induced sensory peripheral neuropathy in CALGB 40101. Clin Cancer Res 18(18):5099–5109. http://www.ncbi.nlm.nih.gov/entrez/query.fcgi?cmd=Retrieve&db=PubMed&dopt=Citation&list_uids=22843789. https://doi.org/10.1158/1078-0432.CCR-12-1590

134. Chen Y, Fang F, Kidwell KM, Vangipuram K, Marcath LA, Gersch CL, Rae JM, Hayes DF, Lavoie Smith EM, Henry NL, Beutler AS, Hertz DL (2020) Genetic variation in Charcot-Marie-Tooth genes contributes to sensitivity to paclitaxel-induced peripheral neuropathy. Pharmacogenomics 23(10):gs-2020

135. Chua KC, Kroetz DL (2017) Genetic advances uncover mechanisms of chemotherapy-induced peripheral neuropathy. Clin Pharmacol Ther 101(4):450–452. https://doi.org/10.1002/cpt.590

136. Bosch TM, Huitema ADR, Doodeman VD, Jansen R, Witteveen E, Smit WM, Jansen RL, van Herpen CM, Soesan M, Beijnen JH, Schellens JHM (2006) Pharmacogenetic screening of CYP3A and ABCB1 in relation to population pharmacokinetics of docetaxel. Clin Cancer Res 12(19):5786–5793. http://clincancerres.aacrjournals.org/content/12/19/5786.abstract. https://doi.org/10.1158/1078-0432.CCR-05-2649

137. Chew SC, Singh O, Chen X, Ramasamy RD, Kulkarni T, Lee EJ, Tan EH, Lim WT, Chowbay B (2011) The effects of CYP3A4, CYP3A5, ABCB1, ABCC2, ABCG2 and SLCO1B3 single nucleotide polymorphisms on the pharmacokinetics and pharmacodynamics of docetaxel in nasopharyngeal carcinoma patients. Cancer Chemother Pharmacol 67(6):1471–1478. https://doi.org/10.1007/s00280-011-1625-9

138. Kus T, Aktas G, Kalender ME, Demiryurek AT, Ulasli M, Oztuzcu S, Sevinc A, Kul S, Camci C (2016) Polymorphism of CYP3A4 and ABCB1 genes increase the risk of neuropathy in breast cancer patients treated with paclitaxel and docetaxel. Onco Targets Ther 9:5073–5080. https://doi.org/10.2147/OTT.S106574. eCollection 2016

139. Sissung TM, Baum CE, Deeken J, Price DK, Aragon-Ching J, Steinberg SM, Dahut W, Sparreboom A, Figg WD (2008) ABCB1 genetic variation influences the toxicity and clinical outcome of patients with androgen-independent prostate cancer treated with docetaxel. Clin Cancer Res 14(14):4543–4549. http://clincancerres.aacrjournals.org/content/14/14/4543.abstract. https://doi.org/10.1158/1078-0432.CCR-07-4230

140. Won HH, Lee J, Park JO, Park YS, Lim HY, Kang WK, Kim JW, Lee SY, Park SH (2012) Polymorphic markers associated with severe oxaliplatin-induced, chronic peripheral neuropathy in colon cancer patients. Cancer 118(11):2828–2836. https://doi.org/10.1002/cncr.26614. Epub 2011 Oct 21

141. Kanai M, Kawaguchi T, Kotaka M, Shinozaki K, Touyama T, Manaka D, Ishigure K, Hasegawa J, Munemoto Y, Matsui T, Takagane A, Ishikawa H, Matsumoto S, Sakamoto J, Saji S, Yoshino T, Ohtsu A, Watanabe T, Matsuda F (2016) Large-scale prospective pharmacogenomics study of oxaliplatin-induced neuropathy in colon cancer patients enrolled in the JFMC41-1001-C2 (JOIN Trial). Ann Oncol 27(6):1143–1148. https://doi.org/10.1093/annonc/mdw074. Epub 2016 Feb 18

142. Oguri T, Mitsuma A, Inada-Inoue M, Morita S, Shibata T, Shimokata T, Sugishita M, Nakayama G, Uehara K, Hasegawa Y, Ando Y (2013) Genetic polymorphisms associated with oxaliplatin-induced peripheral neurotoxicity in Japanese patients with colorectal cancer. Int J Clin Pharmacol Ther 51(6):475–481. https://doi.org/10.5414/CP201851

143. Terrazzino S, Argyriou AA, Cargnin S, Antonacopoulou AG, Briani C, Bruna J, Velasco R, Alberti P, Campagnolo M, Lonardi S, Cortinovis D, Cazzaniga M, Santos C, Kalofonos HP, Canonico PL, Genazzani AA, Cavaletti G (2015) Genetic determinants of chronic oxaliplatin-induced peripheral neurotoxicity: a genome-wide study replication and meta-analysis. J Peripher Nerv Syst 20(1):15–23. https://doi.org/10.1111/jns.12110

144. Dolan ME, El Charif O, Wheeler HE, Gamazon ER, Ardeshir-Rouhani-Fard S, Monahan P, Feldman DR, Hamilton RJ, Vaughn DJ, Beard CJ, Fung C, Kim J, Fossa SD, Hertz DL, Mushiroda T, Kubo M, Einhorn LH, Cox NJ, Travis LB (2017) Clinical and genome-wide analysis of cisplatin-induced peripheral neuropathy in survivors of adult-onset cancer. Clin Cancer Res 23(19):5757–5768. https://doi.org/10.1158/1078-0432.ccr-16-3224

145. Plasschaert SL, Groninger E, Boezen M, Kema I, de Vries EG, Uges D, Veerman AJ, Kamps WA, Vellenga E, de Graaf SS, de Bont ES (2004) Influence of functional polymorphisms of the MDR1 gene on vincristine pharmacokinetics in childhood acute lymphoblastic leukemia. Clin Pharmacol Ther 76(3):220–229. https://doi.org/10.1016/j.clpt.2004.05.007

146. Broyl A, Corthals SL, Jongen JL, van der Holt B, Kuiper R, de Knegt Y, van Duin M, el Jarari L, Bertsch U, Lokhorst HM, Durie BG, Goldschmidt H, Sonneveld P (2010) Mechanisms of peripheral neuropathy associated with bortezomib and vincristine in patients with newly diagnosed multiple myeloma: a prospective analysis of data from the HOVON-65/GMMG-HD4 trial. Lancet Oncol 11(11):1057–1065. https://doi.org/10.1016/S1470-2045(10)70206-0. Epub 2010 Sep 21

147. Campo C, Da Silva Filho MI, Weinhold N, Goldschmidt H, Hemminki K, Merz M, Försti A (2017) Genetic susceptibility to bortezomib-induced peripheral neuroropathy: replication of the reported candidate susceptibility loci. Neurochem Res 42(3):925–931. https://doi.org/10.1007/s11064-016-2007-9. Epub 2016 July 16

148. Corthals SL, Kuiper R, Johnson DC, Sonneveld P, Hajek R, van der Holt B, Magrangeas F, Goldschmidt H, Morgan GJ, Avet-Loiseau H (2011) Genetic factors underlying the risk of bortezomib induced peripheral neuropathy in multiple myeloma patients. Haematologica 96 (11):1728–1732. https://doi.org/10.3324/haematol.2011.041434. Epub 2011 Jul 26

149. Favis R, Sun Y, van de Velde H, Broderick E, Levey L, Meyers M, Mulligan G, Harousseau JL, Richardson PG, Ricci DS (2011) Genetic variation associated with bortezomib-induced peripheral neuropathy. Pharmacogenet Genomics 21(3):121–129. https://doi.org/10.1097/FPC.0b013e3283436b45

150. Campo C, da Silva Filho MI, Weinhold N, Mahmoudpour SH, Goldschmidt H, Hemminki K, Merz M, Försti A (2018) Bortezomib-induced peripheral neuropathy: a genome-wide association study on multiple myeloma patients. Hematol Oncol 36(1):232–237. https://doi.org/10.1002/hon.2391. Epub 2017 Mar 20

151. García-Sanz R, Corchete LA, Alcoceba M, Chillon MC, Jiménez C, Prieto I, García-Álvarez M, Puig N, Rapado I, Barrio S, Oriol A, Blanchard MJ, de la Rubia J, Martínez R, Lahuerta JJ, González Díaz M, Mateos MV, San Miguel JF, Martínez-López J, Sarasquete ME (2017) Prediction of peripheral neuropathy in multiple myeloma patients receiving bortezomib and thalidomide: a genetic study based on a single nucleotide polymorphism array. Hematol Oncol 35(4):746–751. https://doi.org/10.1002/hon.2337. Epub 2016 Sep 8

152. Magrangeas F, Kuiper R, Avet-Loiseau H, Gouraud W, Guérin-Charbonnel C, Ferrer L, Aussem A, Elghazel H, Suhard J, Sakissian H, Attal M, Munshi NC, Sonneveld P, Dumontet C, Moreau P, van Duin M, Campion L, Minvielle S (2016) A genome-wide association study identifies a novel locus for bortezomib-induced peripheral neuropathy in European patients with multiple myeloma. Clin Cancer Res 22(17):4350–4355. https://doi.org/10.1158/1078-0432.CCR-15-3163. Epub 2016 Apr 8

153. Cibeira MT, de Larrea CF, Navarro A, Díaz T, Fuster D, Tovar N, Rosiñol L, Monzó M, Bladé J (2011) Impact on response and survival of DNA repair single nucleotide polymorphisms in relapsed or refractory multiple myeloma patients treated with thalidomide. Leuk Res 35 (9):1178–1183. https://doi.org/10.1016/j.leukres.2011.02.009. Epub 2011 Mar 23

154. Han M, Murugesan A, Bahlis NJ, Song K, White D, Chen C, Seftel MD, Howsen-Jan K, Reece D, Stewart K, Xie Y, Hay AE, Shepherd L, Djurfeldt M, Zhu L, Meyer RM, Chen BE, Reiman T (2016) A pharmacogenetic analysis of the Canadian Cancer Trials Group MY.10 clinical trial of maintenance therapy for multiple myeloma. Blood 128(5):732–735. https://doi.org/10.1182/blood-2016-06-716902. Epub 2016 Jun 23

155. Johnson DC, Corthals SL, Walker BA, Ross FM, Gregory WM, Dickens NJ, Lokhorst HM, Goldschmidt H, Davies FE, Durie BG, Van Ness B, Child JA, Sonneveld P, Morgan GJ (2011) Genetic factors underlying the risk of thalidomide-related neuropathy in patients with multiple myeloma. J Clin Oncol 29(7):797–804. https://doi.org/10.1200/JCO.2010.28.0792. Epub 2011 Jan 18

156. Pachman DR, Qin R, Seisler DK, Smith EM, Beutler AS, Ta LE, Lafky JM, Wagner-Johnston ND, Ruddy KJ, Dakhil S, Staff NP, Grothey A, Loprinzi CL (2015) Clinical course of oxaliplatin-induced neuropathy: results from the randomized phase III trial N08CB (Alliance). J Clin Oncol 33(30):3416–3422. https://doi.org/10.1200/jco.2014.58.8533

157. Attal N, Bouhassira D, Gautron M, Vaillant JN, Mitry E, Lepère C, Rougier P, Guirimand F (2009) Thermal hyperalgesia as a marker of oxaliplatin neurotoxicity: a prospective quantified sensory assessment study. Pain 144(3):245–252. https://doi.org/10.1016/j.pain.2009.03.024. Epub 2009 May 19

158. Loprinzi CL, Reeves BN, Dakhil SR, Sloan JA, Wolf SL, Burger KN, Kamal A, Le-Lindqwister NA, Soori GS, Jaslowski AJ, Novotny PJ, Lachance DH (2011) Natural history of paclitaxel-associated acute pain syndrome: prospective cohort study NCCTG N08C1. J Clin Oncol 29(11):1472–1478. https://doi.org/10.1200/JCO.2010.33.0308

159. Pachman DR, Qin R, Seisler D, Smith EM, Kaggal S, Novotny P, Ruddy KJ, Lafky JM, Ta LE, Beutler AS, Wagner-Johnston ND, Staff NP, Grothey A, Dougherty PM, Cavaletti G, Loprinzi CL (2016) Comparison of oxaliplatin and paclitaxel-induced neuropathy (Alliance

A151505). Support Care Cancer 24(12):5059–5068. https://doi.org/10.1007/s00520-016-3373-1. Epub 2016 Aug 18

160. Kim SH, Kim W, Kim JH, Woo MK, Baek JY, Kim SY, Chung SH, Kim HJ (2018) A prospective study of chronic oxaliplatin-induced neuropathy in patients with colon cancer: long-term outcomes and predictors of severe oxaliplatin-induced neuropathy. J Clin Neurol 14 (1):81–89. https://doi.org/10.3988/jcn.2018.14.1.81

161. Velasco R, Bruna J, Briani C, Argyriou AA, Cavaletti G, Alberti P, Frigeni B, Cacciavillani M, Lonardi S, Cortinovis D, Cazzaniga M, Santos C, Kalofonos HP (2014) Early predictors of oxaliplatin-induced cumulative neuropathy in colorectal cancer patients. J Neurol Neurosurg Psychiatry 85(4):392–398. https://doi.org/10.1136/jnnp-2013-305334. Epub 2013 Jun 29

162. El Chediak A, Haydar AA, Hakim A, Massih SA, Hilal L, Mukherji D, Temraz S, Shamseddine A (2018) Increase in spleen volume as a predictor of oxaliplatin toxicity. Ther Clin Risk Manag 14:653–657. https://doi.org/10.2147/TCRM.S150968. eCollection 2018

163. Cavaletti G, Bogliun G, Marzorati L, Zincone A, Piatti M, Colombo N, Franchi D, La Presa MT, Lissoni A, Buda A, Fei F, Cundari S, Zanna C (2004) Early predictors of peripheral neurotoxicity in cisplatin and paclitaxel combination chemotherapy. Ann Oncol 15 (9):1439–1442. https://doi.org/10.1093/annonc/mdh348

164. Youk J, Kim YS, Lim JA, Shin DY, Koh Y, Lee ST, Kim I (2017) Depletion of nerve growth factor in chemotherapy-induced peripheral neuropathy associated with hematologic malignancies. PLoS One 12(8):e0183491. https://doi.org/10.1371/journal.pone.0183491. eCollection 2017

165. Gaetani L, Blennow K, Calabresi P, Di Filippo M, Parnetti L, Zetterberg H (2019) Neurofilament light chain as a biomarker in neurological disorders. J Neurol Neurosurg Psychiatry 90(8):870–881. https://doi.org/10.1136/jnnp-2018-320106. Epub 2019 Apr 9

166. Meregalli C, Fumagalli G, Alberti P, Canta A, Carozzi VA, Chiorazzi A, Monza L, Pozzi E, Sandelius Å, Blennow K, Zetterberg H, Marmiroli P, Cavaletti G (2018) Neurofilament light chain as disease biomarker in a rodent model of chemotherapy induced peripheral neuropathy. Exp Neurol 307:129–132. https://doi.org/10.1016/j.expneurol.2018.06.005. Epub 2018 Jun 13

167. Meregalli C, Fumagalli G, Alberti P, Canta A, Chiorazzi A, Monza L, Pozzi E, Carozzi VA, Blennow K, Zetterberg H, Cavaletti G, Marmiroli P (2020) Neurofilament light chain: a specific serum biomarker of axonal damage severity in rat models of chemotherapy-induced peripheral neurotoxicity. Arch Toxicol 94(7):2517–2522. https://doi.org/10.1007/s00204-020-02755-w. Epub 2020 Apr 24

168. Sharma MR, Mehrotra S, Gray E, Wu K, Barry WT, Hudis C, Winer EP, Lyss AP, Toppmeyer DL, Moreno-Aspitia A, Lad TE, Velasco M, Overmoyer B, Rugo HS, Ratain MJ, Gobburu JV (2020) Personalized management of chemotherapy-induced peripheral neuropathy based on a patient reported outcome: CALGB 40502 (Alliance). J Clin Pharmacol 60(4):444–452. https://doi.org/10.1002/jcph.1559. Epub 2019 Dec 4

169. Delmotte JB, Beaussier H, Auzeil N, Massicot F, Laprévote O, Raymond E, Coudoré F (2018) Is quantitative sensory testing helpful in the management of oxaliplatin neuropathy? a two-year clinical study. Cancer Treat Res Commun 17:31–36. https://doi.org/10.1016/j.ctarc.2018.10.002. Epub 2018 Oct 10

170. Kleckner IR, Kamen C, Gewandter JS, Mohile NA, Heckler CE, Culakova E, Fung C, Janelsins MC, Asare M, Lin PJ, Reddy PS, Giguere J, Berenberg J, Kesler SR, Mustian KM (2018) Effects of exercise during chemotherapy on chemotherapy-induced peripheral neuropathy: a multicenter, randomized controlled trial. Support Care Cancer 26(4):1019–1028. https://doi.org/10.1007/s00520-017-4013-0

171. Zimmer P, Trebing S, Timmers-Trebing U, Schenk A, Paust R, Bloch W, Rudolph R, Streckmann F, Baumann FT (2018) Eight-week, multimodal exercise counteracts a progress of chemotherapy-induced peripheral neuropathy and improves balance and strength in metastasized colorectal cancer patients: a randomized controlled trial. Support Care Cancer 26(2):615–624. https://doi.org/10.1007/s00520-017-3875-5

172. Hertz DL, Kidwell KM, Vangipuram K, Li F, Pai MP, Burness M, Griggs JJ, Schott AF, Van Poznak C, Hayes DF, Lavoie Smith EM, Henry NL (2018) Paclitaxel plasma concentration after the first infusion predicts treatment-limiting peripheral neuropathy. Clin Cancer Res 24 (15):3602–3610. https://doi.org/10.1158/1078-0432.Ccr-18-0656

173. Vatandoust S, Joshi R, Pittman KB, Esterman A, Broadbridge V, Adams J, Singhal N, Yeend S, Price TJ (2014) A descriptive study of persistent oxaliplatin-induced peripheral neuropathy in patients with colorectal cancer. Support Care Cancer 22(2):513–518. https://doi.org/10.1007/s00520-013-2004-3

174. Zirpoli GR, McCann SE, Sucheston-Campbell LE, Hershman DL, Ciupak G, Davis W, Unger JM, Moore HCF, Stewart JA, Isaacs C, Hobday TJ, Salim M, Hortobagyi GN, Gralow JR, Budd GT, Albain KS, Ambrosone CB (2017) Supplement use and chemotherapy-induced peripheral neuropathy in a cooperative group trial (S0221): the DELCaP study. J Natl Cancer Inst 109(12). https://doi.org/10.1093/jnci/djx098

175. Brami C, Bao T, Deng G (2016) Natural products and complementary therapies for chemotherapy-induced peripheral neuropathy: a systematic review. Crit Rev Oncol Hematol 98:325–334. https://doi.org/10.1016/j.critrevonc.2015.11.014. Epub 2015 Nov 23

176. Walker AF (2007) Potential micronutrient deficiency lacks recognition in diabetes. Br J Gen Pract 57(534):3–4

177. Ambrosone CB, Zirpoli GR, Hutson AD, McCann WE, McCann SE, Barlow WE, Kelly KM, Cannioto R, Sucheston-Campbell LE, Hershman DL, Unger JM, Moore HCF, Stewart JA, Isaacs C, Hobday TJ, Salim M, Hortobagyi GN, Gralow JR, Budd GT, Albain KS (2020) Dietary supplement use during chemotherapy and survival outcomes of patients with breast cancer enrolled in a Cooperative Group Clinical Trial (SWOG S0221). J Clin Oncol 38 (8):804–814. https://doi.org/10.1200/JCO.19.01203. Epub 2019 Dec 19

178. Mongiovi JM, Zirpoli GR, Cannioto R, Sucheston-Campbell LE, Hershman DL, Unger JM, Moore HCF, Stewart JA, Isaacs C, Hobday TJ, Salim M, Hortobagyi GN, Gralow JR, Thomas Budd G, Albain KS, Ambrosone CB, McCann SE (2018) Associations between self-reported diet during treatment and chemotherapy-induced peripheral neuropathy in a cooperative group trial (S0221). Breast Cancer Res 20(1):146. https://doi.org/10.1186/s13058-018-1077-9

179. Kosmidis P, Mylonakis N, Fountzilas G, Samantas E, Athanassiadis A, Pavlidis N, Skarlos D (1997) Paclitaxel (175 mg/m2) plus carboplatin versus paclitaxel (225 mg/m2) plus carboplatin in non-small cell lung cancer: a randomized study. Semin Oncol 24(4 Suppl 12): S12-30–S12-33

180. Winer EP, Berry DA, Woolf S, Duggan D, Kornblith A, Harris LN, Michaelson RA, Kirshner JA, Fleming GF, Perry MC, Graham ML, Sharp SA, Roger K, Henderson IC, Hudis C, Muss H, Norton L (2004) Failure of higher-dose paclitaxel to improve outcome in patients with metastatic breast cancer: Cancer and Leukemia Group B Trial 9342. J Clin Oncol 22 (11):2061–2068. http://jco.ascopubs.org/content/22/11/2061.abstract. https://doi.org/10.1200/JCO.2004.08.048

181. Shulman LN, Cirrincione CT, Berry DA, Becker HP, Perez EA, O'Regan R, Martino S, Atkins JN, Mayer E, Schneider CJ, Kimmick G, Norton L, Muss H, Winer EP, Hudis C (2012) Six cycles of doxorubicin and cyclophosphamide or Paclitaxel are not superior to four cycles as adjuvant chemotherapy for breast cancer in women with zero to three positive axillary nodes: Cancer and Leukemia Group B 40101. J Clin Oncol 30(33):4071–4076. https://doi.org/10.1200/JCO.2011.40.6405. Epub 2012 Jul 23

182. Selvy M, Pereira B, Kerckhove N, Gonneau C, Feydel G, Pétorin C, Vimal-Baguet A, Melnikov S, Kullab S, Hebbar M, Bouché O, Slimano F, Bourgeois V, Lebrun-Ly V, Thuillier F, Mazard T, Tavan D, Benmammar KE, Monange B, Ramdani M, Péré-Vergé D, Huet-Penz F, Bedjaoui A, Genty F, Leyronnas C, Busserolles J, Trevis S, Pinon V, Pezet D, Balayssac D (2020) Long-term prevalence of sensory chemotherapy-induced peripheral neuropathy for 5 years after adjuvant FOLFOX chemotherapy to treat colorectal cancer: a multicenter cross-sectional study. J Clin Med 9(8):E2400. https://doi.org/10.3390/jcm9082400

183. Park SB, Goldstein D, Krishnan AV, Lin CS, Friedlander ML, Cassidy J, Koltzenburg M, Kiernan MC (2013) Chemotherapy-induced peripheral neurotoxicity: a critical analysis. CA Cancer J Clin 63(6):419–437. https://doi.org/10.3322/caac.21204

184. Starobova H, Vetter I (2017) Pathophysiology of chemotherapy-induced peripheral neuropathy. Front Mol Neurosci 10:174. https://doi.org/10.3389/fnmol.2017.00174. eCollection 2017

185. Zajączkowska R, Kocot-Kępska M, Leppert W, Wrzosek A, Mika J, Wordliczek J (2019) Mechanisms of chemotherapy-induced peripheral neuropathy. Int J Mol Sci 20(6):1451. https://doi.org/10.3390/ijms20061451

186. Beijers AJ, Mols F, Tjan-Heijnen VC, Faber CG, van de Poll-Franse LV, Vreugdenhil G (2015) Peripheral neuropathy in colorectal cancer survivors: the influence of oxaliplatin administration. Results from the population-based PROFILES registry. Acta Oncol 54 (4):463–469. https://doi.org/10.3109/0284186X.2014.980912. Epub 2014 Nov 24

187. Bhatnagar B, Gilmore S, Goloubeva O, Pelser C, Medeiros M, Chumsri S, Tkaczuk K, Edelman M, Bao T (2014) Chemotherapy dose reduction due to chemotherapy induced peripheral neuropathy in breast cancer patients receiving chemotherapy in the neoadjuvant or adjuvant settings: a single-center experience. Springerplus 3:366. http://www.ncbi.nlm.nih.gov/entrez/query.fcgi?cmd=Retrieve&db=PubMed&dopt=Citation&list_uids=25089251

188. Simon NB, Danso MA, Alberico TA, Basch E, Bennett AV (2017) The prevalence and pattern of chemotherapy-induced peripheral neuropathy among women with breast cancer receiving care in a large community oncology practice. Qual Life Res 26(10):2763–2772. https://doi.org/10.1007/s11136-017-1635-0. Epub 2017 Jun 29

189. Eckhoff L, Knoop AS, Jensen MB, Ejlertsen B, Ewertz M (2013) Risk of docetaxel-induced peripheral neuropathy among 1,725 Danish patients with early stage breast cancer. Breast Cancer Res Treat 142(1):109–118. https://doi.org/10.1007/s10549-013-2728-2. Epub 2013 Oct 17

190. Molassiotis A, Cheng HL, Lopez V, Au JSK, Chan A, Bandla A, Leung KT, Li YC, Wong KH, Suen LKP, Chan CW, Yorke J, Farrell C, Sundar R (2019) Are we mis-estimating chemotherapy-induced peripheral neuropathy? Analysis of assessment methodologies from a prospective, multinational, longitudinal cohort study of patients receiving neurotoxic chemotherapy. BMC Cancer 19(1):132. https://doi.org/10.1186/s12885-019-5302-4

191. Budd GT, Barlow WE, Moore HC, Hobday TJ, Stewart JA, Isaacs C, Salim M, Cho JK, Rinn KJ, Albain KS, Chew HK, Burton GV, Moore TD, Srkalovic G, McGregor BA, Flaherty LE, Livingston RB, Lew DL, Gralow JR, Hortobagyi GN (2015) SWOG S0221: a phase III trial comparing chemotherapy schedules in high-risk early-stage breast cancer. J Clin Oncol 33 (1):58–64. https://doi.org/10.1200/jco.2014.56.3296

192. Chan JK, Brady MF, Penson RT, Huang H, Birrer MJ, Walker JL, DiSilvestro PA, Rubin SC, Martin LP, Davidson SA, Huh WK, O'Malley DM, Boente MP, Michael H, Monk BJ (2016) Weekly vs. every-3-week paclitaxel and carboplatin for ovarian cancer. N Engl J Med 374 (8):738–748. https://doi.org/10.1056/NEJMoa1505067

193. Pignata S, Scambia G, Katsaros D, Gallo C, Pujade-Lauraine E, De Placido S, Bologna A, Weber B, Raspagliesi F, Panici PB, Cormio G, Sorio R, Cavazzini MG, Ferrandina G, Breda E, Murgia V, Sacco C, Cinieri S, Salutari V, Ricci C, Pisano C, Greggi S, Lauria R, Lorusso D, Marchetti C, Selvaggi L, Signoriello S, Piccirillo MC, Di Maio M, Perrone F (2014) Carboplatin plus paclitaxel once a week versus every 3 weeks in patients with advanced ovarian cancer (MITO-7): a randomised, multicentre, open-label, phase 3 trial. Lancet Oncol 15(4):396–405. https://doi.org/10.1016/S1470-2045(14)70049-X. Epub 2014 Feb 28

194. Schuette W, Blankenburg T, Guschall W, Dittrich I, Schroeder M, Schweisfurth H, Chemaissani A, Schumann C, Dickgreber N, Appel T, Ukena D (2006) Multicenter randomized trial for stage IIIB/IV non-small-cell lung cancer using every-3-week versus weekly paclitaxel/carboplatin. Clin Lung Cancer 7(5):338–343. https://doi.org/10.3816/clc.2006.n.016

195. van der Burg ME, Onstenk W, Boere IA, Look M, Ottevanger PB, de Gooyer D, Kerkhofs LG, Valster FA, Ruit JB, van Reisen AG, Goey SH, van der Torren AM, ten Bokkel Huinink D,

Kok TC, Verweij J, van Doorn HC (2014) Long-term results of a randomised phase III trial of weekly versus three-weekly paclitaxel/platinum induction therapy followed by standard or extended three-weekly paclitaxel/platinum in European patients with advanced epithelial ovarian cancer. Eur J Cancer 50(15):2592–2601. https://doi.org/10.1016/j.ejca.2014.07.015. Epub 2014 Aug 2

196. Gridelli C, Gallo C, Di Maio M, Barletta E, Illiano A, Maione P, Salvagni S, Piantedosi FV, Palazzolo G, Caffo O, Ceribelli A, Falcone A, Mazzanti P, Brancaccio L, Capuano MA, Isa L, Barbera S, Perrone F (2004) A randomised clinical trial of two docetaxel regimens (weekly vs 3 week) in the second-line treatment of non-small-cell lung cancer. The DISTAL 01 study. Br J Cancer 91(12):1996–2004. https://doi.org/10.1038/sj.bjc.6602241

197. Stemmler HJ, Harbeck N, Gröll de Rivera I, Vehling Kaiser U, Rauthe G, Abenhardt W, Artmann A, Sommer H, Meerpohl HG, Kiechle M, Heinemann V (2010) Prospective multi-center randomized phase III study of weekly versus standard docetaxel (D2) for first-line treatment of metastatic breast cancer. Oncology 79(3–4):197–203. https://doi.org/10.1159/000320640. Epub 2011 Mar 1

198. Stemmler HJ, Harbeck N, Gröll de Rivera I, Vehling Kaiser U, Rauthe G, Abenhardt W, Artmann A, Sommer H, Meerpohl HG, Kiechle M, Heinemann V (2010) Prospective multi-center randomized phase III study of weekly versus standard docetaxel plus doxorubicin (D4) for first-line treatment of metastatic breast cancer. Oncology 79(3–4):204–210. https://doi.org/10.1159/000320625. Epub 2011 Mar 1

199. Beijers AJ, Mols F, Vreugdenhil G (2014) A systematic review on chronic oxaliplatin-induced peripheral neuropathy and the relation with oxaliplatin administration. Support Care Cancer 22 (7):1999–2007. https://doi.org/10.1007/s00520-014-2242-z. Epub 2014 Apr 13

200. Ewertz M, Qvortrup C, Eckhoff L (2015) Chemotherapy-induced peripheral neuropathy in patients treated with taxanes and platinum derivatives. Acta Oncol 54(5):587–591. https://doi.org/10.3109/0284186X.2014.995775. Epub 2015 Mar 9

201. Gregg RW, Molepo JM, Monpetit VJ, Mikael NZ, Redmond D, Gadia M, Stewart DJ (1992) Cisplatin neurotoxicity: the relationship between dosage, time, and platinum concentration in neurologic tissues, and morphologic evidence of toxicity. J Clin Oncol 10(5):795–803. https://doi.org/10.1200/jco.1992.10.5.795

202. Gebremedhn EG, Shortland PJ, Mahns DA (2018) The incidence of acute oxaliplatin-induced neuropathy and its impact on treatment in the first cycle: a systematic review. BMC Cancer 18 (1):410. https://doi.org/10.1186/s12885-018-4185-0

203. Grisold W, Cavaletti G, Windebank AJ (2012) Peripheral neuropathies from chemotherapeutics and targeted agents: diagnosis, treatment, and prevention. Neuro Oncol 14 (Suppl 4):iv45–54. https://doi.org/10.1093/neuonc/nos203

204. Weis TM, Marini BL, Nachar VR, Brown AM, Phillips TJ, Brown J, Wilcox RA, Kaminski MS, Devata S, Perissinotti AJ (2020) Impact of a vincristine dose cap on the incidence of neuropathies with DA-EPOCH-R for the treatment of aggressive lymphomas. Leuk Lymphoma 61(5):1126–1132. https://doi.org/10.1080/10428194.2019.1703969. Epub 2019 Dec 26

205. Verstappen CC, Koeppen S, Heimans JJ, Huijgens PC, Scheulen ME, Strumberg D, Kiburg B, Postma TJ (2005) Dose-related vincristine-induced peripheral neuropathy with unexpected off-therapy worsening. Neurology 64(6):1076–1077. https://doi.org/10.1212/01.WNL.0000154642.45474.28

206. Morawska M, Grzasko N, Kostyra M, Wojciechowicz J, Hus M (2015) Therapy-related peripheral neuropathy in multiple myeloma patients. Hematol Oncol 33(4):113–119. https://doi.org/10.1002/hon.2149. Epub 2014 Nov 14

207. Rowinsky EK, Chaudhry V, Forastiere AA, Sartorius SE, Ettinger DS, Grochow LB, Lubejko BG, Cornblath DR, Donehower RC (1993) Phase I and pharmacologic study of paclitaxel and cisplatin with granulocyte colony-stimulating factor: neuromuscular toxicity is dose-limiting. J Clin Oncol 11(10):2010–2020

208. Zhang S, Sun M, Yuan Y, Wang M, She Y, Zhou L, Li C, Chen C, Zhang S (2016) Correlation between paclitaxel Tc > 0.05 and its therapeutic efficacy and severe toxicities in ovarian cancer patients. Cancer Transl Med 2(5):131–136. http://www.cancertm.com/article.asp?issn=2395-3977. https://doi.org/10.4103/2395-3977.192930

209. Xin DS, Zhou L, Li CZ, Zhang SQ, Huang HQ, Qiu GD, Lin LF, She YQ, Zheng JT, Chen C, Fang L, Chen ZS, Zhang SY (2018) TC > 0.05 as a pharmacokinetic parameter of paclitaxel for therapeutic efficacy and toxicity in cancer patients. Recent Pat Anticancer Drug Discov 13 (3):341–347. https://doi.org/10.2174/1574892813666180305170439

210. Huizing MT, Vermorken JB, Rosing H, ten Bokkel Huinink WW, Mandjes I, Pinedo HM, Beijnen JH (1995) Pharmacokinetics of paclitaxel and three major metabolites in patients with advanced breast carcinoma refractory to anthracycline therapy treated with a 3-hour paclitaxel infusion: a European Cancer Centre (ECC) trial. Ann Oncol 6(7):699–704. http://annonc.oxfordjournals.org/content/6/7/699.abstract

211. Mielke S, Sparreboom A, Steinberg SM, Gelderblom H, Unger C, Behringer D, Mross K (2005) Association of paclitaxel pharmacokinetics with the development of peripheral neuropathy in patients with advanced cancer. Clin Cancer Res 11(13):4843–4850. http://clincancerres.aacrjournals.org/content/11/13/4843.abstract. https://doi.org/10.1158/1078-0432.CCR-05-0298

212. Gréen H, Söderkvist P, Rosenberg P, Mirghani RA, Rymark P, Lundqvist EÅ, Peterson C (2009) Pharmacogenetic studies of paclitaxel in the treatment of ovarian cancer. Blackwell, Malden, MA

213. Brown T, Havlin K, Weiss G, Cagnola J, Koeller J, Kuhn J, Rizzo J, Craig J, Phillips J, Von Hoff D (1991) A phase I trial of taxol given by a 6-hour intravenous infusion. J Clin Oncol 9 (7):1261–1267

214. Joerger M, Kraff S, Jaehde U, Hilger RA, Courtney JB, Cline DJ, Jog S, Baburina I, Miller MC, Salamone SJ (2017) Validation of a commercial assay and decision support tool for routine paclitaxel therapeutic drug monitoring (TDM). Ther Drug Monit 39(6):617–624. https://doi.org/10.1097/ftd.0000000000000446

215. Joerger M, von Pawel J, Kraff S, Fischer JR, Eberhardt W, Gauler TC, Mueller L, Reinmuth N, Reck M, Kimmich M, Mayer F, Kopp HG, Behringer DM, Ko YD, Hilger RA, Roessler M, Kloft C, Henrich A, Moritz B, Miller MC, Salamone SJ, Jaehde U (2016) Open-label, randomized study of individualized, pharmacokinetically (PK)-guided dosing of paclitaxel combined with carboplatin or cisplatin in patients with advanced non-small-cell lung cancer (NSCLC). Ann Oncol 27(10):1895–1902. https://doi.org/10.1093/annonc/mdw290

216. Zhang J, Zhou F, Qi H, Ni H, Hu Q, Zhou C, Li Y, Baburina I, Courtney J, Salamone SJ (2019) Randomized study of individualized pharmacokinetically-guided dosing of paclitaxel compared with body-surface area dosing in Chinese patients with advanced non-small cell lung cancer. Br J Clin Pharmacol. https://doi.org/10.1111/bcp.13982

217. Engels FK, Loos WJ, van der Bol JM, de Bruijn P, Mathijssen RH, Verweij J, Mathot RA (2011) Therapeutic drug monitoring for the individualization of docetaxel dosing: a randomized pharmacokinetic study. Clin Cancer Res 17(2):353–362. https://doi.org/10.1158/1078-0432.ccr-10-1636

218. Fukae M, Shiraishi Y, Hirota T, Sasaki Y, Yamahashi M, Takayama K, Nakanishi Y, Ieiri I (2016) Population pharmacokinetic-pharmacodynamic modeling and model-based prediction of docetaxel-induced neutropenia in Japanese patients with non-small cell lung cancer. Cancer Chemother Pharmacol 78(5):1013–1023. https://doi.org/10.1007/s00280-016-3157-9

219. Bruno R, Olivares R, Berille J, Chaikin P, Vivier N, Hammershaimb L, Rhodes GR, Rigas JR (2003) Alpha-1-acid glycoprotein as an independent predictor for treatment effects and a prognostic factor of survival in patients with non-small cell lung cancer treated with docetaxel. Clin Cancer Res 9(3):1077–1082

220. Boer H, Proost JH, Nuver J, Bunskoek S, Gietema JQ, Geubels BM, Altena R, Zwart N, Oosting SF, Vonk JM, Lefrandt JD, Uges DR, Meijer C, de Vries EG, Gietema JA (2015) Long-term exposure to circulating platinum is associated with late effects of treatment in

testicular cancer survivors. Ann Oncol 26(11):2305–2310. https://doi.org/10.1093/annonc/mdv369. Epub 2015 Sep 7

221. Sprauten M, Darrah TH, Peterson DR, Campbell ME, Hannigan RE, Cvancarova M, Beard C, Haugnes HS, Fosså SD, Oldenburg J, Travis LB (2012) Impact of long-term serum platinum concentrations on neuro- and ototoxicity in Cisplatin-treated survivors of testicular cancer. J Clin Oncol 30(3):300–307. https://doi.org/10.1200/JCO.2011.37.4025. Epub 2011 Dec 19

222. Trendowski MR, El-Charif O, Ratain MJ, Monahan P, Mu Z, Wheeler HE, Dinh PC Jr, Feldman DR, Ardeshir-Rouhani-Fard S, Hamilton RJ, Vaughn DJ, Fung C, Kollmannsberger C, Mushiroda T, Kubo M, Hannigan R, Strathmann F, Einhorn LH, Fossa SD, Travis LB, Dolan ME (2019) Clinical and genome-wide analysis of serum platinum levels after cisplatin-based chemotherapy. Clin Cancer Res 25(19):5913–5924. https://doi.org/10.1158/1078-0432.CCR-19-0113. Epub 2019 Jul 11

223. Chalret du Rieu Q, White-Koning M, Picaud L, Lochon I, Marsili S, Gladieff L, Chatelut E, Ferron G (2014) Population pharmacokinetics of peritoneal, plasma ultrafiltrated and protein-bound oxaliplatin concentrations in patients with disseminated peritoneal cancer after intra-peritoneal hyperthermic chemoperfusion of oxaliplatin following cytoreductive surgery: correlation between oxaliplatin exposure and thrombocytopenia. Cancer Chemother Pharmacol 74(3):571–582. https://doi.org/10.1007/s00280-014-2525-6

224. Ishibashi K, Okada N, Miyazaki T, Sano M, Ishida H (2010) Effect of calcium and magnesium on neurotoxicity and blood platinum concentrations in patients receiving mFOLFOX6 therapy: a prospective randomized study. Int J Clin Oncol 15(1):82–87. https://doi.org/10.1007/s10147-009-0015-3

225. Shord SS, Bernard SA, Lindley C, Blodgett A, Mehta V, Churchel MA, Poole M, Pescatore SL, Luo FR, Chaney SG (2002) Oxaliplatin biotransformation and pharmacokinetics: a pilot study to determine the possible relationship to neurotoxicity. Anticancer Res 22(4):2301–2309

226. Lavoie Smith EM, Li L, Hutchinson RJ, Ho R, Burnette WB, Wells E, Bridges C, Renbarger J (2013) Measuring vincristine-induced peripheral neuropathy in children with acute lymphoblastic leukemia. Cancer Nurs 36(5):E49–E60. https://doi.org/10.1097/NCC.0b013e318299ad23

227. Van den Berg HW, Desai ZR, Wilson R, Kennedy G, Bridges JM, Shanks RG (1982) The pharmacokinetics of vincristine in man: reduced drug clearance associated with raised serum alkaline phosphatase and dose-limited elimination. Cancer Chemother Pharmacol 8(2):215–219

228. Crom WR, de Graaf SS, Synold T, Uges DR, Bloemhof H, Rivera G, Christensen ML, Mahmoud H, Evans WE (1994) Pharmacokinetics of vincristine in children and adolescents with acute lymphocytic leukemia. J Pediatr 125(4):642–649. https://doi.org/10.1016/s0022-3476(94)70027-3

229. Moore AS, Norris R, Price G, Nguyen T, Ni M, George R, van Breda K, Duley J, Charles B, Pinkerton R (2011) Vincristine pharmacodynamics and pharmacogenetics in children with cancer: a limited-sampling, population modelling approach. J Paediatr Child Health 47(12):875–882. https://doi.org/10.1111/j.1440-1754.2011.02103.x

230. Chong CD, Logothetis CJ, Savaraj N, Fritsche IIA, Gietner AM, Samuels ML (1988) The correlation of vinblastine pharmacokinetics to toxicity in testicular cancer patients. J Clin Pharmacol 28(8):714–718

231. Li J, Sausville EA, Klein PJ, Morgenstern D, Leamon CP, Messmann RA, LoRusso P (2009) Clinical pharmacokinetics and exposure-toxicity relationship of a folate-Vinca alkaloid conjugate EC145 in cancer patients. J Clin Pharmacol 49(12):1467–1476. https://doi.org/10.1177/0091270009339740

232. Lee SE, Choi K, Han S, Lee J, Hong T, Park GJ, Yim DS, Min CK (2017) Bortezomib pharmacokinetics in tumor response and peripheral neuropathy in multiple myeloma patients receiving bortezomib-containing therapy. Anticancer Drugs 28(6):660–668. https://doi.org/10.1097/CAD.0000000000000506

Evaluation of Chemotherapy-Induced Peripheral Neuropathy

Youmin Cho, Kathryn J. Ruddy, and Ellen M. Lavoie Smith

Abstract

Measurement of chemotherapy-induced neuropathy can be based on patient report, clinician report, or objective assessment of nerve function. In this chapter, we discuss patient-reported outcomes (PROs), including the European Organization for Research and Treatment of Cancer Quality of Life-Chemotherapy-Induced Peripheral Neuropathy questionnaire (EORTC QLQ-CIPN20), the Functional Assessment of Cancer Therapy/Gynecological Cancer Group-Neurotoxicity questionnaire (FACT/GOG-Ntx), and the PRO-Common Terminology Criteria for Adverse Events (PRO-CTCAE). In addition, we describe clinician-reported scales: the CTCAE, the Eastern Cooperative Oncology Group (ECOG) criteria, the World Health Organization (WHO) neurotoxicity scale, and the Ajani scale. Two scales that are specifically used for oxaliplatin-induced neurotoxicity are also discussed: the Oxaliplatin Neurological Toxicity Scale (ONTS] and the Neurotoxicity Criteria of Debiopharm (DEB-NTC). We also explain how nerve conduction studies assess peripheral nerve action potential amplitudes and velocity, and we describe the use of quantitative sensory testing, in which patients report whether they can detect sensations of vibration, mechanical stimuli, and warmth/cold applied to the skin at different locations. We

Y. Cho
University of Michigan School of Nursing, Ann Arbor, MI, USA
e-mail: youmcho@umich.edu

K. J. Ruddy
Mayo Clinic, Rochester, MN, USA
e-mail: Ruddy.Kathryn@mayo.edu

E. M. Lavoie Smith (✉)
University of Alabama at Birmingham School of Nursing, Birmingham, AL, USA
e-mail: esmith3@uab.edu

© The Author(s), under exclusive license to Springer Nature Switzerland AG 2021
M. Lustberg, C. Loprinzi (eds.), *Diagnosis, Management and Emerging Strategies for Chemotherapy-Induced Neuropathy*,
https://doi.org/10.1007/978-3-030-78663-2_3

discuss how these two types of studies can augment patient- and clinician-reported assessments, in part because they can detect abnormalities before they become clinically evident. Further, we highlight important biomarkers of CIPN, discuss options for nerve imaging, and make recommendations for clinical practice.

Keywords

CIPN · Chemotherapy · Neuropathy · Patient reported outcomes

3.1   Introduction

Quantitative and qualitative measures of chemotherapy-induced peripheral neuropathy (CIPN) are critical to research that focuses on improving outcomes for patients during and after receipt of neurotoxic chemotherapy. In addition, CIPN assessment will help oncology care teams better understand and address the symptoms experienced by their patients, resulting in higher-quality clinical care. Patient-reported, clinician-reported, and objective measures of nerve function are available. In this chapter, we describe a variety of clinical grading scales used by clinicians to report neuropathy severity and functional interference, many instruments that collect patient-reported data on neuropathy symptoms, and a few objective measures of nerve function and appearance. These measures nearly universally assess changes in sensory nerve function, the most common manifestation of CIPN; some also assess motor and autonomic nerve function.

3.2   Clinical Grading Scales

Clinical grading scales for assessing chemotherapy-related toxicities reflect the severity of treatment-related symptoms [1]. In a clinical setting, grade 2 or higher neuropathy indicates severe neurotoxicity that often results in chemotherapy dose reduction [1, 2]. Several grading scales are used to quantify CIPN, most frequently the Common Toxicity Criteria (CTC) scales, including the National Cancer Institute-Common Terminology Criteria for Adverse Events (NCI-CTCAE, previously NCI-CTC), the Eastern Cooperative Oncology Group (ECOG) criteria, the World Health Organization (WHO) neurotoxicity scale, and the Ajani scale (Tables 3.1, 3.2, and 3.3) [3].

Additionally, two scales are available specifically to assess oxaliplatin-induced CIPN: the Oxaliplatin Neurological Toxicity Scale (ONTS) and the Neurotoxicity Criteria of Debiopharm (DEB-NTC) (Table 3.4) [3, 4]. However, since specific grading scale scores are based not on objective measurements (e.g., deep tendon reflex or sensory examination) but rather on clinicians' scoring of patients' reported symptoms, reliance on clinical grading scales in research and practice has been subject to recent scrutiny for two main reasons. First, accuracy of the assessment

Table 3.1 Clinical grading scales for chemotherapy-induced peripheral neuropathy (CIPN)—sensory

	Grade 1	Grade 2	Grade 3	Grade 4	Grade 5
NCI-CTCAE (ver. 5.0) [9]	Asymptomatic	Moderate symptoms; limiting instrument ADL	Severe symptoms; limiting self-care ADL	Life-threatening consequences; urgent intervention indicated	–
	Grade 0	Grade 1	Grade 2	Grade 3	Grade 4
ECOG-sensory [12]	None or no change	Mild paresthesias; loss of deep tendon reflexes	Mild or moderate objective sensory loss; moderate paresthesias	Severe objective sensory loss or paresthesia that interferes with function	–
ECOG-vision [12]	None or no change	–	–	Symptomatic subtotal loss of vision	Blindness
ECOG-hearing [12]	None or no change	Asymptomatic, hearing loss on audiometry only	Tinnitus	Hearing loss interfering with function but correctable with hearing aid	Deafness, not correctable
WHO [13]	None	Paresthesias	Severe paresthesias	Intolerable paresthesias	–
Ajani scale [14]	–	Paresthesia; decreased deep tendon reflexes	Mild objective abnormality; absence of deep tendon reflexes; mild to moderate functional abnormality	Severe paresthesia; moderate objective abnormality; severe functional abnormality	Complete sensory loss; loss of function

NCI-CTCAE, National Cancer Institute-Common Terminology Criteria for Adverse Events; ADL, Activities of Daily Living; ECOG, Eastern Cooperative Oncology Group; WHO, World Health Organization

Table 3.2 Clinical grading scales for chemotherapy-induced peripheral neuropathy (CIPN)—motor

	Grade 1	Grade 2	Grade 3	Grade 4	Grade 5
NCI-CTCAE (ver. 5.0) [9]	Asymptomatic; clinical or diagnostic observations only	Moderate symptoms; limiting instrument ADL	Severe symptoms; limiting self-care ADL	Life-threatening consequences; urgent intervention indicated	Death
ECOG-motor [12]	Grade 0	Grade 1	Grade 2	Grade 3	Grade 4
	None or no change	Subjective weakness; no objective findings	Mild objective weakness without significant impairment of function	Objective weakness with impairment of function	Paralysis
WHO [13]	None	Decreased tendon reflexes	Mild weakness	Marked motor loss	Paralysis
Ajani [14]	–	Mild or transient muscle weakness	Persistent moderate weakness but ambulatory	Unable to ambulate	Complete paralysis

NCI-CTCAE, National Cancer Institute-Common Terminology Criteria for Adverse Events; ADL, Activities of Daily Living; ECOG, Eastern Cooperative Oncology Group; WHO, World Health Organization

Table 3.3 Clinical grading scales for chemotherapy-induced peripheral neuropathy (CIPN)—autonomic

	Grade 0	Grade 1	Grade 2	Grade 3	Grade 4
ECOG-constipation [12]	None or no change	Mild	Moderate	Severe	Ileus > 96 h
WHO-constipation [13]	None	Mild	Moderate	Abdominal distension	Distention and vomiting
Ajani-bladder dysfunction [14]	–	–	Bladder dysfunction not requiring catheterization	Bladder dysfunction requiring temporary catheterization	Bladder dysfunction requiring permanent indwelling catheter
Ajani-constipation [14]	–	Mild constipation requires no specific therapy	Constipation requiring occasional specific oral therapy	Constipation requiring daily specific therapy (oral or enema)	Abdominal distention, impaction, and vomiting (requires inpatient management)
Ajani-general [14]	–	Abnormal sweating	Impotence	Asymptomatic arrhythmias (no therapy required), orthostatic hypotension	Symptomatic arrhythmias (requiring therapy), orthostatic hypotension (necessitating hospitalization)

ECOG, Eastern Cooperative Oncology Group; WHO, World Health Organization

Table 3.4 Clinical grading scales for oxaliplatin-induced peripheral neuropathy (OIPN)

	Grade 1	Grade 2	Grade 3	Grade 4
ONTS [15]	Paresthesia and/or dysesthesia with compete regression within 7 days	Paresthesia and/or dysesthesia with compete regression within 14 days	Paresthesia and/or dysesthesia with incomplete regression between courses	Paresthesia and/or dysesthesia with functional impairment
DEB-NTS [18]	Paresthesias or dysesthesias within 7 days	Paresthesias or dysesthesias more than 7 days	Functional impairment interfering with activities of daily living	–

ONTS, Oxaliplatin Neurological Toxicity Scale; DEB-NTS, Neurotoxicity Criteria of Debiopharm

requires agreement between clinicians' ratings and patients' symptom experiences [5, 6], and second, a considerable number of studies have demonstrated poor grading scale reliability and validity [1, 4, 7, 8].

3.2.1   NCI-CTC/CTCAE

The NCI-CTCAE (previously NCI-CTC) is commonly used to report cancer treatment-induced adverse event (AE) severity [9]. It was initially developed in 1983 by the cooperative oncology groups in North America and Canada, and is regularly updated [1, 4]. The most recent version, 5.0, was issued in November 2017, with the next updated version 6.0 being expected in Fall 2022 [9]. Each AE term has a 5-point grading scale that indicates the severity of the AE. NCI-CTCAE includes two AE terms for CIPN: peripheral motor neuropathy and peripheral sensory neuropathy.

The NCI-CTCAE is the most widely used CIPN grading system despite its numerous limitations. In a cross-sectional study ($N = 37$) comparing inter-observer reliability of the NCI-CTC, ECOG, WHO, and Ajani scales, the NCI-CTC was found to have the lowest agreement coefficient between two observers for all grades (1–5) and for the dichotomy (1–3 vs. 4) (45.9% and 81.1%, respectively) [1]. Empirical evidence supports the NCI-CTCAE sensory neuropathy scale's construct validity based on low to moderate correlation with the Total Neuropathy Score (TNS©) and the Neuropathic Pain Scale for Chemotherapy-Induced Neuropathy (NPS-CIN) ($r = 0.22$–0.63, $p = 0.05$ to <0.001), but motor scale scores did not correlate [8]. Further, the NCI-CTC grades were low, ranging from 0 to 2, even though CIPN symptoms were clinically relevant, suggesting that the NCI-CTC scales are insensitive to minor changes in CIPN severity (floor effect) [10]. Similarly, the NCI-CTC sensory grades did not correlate with Total Neuropathy Score (TNS©) assessments of objective pin sensibility ($rs = 0.171$), vibration sensibility ($rs = 0.217$), and deep tendon reflex ($rs = 0.217$), nor did NCI-CTC motor neuropathy grades correlate with objective TNS© muscle strength scores ($rs = 0.080$) [7].

Despite its pitfalls, the NCI-CTCAE is preferred by many clinicians because it is easy and efficient to use in busy clinical settings [4]. Also, because it provides grading rubrics for numerous adverse effects, such as neutropenia, one scale can be used to grade all cancer-associated AEs. However, NCI-CTCAE scores should be interpreted cautiously given the inter-observer disagreements, poor sensitivity due to floor effects, and the suboptimal construct validity findings from several studies. To address those limitations of the NCI-CTCAE scoring system, NCI developed the Patient-Reported Outcomes version of the CTCAE (PRO-CTCAE), which will be described later in this chapter.

3.2.2 ECOG Common Toxicity Criteria

The ECOG Common Toxicity Criteria were developed by the Eastern Cooperative Oncology Group (ECOG) in 1974 [11]. The ECOG criteria have been used in all ECOG studies and publications [11]. The ECOG criteria are based on five CIPN-related toxicities: sensory, vision, hearing, motor, and constipation [12]. Each criterion is scored from 0 to 4: high scores reflect worse CIPN. A revised version of the ECOG neurotoxicity criteria addresses more parameters (e.g., autonomic symptoms) and the concept of "disabling" sensory loss [4]. The inter-observer agreement of the ECOG criteria was the highest for all grades (0–4) (Intraclass Correlation Coefficient [ICC] = 0.75) in comparison to NCI-CTC (ICC = 0.58), WHO (ICC = 0.55), and Ajani scales (ICC = 0.37) [1]. Limited evidence describes the ECOG scale's validity, sensitivity, and responsiveness to change over time.

3.2.3 WHO Recommendations

The WHO Recommendations, developed in 1979, has two CIPN-related criteria: peripheral neuropathic symptoms and constipation [13]. Clinicians rate peripheral neuropathy based on the presence and severity of paresthesias and muscle weakness. Each criterion is scored from 0 to 4. Psychometric properties have not been extensively tested [4]. These grading criteria are rarely used for the assessment of CIPN.

3.2.4 Ajani Scale

The Ajani scale was developed in 1990 by the Chemotherapy Working Group and the Departments of Medical Specialties and Neuro-oncology in the Houston Cancer Center [14]. The prototype of this scale was based partly on the WHO criteria [14]. One way in which the Ajani scale differs from the other clinical grading scales is that it incorporates clinical significance within each criterion [14]. For assessment of CIPN, the Ajani scale evaluates autonomic nervous system toxicity (i.e., bladder dysfunction, constipation, sweating, impotence, and arrhythmias), and motor and sensory deficits [14]. Each criterion is scored from 1 to 4; high scores reflect worse

CIPN. When comparing to the NCI-CTC, ECOG, and WHO scales, the inter-observer agreement was the lowest in all grades (ICC = 0.37) [1]. Published evidence regarding other psychometric properties is limited.

3.2.5 Oxaliplatin-Induced Peripheral Neuropathy (OIPN) Assessment

In addition to the general grading scales that quantify all types of CIPN (e.g., NCI-CTCAE), two specialized scales have been developed to evaluate oxaliplatin-induced peripheral neuropathy (OIPN), which has unique, acute, cold-induced manifestations [3].

3.2.5.1 Oxaliplatin Neurological Toxicity Scale (ONTS)

The Oxaliplatin Neurological Toxicity Scale (ONTS) was developed in 1993 for use in a randomized controlled trial of an oxaliplatin chemotherapy regimen [15]. Although this 4-point scale evaluates only peripheral sensory neuropathy, it also quantifies symptom duration [16].

One study showed that the ONTS was more sensitive than the NCI-CTC ($N = 114$) [16]. More specifically, of 53 patients with severe grade 3–4 CIPN based on the ONTS, 23, 18, and 12 received a NCI-CTC grade of 1, 2, and 3, respectively, and no patient received an NCI-CTC grade 4 [16]. Furthermore, the ONTS was responsive to change over time [16]. Beyond the information presented in this one paper, no additional published evidence supports its general validity or reliability.

3.2.5.2 The Neurotoxicity Criteria of Debiopharm (DEB-NTC)

The Neurotoxicity Criteria of Debiopharm (DEB-NTC) is based on sensory CIPN symptom duration [17], but also addresses CIPN-associated functional deficits [17, 18]. However, the concordance rate between grade 0–2 NCI-CTCAE grades and the DEB-NTC was low: 48.8% and 47.3% in oxaliplatin- and irinotecan-treated patients, respectively ($\kappa = 0.26$ and 0.18, respectively, $p < 0.001$) [17]. Low concordance/agreement between NCI-CTCAE and DEB-NTC grades does not suggest that the DEB-NTC lacks convergent validity, but that, again, the NCI-CTCAE lacks sensitivity to detect the full range of CIPN severity. The DEB-NTC demonstrated earlier detection of mild OIPN than did the NCI-CTCAE ver. 3.0 [17]. In particular, the DEB-NTC was able to detect grade 1–2 OIPN from a significantly lower cumulative oxaliplatin dose than was the NCI-CTCAE ($p < 0.001$) [17].

3.3   Composite Scales

Several composite measures, which combine scores from several subjective and objective components of a comprehensive CIPN examination, have been developed for diabetic and inflammatory neuropathy; however, only two have been validated to assess CIPN: the Total Neuropathy Score (TNS©) [19] and the modified Inflammatory Neuropathy Cause and Treatment Group Sensory Sum Score (mISS) [20].

3.3.1   Total Neuropathy Score (TNS©)

Originally developed in 1994 by a team of neurologists at the Johns Hopkins University, the TNS© is the most commonly used and best validated of the two composite scales [19].

Several abbreviated TNS© variants have since been developed and tested for use with adult and pediatric populations [3, 21–25]. Each TNS© item is scored from 0 to 4 and summed; high scores reflect more severe CIPN. Scores reflect distal to proximal extension of CIPN signs and symptoms. The more proximal the extension, the higher the score. For example, a patient with numbness only in the toes would receive a lower sensory symptom score than someone with numbness extending from the toes to the ankles.

The original 11-item TNS© contains rubrics for assessing subjective sensory, motor, and autonomic symptoms. When assessing subjective sensory CIPN, the score reflects the distal to proximal extension of three sensory symptoms—numbness, tingling, neuropathic pain—but only the worst of the three scores is included in the total score. The original TNS© also provides scores for the following physical examination findings: pinprick sensation, vibration sensation threshold using a 128 Hz tuning fork, deep tendon reflexes, and motor strength, which is assessed in the toes, ankles, hips, fingers, thumbs, wrist, and arms. For all physical examination components, the assessor conducts bilateral assessments; if the findings are asymmetrical, the higher score from the two sides is used for the final summed score.

Nerve conduction study (NCS) scores, based on peripheral nerve action potential amplitudes for the right sural sensory and peroneal motor nerves, are included in some TNS© variants; low TNS© scores reflect action potentials that are normal or $\geq 96\%$ of what would be expected for the patient's age. Quantitative sensory testing (QST) scores are also included in some of the variants and provide additional assessments of vibration and thermal sensation thresholds based on norm-adjusted, percentile-based values. Low (normal) QST threshold scores mean that the patient can detect very subtle vibratory or thermal sensations.

The total TNS© score varies based on the number of items in the variant (Table 3.5). Scores obtained using the original 11-item TNS© range from 0 to 44 [24]. Scores ≥ 5 indicate CIPN [26]. The TNSr© (reduced) variant excludes QST, and subjective motor and autonomic items (score range 0–28) [24]. The mTNS© (modified) excludes tuning fork vibratory threshold assessments, thermal QST, NCS, and subjective autonomic scores (score range 0–24) [24]. The clinical

Table 3.5 Total neuropathy score variants (modified with permission) [24]

	TNS©	TNS©r	mTNS©	TNS©c	5-item TNS©	TNS©-SF
Pin Prick	√	√	√	√	√	
Vibration via 124 Hz or Rydel–Seiffer tuning fork	√	√		√	√	√
Vibration threshold via Quantitative Sensory Testing	√		√			
Thermal threshold via Quantitative Sensory Testing	√					
Nerve Conduction Studies—Sensory (Sural Nerve Conduction Amplitude)	√	√				
Nerve Conduction studies—Motor (Peroneal Nerve Conduction Amplitude)	√	√				
Deep Tendon Reflexes	√	√	√	√	√	
Strength	√	√	√	√	√	
Subjective Report					√	
Worst Sensory Score of 3 symptoms (numbness, tingling, neuropathic pain)	√	√	√	√		
Numbness						√
Tingling						√
Neuropathic pain						√
Motor (e.g., walking on toes/heels, climbing stairs, buttoning, writing, combing hair)	√		√	√		
Autonomic (fainting, impotence, bloating, constipation, loss of bowel and bladder control)	√			√		
Total Score Range	0–44	0–28	0–24	0–28	0–20	0–16

TNS, Total Neuropathy Score; TNSr, TNS-reduced; mTNS, modified TNS; TNSc, TNS-clinical; TNS-SF, TNS-Short Form

variant (TNSc©), designed for use by non-neurologists in oncology practice settings, includes items that quantify pinprick sensation, tuning fork-based vibration threshold, deep tendon reflexes, strength, and all three subjective symptom categories (sensory, motor, autonomic) (score range 0–28) [24]. The TNS© items most often excluded from the various variants are those that require specialized tests/ training (QST, NCS), deep tendon reflexes because they can be difficult to assess, motor symptoms and examination findings because motor neuropathy occurs less often, and autonomic symptoms (e.g., constipation, orthostatic hypotension, impotence), which can be difficult to attribute to CIPN.

While extensive evidence supports the reliability and validity of the TNS© [10, 19, 27–29], TNSr© [10, 26, 27, 30–32], TNSc© [27, 28, 30], and mTNS© [29], recent psychometric analyses suggest that the TNS© can be further reduced, and that minor revisions in the scoring procedures will improve its performance [8, 22, 33]. For assessing taxane- or platinum-induced CIPN, empirical evidence supports the reliability, validity, and sensitivity of a 5-item TNS© that includes the worst of the three subjective symptom scores, and the motor strength, pinprick sensation, vibration threshold, and tendon reflex scores (score range 0–20). An even shorter variant, the TNS©-SF (short form), is more internally consistent ($\alpha = 0.80$) than the 5-item version ($\alpha = 0.56$), and is also valid and sensitive [34, 35]. The TNS©-SF has just four items: all three subjective symptom scores and the tuning fork-assessed vibration threshold score (score range 0–16). Of all the variants, the TNS©-SF is the most clinically feasible for use by non-neurologists because it excludes QST, NCS, and deep tendon reflex assessments, all of which require specialized clinical training and experience to obtain accurate and reliable scores. Individuals who do not have extensive neurology training can easily learn how to obtain reliable tuning fork assessments using either a simple 128 Hz tuning fork, or the graduated Rydel–Seiffer (Fig. 3.1) [36]. Another advantage of the

Fig. 3.1 Rydel–Seiffer tuning fork (toe placement)

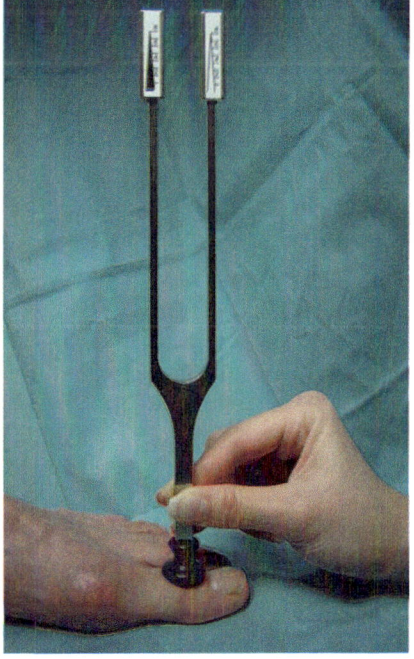

TNS©-SF is that all subjective symptom item scores are included in the total score, not just the worst of the three. This approach allows quantification of non-painful (numbness and tingling) and painful CIPN, and may prompt the clinician to offer evidence-based treatment (i.e., duloxetine) for painful CIPN [37].

3.3.2 Inflammatory Neuropathy Cause and Treatment Group Sensory Sum Score (mISS)

The ISS quantifies solely sensory neuropathy through five assessments: pinprick and vibration sensibility in the arms and legs, and two-point discrimination at the ventral index finger [20]. Items are scored 0–4 distally to proximally, and the five scores are summed (total score range = 0–20). Normal sensation in the most distal location (i.e., index finger or big toe) is scored as "0," which assumes that findings more proximal will also be normal. Higher scores are associated with more proximal abnormalities, such as at the wrist, elbow shoulder, ankle, knee, or groin, and reflect more severe sensory neuropathy. The modified version (mISS) includes light touch and joint position items [38].

The ISS was initially tested in patients with inflammatory neuropathy (e.g., Guillain–Barré syndrome) and found to be valid, reliable, and responsive to change over time [20].

Specifically, construct validity was supported based on moderately strong and statistically significant correlations between scores from the mISS and upper and lower extremity functional measures. Inter-rater reliability was demonstrated based on strong correlations between scores obtained by different raters, and responsiveness assessments were based on observed changes in scores following intravenous immunoglobulin treatments. The mISS's psychometric properties have also been tested in patients who had received taxanes, platinums, thalidomide, vincristine, or bortezomib [30]. Published evidence supports the mISS's strong inter-rater/equivalence reliability ($r = 0.84$) and construct validity because scores were strongly correlated with the TNSc© ($r = 0.72$–0.76) [30].

While the mISS is a reliable and valid measure based on psychometric data from patients who were receiving many different neurotoxic chemotherapeutic agents, the mISS is not clinically feasible for use by non-neurologists in busy oncology clinical settings. The assessor must conduct neurological assessments at multiple anatomic sites, a time-consuming task.

Further, several of the mISS assessments require specialized training (e.g., joint position, two-point discrimination). Lastly, the mISS only provides a sensory neuropathy score and should not be used as the sole CIPN measure when evaluating patients receiving drugs that also cause significant motor neuropathy (e.g., vincristine).

3.4 Patient-Reported Outcome Measures

Patient-reported outcomes (PROs) are described by patients directly, without interpretation by clinicians or anyone else [39]. Clinicians commonly assess CIPN via clinical grading scales (e.g., NCI-CTCAE), which underestimate symptoms and are too insensitive to detect subtle differences in mild symptoms [5, 40]. Since CIPN symptoms are subjective, clinician-based grading scales cannot measure patients' experiences of symptoms. Therefore, an essential component of a comprehensive symptom management strategy is using valid PRO measures to collect patients' self-reported symptom information [6, 40].

The European Organization for Research and Treatment of Cancer Quality of Life-Chemotherapy-Induced Peripheral Neuropathy questionnaire (EORTC QLQ-CIPN20) and the Functional Assessment of Cancer Therapy/Gynecological Cancer Group-Neurotoxicity questionnaire (FACT/GOG-Ntx) are the most commonly used PRO measures. In addition, the National Cancer Institute recently issued a PRO version of the NCI-CTCAE scale, the PRO-CTCAE. These and other PRO measures are described in this section.

3.4.1 EORTC QLQ-CIPN20

The EORTC Quality of Life group developed the Quality of Life Questionnaire-Core 30 (EORTC QLQ-C30) in 2005 [41]. The EORTC QLQ-CIPN20 questionnaire, a supplemental module of the EORTC QLQ-C30 and currently available in 42 languages [42], provides a comprehensive assessment of patients' CIPN-associated symptom experiences and functional limitations [43]. This 20-item questionnaire contains three subscales: sensory (9 items), motor (8 items), and autonomic (3 items) [43]. Each item is scored using a 4-point Likert scale (1 = "not at all," 2 = "a little," 3 = "quite a bit," and 4 = "very much."), with a 7-day recall period. Subscale scores are summed, and higher scores reflect worse CIPN. Although the scale was developed initially to include three subscales (i.e., sensory, motor, autonomic), these were not empirically tested. Recent evidence based on confirmatory factor analysis findings from several studies have revealed an unstable factor structure and poor correlation among some items within the sensory, motor, and autonomic subscales [44–46]. Therefore, a summed score of all items is now recommended, rather than calculating three separate subscale scores [44].

The psychometric properties of the EORTC QLQ-CIPN20 have been extensively evaluated; the tool is reliable, valid, sensitive, and responsive [44–51]. The internal consistency reliability was acceptable at the initial development stage and excellent alpha coefficients ($\alpha \geq 0.80$) have been confirmed in subsequent studies [43, 45, 51]. The EORTC QLQ-CIPN20 demonstrated a significant association with the NCI-CTCAE scale ($p < 0.0001$), which supports strong convergent validity of this tool [47]. However, some of the autonomic subscale items, such as those assessing dizziness, blurred vision, hearing loss, and erectile dysfunction, demonstrated weak correlations with other items and total scores (item-item correlations $r \leq 0.30$, item-total score correlations $r \leq 0.40$) [45]. Furthermore, results of a secondary data analysis ($N = 1155$) revealed poor internal consistency for the autonomic subscale (Cronbach's α coefficients $= 0.62$ for male, 0.39 for female) [46].

Because of these suboptimal findings, a 16-item modified version was developed. It excludes the dizziness, blurred vision, hearing loss, and erectile dysfunction items, and was found to be reliable, valid, and sensitive [45, 49]. The EORTC QLQ-CIPN15 is another reduced-item version that deletes an additional problematic item about driving ability, based on data obtained following cognitive interviews with 25 patients [49]. Empirical evidence supports the 15-item version's strong reliability, validity, sensitivity, and responsiveness to change [48].

3.4.2 FACT/GOG-Ntx

The Functional Assessment of Cancer Therapy (general version) (FACT-G) is a 27-item questionnaire that measures health-related quality of life in patients with cancer and chronic illnesses [52]. The FACT/Gynecologic Oncology Group-Neurotoxicity questionnaire (FACT/GOG-Ntx) includes the 27 FACT-G items and an additional 11-item subscale that evaluates CIPN symptoms and related concerns [52], and is currently available in 45 languages [53]. Development of the Ntx subscale was based on the Gynecologic Oncology Group (GOG-PN) peripheral neuropathy scale [54]. Each item in the FACT/GOG-Ntx is scored using a 5-point Likert scale (0 = "not at all" to 4 = "very much") and summed, with a 7-day recall period. Higher scores reflect worse CIPN severity.

Empirical evidence indicates that the FACT/GOG-Ntx is a reliable, valid, and responsive tool for CIPN assessment, is internally consistent across all evaluation points up to 12 months after initial treatment (Cronbach's α coefficients > 0.80) [52], and has moderate to high internal consistency [55, 56]. In a recent study ($N = 343$), item-total score correlations and item-item correlations were moderate to strong ($r = 0.66$–0.79, 0.34–0.73, respectively) [55]. Contrasting group validity was confirmed by clinically significant differences in the FACT/GOG-Ntx scores between patients with and without known neurotoxic chemotherapy ($p < 0.05$ at baseline and 3- and 6-month follow-up) [52]. Further, the FACT/GOG-Ntx was able to differentiate patients with NCI-CTC grade ≥ 1 from those with NCI-CTC score < 1 (the Area Under the Curve [AUC] in the Receiver Operating Curve [ROC] $= 0.81$) [55]. Scores from objective measures (i.e., pin test, pin sensitivity, vibration, and cold test) were significantly correlated with FACT/GOG-Ntx scores over time, providing evidence of good concurrent validity [52], although a recent study showed low to moderate correlation with monofilament tests. When compared to the EORTC-QLQ CIPN20, satisfactory convergent validity was supported based on high correlations ($r = 0.79$–0.93, $p < 0.01$) [56]. The tool demonstrated moderate to high responsiveness to change over time in multiple studies [52, 55, 56]. The FACT/GOG-Ntx 4, a shorter version with only four sensory-specific items, is reliable and valid despite its reduced length [55].

3.4.3   PRO-CTCAE

Since substantial evidence indicates that the NCI-CTCAE grading scale is less effective in detecting patients' symptomatic adverse events than PRO measures, the NCI developed the PRO-CTCAE (Patient-Reported Outcomes version of Common Terminology Criteria of Adverse Events) [57, 58] to complement the clinician-graded CTCAE. Currently, 78 adverse events listed in the CTCAE are available in the PRO-CTCAE [58], including two items relevant to CIPN: the severity of numbness and tingling in hands or feet, and the interference of those symptoms in daily activities [3]. Each item is scored on a 5-point scale ("none" to "very severe" for the severity, and "not at all" to "very much" for the interference), with a 7-day recall period.

The PRO-CTCAE has demonstrated good reliability, validity, and responsiveness [59]. In one study ($N = 975$), all PRO-CTCAE items were strongly correlated with the conceptually related EORTC QLQ-C30 items [59]. The two CIPN-related items exhibited moderate to strong test–retest reliability (ICC = 0.80 for the severity and 0.55 for the interference) [59]. Most PRO-CTCAE items were able to distinguish subgroups based on low and high ECOG performance status, cancer type, treatment, or other clinically related factors [59]. In a recent comparison between the PRO-CTCAE and the EORTC-QLQ CIPN20, the correlation between the two CIPN-related PRO-CTCAE items and the sensory and motor subscale scores were moderate to high (PRO-CTCAE severity: sensory $r = 0.76$, motor $r = 0.55$; PRO-CTCAE interference: sensory $r = 0.78$, motor $r = 0.77$) [51]. However, the correlation with autonomic subscale scores in the EORTC-QLQ CIPN20 was low (severity: $r = 0.14$, interference: $r = 0.28$) [51]. The two CIPN-relevant items were responsive to change based on effect size data [59]. Lastly, the PRO-CTCAE could detect CIPN symptoms in early or mid-treatment (11–19%), which were not detected by the NCI-CTCAE grading scale during the same period [40].

A large, multi-site national oncology clinical trial ($N = 153$) in the United States demonstrated that it is feasible to implement PRO-CTCAE in the research settings [60]. Patient compliance ranged from 72 to 86% with a median time of 15 min (range, 0–60 min) taken per patient to complete PRO-CTCAE [60]. The median time needed for clinical research professionals to learn about the system was 60 min (range, 30–240 min), and it took 10 min to teach patients (range, 2–60 min) [60]. These results suggest that PRO-CTCAE can be adopted in clinical research settings with minimal workflow disruption for researchers and burden for patients.

3.4.4   PRO Measures for the Assessment of Functional Limitations

Although substantial evidence indicates that most PRO measures are valid to evaluate CIPN-related symptoms, some measures do not capture CIPN-associated interference with patients' daily activities [3]. To address this gap, several PRO measures are available for the specific assessment of functional limitations related to CIPN.

3.4.4.1 Patient Neurotoxicity Questionnaire (PNQ)

The PNQ was developed by BioNumerik Pharmaceuticals to quantify the severity of CIPN-related symptoms [61]. This 3-item scale evaluates sensory, motor, and functional loss with an A–E scale: score A = no symptoms, E = severe symptoms [61]. In addition, patients who score their symptoms as D or E are required to select the type of activity interference experienced [61]. The PNQ distinguishes between the absence (score $\leq$ C) and presence (score $>$C) of the neuropathic symptoms that interfere with patients' daily activities [61–63]. A phase III trial ($N = 300$) of adjuvant taxane chemotherapy in patients with an operable breast cancer supports the measure's validity and responsiveness [63, 64]. In this trial, the PNQ and the FACT/GOG-Ntx scores were obtained from patients, while NCI-CTC grades were obtained by physician raters [64]. PNQ scores encompassed the full score range, whereas NCI-CTC scores were 0 or 1 [64]. In particular, patients who reported maximum sensory and motor neuropathic symptom scores often received physician-rated NCI-CTC grades of 0 or 1 (weighted κ coefficients = 0.16 and 0.02 for sensory and motor scores, respectively) [64]. Despite the low correlations with the psychometrically weak NCI-CTC scale, PNQ scores were moderately correlated with FACT/GOG-Ntx scores ($r = 0.66$ and 0.51 for sensory and motor subscales, respectively) [64], providing evidence of the PNQ's convergent validity. Lastly, PNQ sensory scores were more responsive to change over time than the motor scores (Cohen's $d = 0.79$ and 0.38 for sensory and motor scores, respectively, $p < 0.0001$) [64]. However, further evaluations to identify the tool's reliability are needed.

3.4.4.2 The Rasch-Built Overall Disability Scale for CIPN (CIPN R-ODS)

The CIPN R-ODS is an interval-weighted scale to assess activity limitations and disability associated with CIPN symptoms [65]. The preliminary version of the CIPN R-ODS included 146 items of activity and participation outcome items selected from the International Classification of Functioning, Disability, and Health (ICF) [65]. In order to select the final items, the researchers conducted a Rasch analysis to convert ordinal-level items to an interval-level scale based on the patient's ability to perform each task (task difficulty) [65]. After the Rasch model-fitting test, 28 items were selected [65]. Each item is scored as 0 (impossible to perform), 1 (performed but with difficulty), and 2 (easily performed without difficulty). The scores are summed, and a high score means that the patient is less disabled and has more ability to complete difficult daily activity tasks [65]. Empirical evidence supports the internal consistency reliability (Personal Separation Index = 0.92) and convergent validity with a strong correlation with NCI-CTC scores [65]. However, it is important to keep in mind that this evidence does not fully support the CIPN R-ODS's validity because it was compared to a weak CIPN measure (NCI-CTC).

3.4.4.3 Chemotherapy-Induced Neurotoxicity Questionnaire (CINQ)

The CINQ was developed to assess the impact of CIPN symptoms on patients' quality of life and daily activities [66]. It assesses paresthesias and dysesthesias in the following body parts: arms, hands, fingers, legs, feet, toes, face, and mouth

[66]. In addition, the CINQ assesses motor (e.g., unbuttoning a blouse, opening a jar, less strength in legs) and autonomic (e.g., bladder control, erectile dysfunction, dry vagina) symptoms [67]. Patients rate the presence and severity of symptoms and the associated impact on daily activity using a 0–5 scale (higher scores reflect more severe symptoms). Convergent validity has been evaluated through the correlation with the FACT/GOG-Ntx: the Ntx subscale and the CINQ showed a strong negative correlation ($r = -0.74$, $p \leq 0.001$) [66]. Further, a strong negative correlation ($r = -0.73$, $p \leq 0.001$) with the FACT/GOG was demonstrated, which indicates that severe CIPN symptoms were related to a lower quality of life [66]. When compared to the Semmes–Weinstein filament tests, a weak but statistically significant correlation was demonstrated ($\kappa = 0.32$, $p < 0.001$) [68]. The CINQ was able to distinguish between chemotherapy-receiving and chemotherapy-naïve patients in multiple studies, providing evidence of contrasting group validity [66–68]. However, no evidence of this tool's reliability has been published.

3.4.4.4 Chemotherapy-Induced Peripheral Neuropathy Assessment Tool (CIPNAT)

The CIPNAT assesses CIPN symptom patterns and characteristics (i.e., symptom occurrence, severity, distress, and frequency), and related performance (interference with activities) [69]. The CIPNAT has 36 neuropathy symptom-related items and 14 interference-related items. Empirical evidence supports its reliability based on high test–retest reliability scores ($r = 0.93$, $p < 0.001$) and strong internal consistency (Cronbach's α coefficient = 0.95) [69]. Strong convergent validity has been demonstrated through comparison to the FACT/GOG-Ntx ($r = 0.83$, $p < 0.001$) [69]. However, the length of the questionnaire (average time to complete = 15 min) may limit its use in clinical practice settings. Reliable and valid Turkish and Arabic versions of the CIPNAT are available [70–72].

3.4.5 Miscellaneous PRO Measures

Other PRO measures for CIPN-related symptom evaluation are described below. Table 3.6 lists symptoms addressed and psychometric properties of each measure.

3.4.5.1 Treatment-Induced Neuropathy Assessment Scale (TNAS)

The TNAS also assesses the severity of CIPN-related symptoms [73]. Version 1.0. contains 11 items; additional two items covering new domains identified as important in cognitive interviewing are included in version 2.0. [73]. Patients rate the severity of their neuropathic symptoms using a 0–10 scale (0 = the symptom is not present, 10 = the symptom is as bad as you can imagine), with a 24-h recall period [73]. It is an efficient method of CIPN assessment based on the short average time to complete the questionnaire (<2 min) [73]. A preliminary study supports its reliability, validity, and sensitivity when tested in a mixed population of patients with multiple myeloma ($n = 223$) and colorectal cancer ($n = 186$) [73]. When tested with multiple myeloma and colorectal cancer patients, good internal consistency

Table 3.6 Patient-reported outcome measures—symptoms addressed and psychometric properties

	Sensory	Motor	Autonomic	Functional limitations	Psychometric properties
EORTC-QLQ CIPN20	✓	✓	✓	✓	*Reliability*: high Cronbach's α coefficients [43, 45, 46, 51], good test–retest reliability [47], weak item–item or item-total score correlations, and low Cronbach's α coefficients in autonomic subscales [45, 46] *Validity*: strong convergent validity with NCI-CTC/CTCAE [47] *Responsiveness*: Yes [45, 47]
FACT/GOG-Ntx	✓	✓		✓	*Reliability*: high Cronbach's α coefficients [52, 55, 56] *Validity*: strong convergent validity with NCI-CTC/CTCAE and objective test results [52, 55, 56] *Responsiveness*: Yes [52, 55, 56] Sensitivity: high [52, 55]
PRO-CTCAE	✓			✓	*Reliability*: good test–retest reliability *Validity*: strong convergent validity with the EORTC-QLQ-C30 *Responsiveness*: Yes [59]
Patient Neurotoxicity Questionnaire (PNQ)	✓	✓		✓	*Validity*: poor correlation with the FACT-G, moderate correlation with the FACT/GOG-Ntx [64] *Responsiveness*: Yes [63]
The Rasch-built Overall Disability Scale for CIPN (CIPN R-ODS)				✓	*Reliability*: high Personal Separation Index *Validity*: strong convergent validity with NCI-CTC, but showed ceiling effect (9.4%) [65]
Chemotherapy-Induced Neurotoxicity Questionnaire (CINQ)	✓			✓	*Reliability*: strong internal consistency reliability through the high Personal Separation Index *Validity*: strong criterion validity with FACT/GOG-Ntx [66]
Chemotherapy-Induced Peripheral Neuropathy Assessment Tool (CIPNAT)	✓	✓		✓	*Reliability*: high test–retest reliability, high Cronbach's α coefficients *Validity*: strong convergent validity with the FACT/GOG- Ntx [69]

Tool				Reliability/Validity
Treatment-induced Neuropathy Assessment Scale (TNAS)	✓	✓		*Reliability*: high Cronbach's α coefficients *Validity*: moderate convergent validity with the EORTC-QLQ CIPN20 [73]
Neuropathy Screening Questionnaire (NSQ)	✓			*Validity*: moderate correlation with the EORTC-QLQ CIPN20 Responsible *sensitivity* and *specificity* [51]
The Scale for Chemotherapy-Induced long-term Neurotoxicity (SCIN)	✓		✓	*Reliability*: moderate to high Cronbach's α coefficients *Validity*: discriminant validity, correlation between ototoxicity subscale and audiometry results [76]
CIPN Self-Check Sheet	✓	✓	✓	*Reliability*: high inter-rater reliability [77]
Indication for Common Toxicity Criteria Grading of Peripheral Neuropathy Questionnaire (ICPNQ)	✓	✓		*Reliability*: high Cronbach's α coefficients *Validity*: strong convergent validity, divergent validity [78]
Oxaliplatin Associated Neurotoxicity Questionnaire (OANQ)	✓	✓		*Reliability*: high Cronbach's α coefficients, strong test–retest reliability [80]

EORTC-QLQ CIPN20, European Organization for Research and Treatment of Cancer Quality of Life Questionnaire Chemotherapy-Induced Peripheral Neuropathy 20; FACT/GOG-Ntx, Functional Assessment of Cancer Therapy/Gynecologic Oncology Group-Neurotoxicity questionnaire; PRO-CTCAE, Patient-Reported Outcomes version of Common Terminology Criteria for Adverse Events

reliability was demonstrated by high Cronbach's α coefficients (0.86 and 0.87, respectively) [73]. Version 1.0 was able to detect changes in CIPN symptoms during the course of treatment [73]. The TNAS version 2.0 moderately correlated with the EORTC-QLQ CIPN20 ($r = 0.46$–0.64) [73], but its sensitivity was not tested [73]. A recent qualitative study yielded a modified 9-item measure, TNAS version 3.0 [74, 75], which also demonstrated favorable reliability (Cronbach's α coefficient $= 0.90$) and validity based on moderately strong score correlations with two of the three EORTC QLQ-CIPN20 subscale scores ($r = 0.69$, sensory; $r = 0.70$, motor; $r = 0.32$, autonomic) [75].

3.4.5.2 Neuropathy Screening Questionnaire (NSQ)

The NSQ is a 10-item electronic PRO measure developed for use within the Carevive® Cancer Care Planning System [51]. Patients first report the presence of numbness and tingling in their hands or feet within the past 7 days, then rate the severity of the symptom(s) that they indicated as "Yes" [51]. Each item is scored from 0 to 10: higher scores reflect worse symptom severity. Convergent validity was supported based on moderately strong score correlations with EORTC QLQ-CIPN20 scores ($r = 0.67$, $p < 0.001$) [51]. In addition, the NSQ was sensitive (0.67) and specific (1.0) [51]. However, the study sample size was small ($N = 25$), and no additional psychometric data have been published.

3.4.5.3 The Scale for Chemotherapy-Induced Long-Term Neurotoxicity (SCIN)

The SCIN contains 6 items that assess paresthesias, Raynaud's symptoms, and ototoxicity [76]. Each item is scored on a 4-point Likert scale ($0 =$ "not at all" to $3 =$ "very much") [76]. Scores are summed and high scores reflect worse CIPN. Data from a cross-sectional study with testicular cancer survivors ($N = 684$) support the SCIN's internal consistency (Cronbach's α coefficient $= 0.72$) [76]. The ototoxicity subscale was significantly correlated with the audiometry results ($p < 0.00001$) [76]. Because acceptable psychometric properties were confirmed with cancer survivors (i.e., survivors who had been treated at least 4 years prior), the use of the SCIN can be used to evaluate chronic CIPN [76].

3.4.5.4 The Chemotherapy-Induced Peripheral Neuropathy Self-Check Sheet

The CIPN self-check sheet was developed (2015) and primarily tested in Japan [77]. The CIPN self-check sheet addresses four main categories: upper limbs (6 items), lower limbs (5 items), pain (2 items), and limitations in daily activities (1 item) [77]. Each item is scored by Yes/No indicating the presence of symptoms (e.g., dullness, difficulty discriminating temperature, pain) and the limitation in daily activities, except for the pain subscale (10-point scale, higher scores reflect severe pain). The validity of the tool was demonstrated through the cross-classification method comparing patients' answers to the CIPN self-check sheet and clinicians' physical exam and free-style interview [77]. Inter-rater reliability was higher in the CIPN self-check sheet than clinicians' physical exam and interview

($\kappa = 0.988$ vs. $0.501, p < 0.01$) [77]. However, no further evidence of psychometric properties has been published.

3.4.5.5 Indication for Common Toxicity Criteria Grading of Peripheral Neuropathy Questionnaire (ICPNQ)

The ICPNQ was developed to monitor the consistency and accuracy of the NCI-CTCAE grades in multiple myeloma patients [78]. The ICPNQ assesses sensory (5-item), autonomic (9-item), and motor (3-item) symptoms [78]. Patients select "Yes" if a symptom was present within the past seven days. In addition, each item in the sensory category requires patients to select the place (e.g., toes, fingers) where they have symptoms. The motor symptoms category covers a loss of muscle strength in the arms and legs, and limitations in self-care and instrumental activities of daily living. Psychometric evaluation in multiple myeloma patients ($N = 156$) demonstrated favorable reliability for the sensory and motor scales, but not for the autonomic scale (Cronbach's α coefficient $= 0.84$, 0.74, and 0.61, respectively) [78]. When compared to the EORTC-QLQ CIPN20, correlations were moderate to high ($r = 0.40$–0.72, $p < 0.001$), which supports good validity [78]. The ICPNQ has only been tested in multiple myeloma patients in a single cross-sectional study.

3.4.5.6 Oxaliplatin Associated Neurotoxicity Questionnaire (OANQ)

The OANQ was initially used in a Phase I clinical trial of an oxaliplatin- and capecitabine-containing chemotherapy regimen [79]. This 19-item questionnaire evaluates oxaliplatin-specific peripheral neurotoxic symptoms in three main areas: upper extremities, lower extremities, and oral/facial [79]. Patients answer Yes/No to whether each symptom exists, then score the symptom severity on a 1–4 scale if they indicated Yes; higher scores reflect worse CIPN symptoms. In addition, patients are required to score each symptom's interference with their daily activities, on a 1–4 scale; high scores reflect extreme interference. Results from a small pilot study ($N = 23$) support the OANQ's strong internal consistency (Cronbach's α coefficients $= 0.840$–0.935) and test–retest reliability based on overall excellent reproducibility (ICC > 0.75 in 83%, weighted $\kappa > 0.80$ in 59% of all items) [80]. However, the sample size in this study was small and no additional evidence of validity or responsiveness has been published.

3.5 Pain Scales

Although most PRO measures include one or two items about CIPN-related pain, they do not provide a comprehensive assessment of CIPN pain characteristics [3, 81]. More specifically, CIPN PRO measures neither characterize nor quantify the distinct types of painful CIPN, such as painful numbness or painful tingling [81]. Therefore, several general pain scales are currently used for quantifying CIPN-associated painful symptoms. In addition, several scales exist that assess neuropathic pain symptoms experienced by patients with CIPN. Table 3.7 summarizes the CIPN-related pain characteristics quantified by each measure, and the psychometric

Table 3.7 Pain scales–neuropathic pain characteristics addressed and psychometric properties evaluated in chemotherapy-induced peripheral neuropathy (CIPN) population

	Pain presence	Pain severity	Interference with daily activities	Psychometric evaluation in CIPN population
BPI Short Form (BPI-SF)	√	√	√	Contrasting group/construct validity [84]
Pain Intensity Numerical Rating Scale (PI-NRS)	√	√		Test–retest reliability [85]
Visual Analogue Scale (VAS)	√	√		Test–retest reliability Responsiveness [8]
Neuropathic Pain Scale for chemotherapy-induced neuropathy (NPS-CIN)	√	√		Internal consistency validity Contrasting group validity Convergent validity (with NCI-CTC) [8]
Neuropathic Pain Symptom Inventory (NPSI)	√	√		Chinese version—internal consistency validity, internal consistency reliability [91]
The PROMIS-Pain Quality Neuro (PROMIS-PQ Neuro)	√	√		–
The Douleur Neuropathique 4 (DN4)	√			Sensitivity, specificity [94]
The Leeds Assessment of Neuropathic Symptoms and Signs (LANSS)	√			Sensitivity, specificity [94]
ID Pain	√			Convergent validity, sensitivity, specificity [96]

BPI, Brief Pain Inventory; NCI-CTC, National Cancer Institute-Common Toxicity Criteria; PROMIS, Patient-Reported Outcome Measurement Information System

properties of the measure based on studies that were conducted in a CIPN population.

3.5.1 General Pain Scales

Published empirical evidence suggests that several general pain scales can be used to assess CIPN-related pain.

3.5.1.1 Brief Pain Index (BPI)

The long-form BPI is a well-validated tool that can be used to assess pain history, intensity, location, and quality of pain [82]. The BPI Short Form (BPI-SF) is recommended for use in clinical trials due to its conciseness and availability in multiple languages [83]. The BPI-SF has two main sections: pain severity (6-item) and pain interference (1 item). The pain severity items measure "worst," "least," "average," and "right now (current)" pain [83]. Each pain severity item is scored from 0 = "no pain" to 10 = "pain as bad as you can imagine." In addition, patients are required to mark the place(s) where they have pain on a body map. The pain interference items measure the extent of interference with seven daily activities (i.e., general activity, walking, work, mood, enjoyment of life, relations with others, and sleep) [83]. Each pain interference item is scored from 0 = "does not interfere" to 10 = "completely interferes." Two additional items ask about the presence of abnormal pain and the percentage of relief provided by medications or treatments. However, the usefulness and psychometrics of the two items have not been tested [83]. Although no studies directly evaluate the BPI's psychometric properties when used to assess CIPN pain, one study found that 59% of patients with CIPN who received duloxetine reported a decrease in BPI scores, compared to 38% of the placebo group patients [84]; this provides empirical evidence of contrasting group/construct validity.

3.5.1.2 Pain Intensity Numerical Rating Scale (PI-NRS)

The PI-NRS is an 11-point numerical pain rating scale. Pain severity is scored from 0 (no pain) to 10 (worst possible pain), with a 24-h recall period [85]. A psychometric evaluation with stable CIPN patients ($N = 281$) exhibited favorable test–retest reliability (0.768) [86]. Other evidence that supports the PI-NRS' reliability, validity, and responsiveness in evaluating CIPN-related pain has not been published to date.

3.5.1.3 Visual Analogue Scale (VAS)

The VAS is a widely used pain measurement scale that evaluates a characteristic or attitude toward pain from patients' perspectives [87, 88]. The VAS measures pain as a continuous spectrum [87]. A horizontal 100 mm line has anchors representing the complete absence of a symptom on one end of the line and extreme symptom severity on the other end [88]. Patients draw a mark on the line to indicate the degree of symptom severity experienced, and the score is the number of millimeters to the patient's mark. In a study with stable CIPN patients ($N = 281$), the VAS demonstrated good test–retest reliability (0.724) [86]. A study with patients receiving paclitaxel- ($n = 59$) and docetaxel-containing ($n = 34$) regimens used the VAS to evaluate CIPN-related pain and numbness separately [89]. Results showed that the patterns of pain and numbness were slightly different in the two groups, and that the VAS could appropriately recognize the change in neuropathic symptom severity over time [89]. Further, the two different chemotherapy regimens caused significantly different CIPN pain and numbness change patterns over time [89].

3.5.2 Neuropathic Pain Scales

Painful CIPN is classified as neuropathic pain because it arises from damage to peripheral or central nervous system tissue; therefore, neuropathic pain scales can be used to quantify painful CIPN.

3.5.2.1 Neuropathic Pain Scale for Chemotherapy-Induced Neuropathy (NPS-CIN)

The NPS-CIN, a 6-item pain scale that assesses intense, unpleasant, sharp, deep, numb, and tingling pain qualities within a 24-h recall period [8, 22], was modified from existing pain scales: the Neuropathic Pain Scale (NPS) and the Pain Quality Assessment scale [8, 22]. Each item of the NPS-CIN is scored using a 5-point scale (0 = "not at all" to 4 = "excruciating"), and the scores are then summed to obtain a total score. Empirical evidence supports the measure's strong internal consistency reliability ($\alpha = 0.96$) [22]. Moreover, because NPS-CIN scores were significantly higher in patients with diabetes—a risk factor for developing more severe CIPN—than in those without this diagnosis, this provides evidence of contrasting group validity [8]. Although the NPS-CIN scores were moderately correlated with the NCI-CTC sensory scores ($r = 0.63$) [8], a correlation with the weak NCI-CTC should be interpreted cautiously, as stated previously.

3.5.2.2 Neuropathic Pain Symptom Inventory (NPSI)

The NPSI is a 12-item questionnaire that contains four pain characteristics: spontaneous, paroxysmal, evoked, and dysesthesia/paresthesia [90]. Ten items quantify various neuropathic pain characteristics and are scored using a 0–10 scale (higher scores reflect severe pain) with a 24-h recall period, and two items assess the duration of spontaneous pain and paroxysmal pain [90]. Empirical evidence supports good reliability and validity [90]. In a psychometric evaluation for CIPN conducted in China with the Chinese version of the NPSI (C-NPSI) ($N = 106$) [91], the C-NPSI demonstrated high internal consistency reliability (Cronbach's α coefficient = 0.9) [91]. Item-item correlations and item-total score correlations ranged from 0.082 to 0.429, supporting a weak but statistically significant positive correlation ($p < 0.05$) [91]. However, convergent validity was not tested via score comparisons with reliable and valid CIPN measurements; the C-NPSI scores were only compared to scores from a mood status measurement.

3.5.2.3 PROMIS-PQ Neuro

The Patient-Reported Outcome Measurement Information System (PROMIS) is a comprehensive measurement system of PRO measures related to numerous common medical conditions [92]. The PROMIS-Pain Quality Neuro (PROMIS-PQ Neuro) scale, derived from the PROMIS pain quality items, specifically assesses neuropathic pain symptoms [92]. This 5-item measure addresses numbness, tingling, pinprick pain, stinging, and electrical (shock-like) symptoms with a 7-day recall period [92]. While acceptable reliability and validity were demonstrated in all neuropathic pain types, including CIPN ($n = 134$), no other studies have evaluated its psychometric properties specifically for CIPN assessment.

3.5.2.4 The Douleur Neuropathique 4 (DN4)

The DN4 was developed by The French Neuropathic Pain Group to address the difference in chronic neuropathic pain induced by neurological (peripheral or central) and somatic tissue injuries [93]. The DN4 assesses four main categories:

description of pain, paresthesia/dysesthesia, sensory deficits, and evoked pain [93]. Patients identify whether they have paresthesia/dysesthesia and pain, and clinicians use physical examination techniques to assess hypoesthesia to touch and/or pinprick, and brush-induced pain [93]. In a cross-sectional study with $N = 358$ cancer patients undergoing active chemotherapy, the DN4 exhibited overall moderate to high sensitivity (87.5%) and specificity (88.4%) [94]. The DN4 was more sensitive to detect mild neuropathic pain than the Leeds Assessment of Neuropathic Symptoms and Signs (LANSS) [94]. However, no further published evidence supports the tool's reliability and validity for CIPN-related pain assessment.

3.5.2.5 The Leeds Assessment of Neuropathic Symptoms and Signs (LANSS)

The LANSS was developed to distinguish neuropathic from nociceptive pain [95]. The scale contains seven screening questions scored by "Yes" (1–5) or "No" (0) [95]. The LANSS uses several descriptors of neuropathic pain (e.g., pricking, tingling, electric shocks), so that the tool can distinguish neuropathic and non-neuropathic pain [95]. If the total score is less than 12, the pain is not neuropathic [95]. When compared to the DN4, the LANSS demonstrated higher specificity (93.4%) but lower sensitivity (65.8%) [94]. No other published evidence supporting the tool's reliability and validity for CIPN-related pain assessment is available.

3.5.2.6 ID Pain

The ID Pain is a 6-item scale that also differentiates neuropathic from nociceptive pain [96]. The ID Pain uses 6 descriptors of neuropathic pain (e.g., pain feels like pin and needles, hot/burning, electric shocks) to help distinguish nociceptive pain and neuropathic pain [96]. Each item is scored by "Yes" (1, except for the question asking about limitations of joints $= -1$) or "No" (0); higher total scores reflect pain with more neuropathic components [96]. Cut points that delineate the presence of neuropathic pain are as follows: very likely neuropathic pain (score 4 or 5), likely neuropathic pain (score 2 or 3), possible neuropathic pain (score 1), and unlikely (score 0 and –1) [96]. In a cross-sectional study with breast cancer patients receiving taxane-containing chemotherapy ($N = 240$), the ID Pain scores were significantly correlated with a clinical diagnosis of neuropathic pain (positive neuropathic pain $\geq$ ID Pain score 2, $rs = 0.41$; $p < 0.0001$) [96].

Further, the ID Pain scores were significantly correlated with the LANSS ($r = 0.58$; $p < 0.005$) [96]. Lastly, the ID Pain demonstrated high specificity (86% with clinical diagnosis, 93.5% with the LANSS) and moderate sensitivity (50% with clinical diagnosis, 67% with the LANSS) [96].

3.6   Functional Tests

Patients with CIPN commonly report functional deficits that increase fall risk and negatively affect overall quality of life [97, 98]. Thus, upper and lower extremity functional ability should be evaluated in clinical practice and research settings using previously described PRO surveys; the following functional tests may also be useful outcome measures in clinical trials [99].

3.6.1   Postural Stability Tests

Loss of sensation in the plantar surfaces of the feet, foot drop, and reduced lower extremity muscle strength are physical manifestations of CIPN that increase fall risk. Although complex testing protocols and specialized equipment (e.g., forceplate) have been used in research settings to assess postural stability, empirical evidence supports the reliability and validity of two clinical measures that are easier to use: the Timed Up & Go (TUG) test and the Fullerton Advanced Balance Scale (FABS).

3.6.1.1 The TUG (Timed Up and Go) Test

The TUG test is a reliable and valid method for assessing mobility and functional stability when used with older frail individuals [100], community dwelling elders [101], and patients with diabetic neuropathy [101]. Published data suggest that the TUG has good sensitivity (87%) and specificity (87%) to detect increased fall risk [101]. The assessor uses a stopwatch to record the number of seconds needed for the patient to rise from a chair, walk three meters, turn around, and return to and sit down on the chair [100]. The patient is instructed to "GO," and timing begins when the buttocks leave the chair.

3.6.1.2 FABS (Fullerton Advanced Balance Scale)

The FABS quantifies the patient's ability to perform 10 tasks that require balance. The assessor scores each item from 0 to 4 and sums the scores (maximum score = 40): high scores reflect more impairment and scores above 22 are predictive of increased fall risk in independently-living older adults [102, 103]. Individuals complete the following tasks: (1) stand with their feet together and eyes closed, (2) lean forward to reach for an object, (3) walk in a tight circle (both directions), (4) step onto and over a 6-inch bench, (5) tandem walk (heel-to-toe), (6) stand on one leg, (7) stand on a foam pad with eyes closed, (8) jump for distance, (9) walk with head turns, and (10) attempt recovery from an unexpected loss of balance [102]. Empirical evidence supports the FABS's validity, reliability, and sensitivity when used with older patients and those with Parkinson's disease or breast cancer [29, 102–105]. Although testing takes minimal time (approximately 10 min), special testing equipment is required, including a 6-in. high bench and foam pads. Therefore, use of the FABS outside of a research setting may be challenging.

3.6.2 6-Minute Walk Test (6MWT)

This test is easy to administer and can be used to assess walking ability in patients with CIPN [106]. Patients walk between two markers set 15 m apart as many times as possible over six minutes. In a diverse sample ($N = 100$) of patients who had received a wide assortment of neurotoxic chemotherapeutic agents, lower walking distance scores were associated with worse patient-reported CIPN-R-ODS disability scores ($r = 0.63$) and CIPN based on TNSc ($r = 0.48$) and EORTC CIPN20 ($r = 0.50$) scores [106].

3.6.3 Grooved Pegboard Test

In addition to lower extremity function, CIPN can compromise the ability to perform everyday tasks that require fine motor skills, such as buttoning/zipping clothing, writing, or holding utensils. The Grooved Pegboard Test, also called the nine-hole peg test, can quantify fine motor skills in patients with neuropathy. Patients are timed as they fill in nine pegboard slots with pegs, and then remove the pegs, one at a time. Although the pegboard test has not been validated for assessing CIPN-associated functional deficits, empirical evidence supports its construct validity based on data obtained from patients with inherited peripheral neuropathy (Charcot–Marie–Tooth Disease). Patients had significantly slower dominant hand pegboard speed than did healthy controls ($p < 0.001$) [107]. This test is most feasible for use in research studies.

3.7 Electrophysiological Assessments

3.7.1 Nerve Conduction Studies (NCS)

Nerve conduction studies (NCS) have long been considered the gold-standard approach for evaluating peripheral nerve action potential amplitude and conduction velocity in patients with peripheral neuropathy from diverse causes (e.g., immune-mediated polyneuropathies, diabetes, toxic neuropathy) [108–110]. These tests provide information about the physiological function of large myelinated nerve fibers (Aβ), but cannot quantify small-fiber neuropathy resulting from damage to thinly myelinated (Aδ) or unmyelinated (c) fibers [111].

Neurotoxic chemotherapeutic agents cause two main types of neuropathy that are distinguishable with NCS, neuronopathy, and axonopathy. Drugs that target the nerve cell bodies within the dorsal root ganglion, mainly platinum-based drugs, cause a non-length dependent neuronopathy, meaning that long nerve fibers—those extending to the toes—and shorter nerve fibers—those extending to the fingers—are affected simultaneously. Neuronopathy due to platinum drugs is evidenced by diminished or absent sensory action potentials in the sural (lower extremity) and radial (upper extremity) nerves [108].

Taxanes, vincristine, thalidomide, and bortezomib cause length-dependent axonopathy from damage to sensory and motor nerve axons [108]. With this type of neuropathy, signs and symptoms typically emerge in the lower extremities first and, as neuropathy worsens, extend more proximally to the upper extremities as well. Nerve conduction studies reveal absent or diminished sural (sensory) nerve action potentials, and low distal compound motor action potential (dCMAP) in the peroneal (lower extremity) nerves first, and eventually progress to the shorter median and radial (upper extremity) nerves [108]. These findings reveal the damage that occurs before patients become symptomatic.

Neurotoxic chemotherapy can also cause demyelination, which results in diminished nerve conduction velocity. Reduced motor nerve conduction velocity is evidenced by F-wave abnormalities [108]. Motor nerve conduction abnormalities are less common than sensory abnormalities, and are most often associated with taxane- or vincristine-induced CIPN [112]. By identifying muscle tissue denervation, electromyography (EMG) can also detect motor neuropathy [109].

Although empirical evidence suggests that NCS obtained at baseline and midway through a course of neurotoxic chemotherapy can serve as an early biomarker to predict later severe neuropathy [113], NCS are rarely used to assess CIPN in non-research clinical settings. The tests are uncomfortable, costly, and inconvenient, due to the need for a referral to neurology subspecialty services. Further, the information provided by NCS usually does not provide new information that further informs clinical decision-making beyond what an oncology clinician can determine from simpler clinical assessments (i.e., monofilament and tuning fork-based vibration testing) [114]. Moreover, NCS provide no information about painful small-fiber neuropathy. For this reason, and based on the premise that symptomatic relief, not improved physiological function, is the desired outcome, a recent trend is reliance on PROs, not NCS, to demonstrate intervention efficacy in intervention studies [99, 115]. However, NCS provide important diagnostic information that can identify the specific type of polyneuropathy in complex cases when chemotherapy may not be the sole cause of the neuropathy. Further, NCS can reveal the mechanism of action of CIPN preventative interventions; therefore, they may be important outcome variables in physiological experiments.

3.7.2 Autonomic Function Tests

Autonomic neuropathy arises from damage to peripheral nerves that control involuntary functions, such as vascular diameter, which affects blood pressure, diaphoresis, gastrointestinal and sexual function, and urination [116]. Orthostatic hypotension (due to impaired neurological control of vascular diameter), constipation, erectile dysfunction, and urinary incontinence are the typical manifestations of autonomic CIPN. Tests such as the quantitative sudomotor axon reflex test (QSART), skin sweat testing, and heart variability assessments quantify autonomic neuropathy [111, 116]. For CIPN, however, these tests are not recommended because the findings are no more informative than other assessment approaches and generally cannot be directly attributed to CIPN rather than other causes (e.g., comorbidity, medication use, diet) [108].

3.7.3 Miscellaneous Electrophysiological Assessments

Other tests, such as nerve excitability studies and microneurography, are available to quantify peripheral nerve physiological function [108]. Nerve excitability studies provide information about axon excitability; microneurography is an invasive procedure: needles are inserted into the nerve to examine nociceptor function in patients with pain disorders [108]. However, these tests are not used in clinical practice due to their complexity and the need for highly trained examiners and specialized equipment. For now, these approaches are most appropriate for use in research settings when very detailed physiological data are needed to identify CIPN mechanisms.

3.8 Quantitative Sensory Testing (QST)

Quantitative sensory testing (QST) can be used to assess small and large fiber dysfunction due to treatment with bortezomib, vincristine, taxanes, and platinums [108, 111]. Vibration, mechanical stimuli, and varying degrees of warmth/cold are transmitted through a probe or other device that is applied to the skin at different locations (e.g., fingers, thenar, face, foot dorsum), and patients report whether they can detect these sensations. QST is used most often to provide information regarding specific nerve fiber dysfunction via objective (equipment-generated) and subjective (patient-reported) assessments of mechanical detection and vibration thresholds (Aβ fibers), mechanical pain and cold detection thresholds (Aδ fibers), warmth detection and heat pain thresholds (C fibers), and cold pain thresholds (Aδ and C fibers) [111].

Several methods for conducting QST are described in detail elsewhere [117], a few of which we will describe here. Vibration threshold is often measured using a simple Rydel–Seiffer tuning fork (Fig. 3.1), which is placed on bony prominences bilaterally at the toe, ankle, knee, hip, finger, wrist, and elbow (Fig. 3.2) [117, 118]. The assessor activates the tuning fork, and obtains a numerical reading from the standardized fork markings at the point when the patient reports no longer feeling the vibration, and the findings are interpreted according to age-based norms [36]. Mechanical detection threshold is determined by applying varying-sized von Frey filaments, using progressively larger (ascending) and then smaller (descending) filament sizes. Each filament is applied to the skin for two seconds and the filament size (size=grams) that cannot be detected by the patient with each of the five ascending and five descending testing sequences is recorded and averaged [117, 118]. Mechanical pain threshold tests involve application to the skin of blunt needles at varying weights in five ascending and descending series, resulting in an average weight that elicits pain [117, 118]. For temperature threshold and pain testing, a probe delivers the sensation. The baseline temperature is typically set at

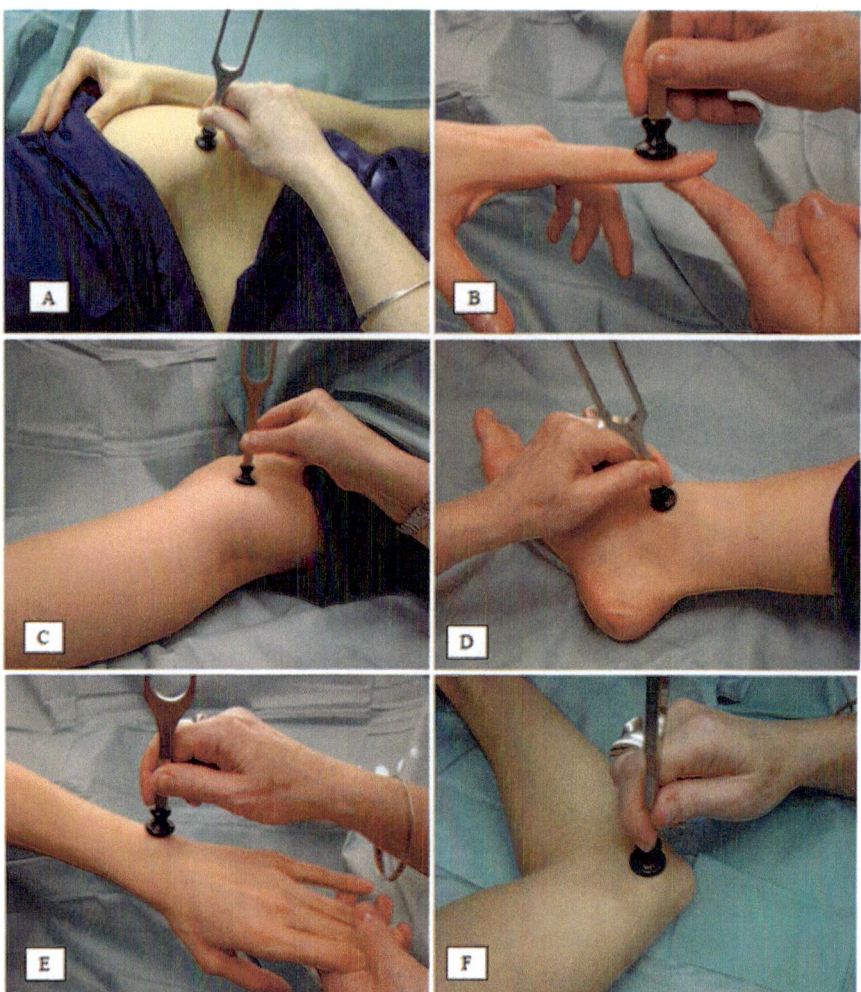

Fig. 3.2 Rydel–Seiffer tuning fork placement. (**a**) Hip. (**b**) Finger. (**c**) Knee. (**d**) Ankle. (**e**) Wrist. (**f**) Elbow

32° centigrade and slowly increased by 1°every second (maximum = 50°) until the patient reports feeling a change in temperature or intolerable discomfort [117–120].

Like any CIPN assessment approach, QST has its advantages and disadvantages. Unlike NCS methods for quantifying CIPN, QST is non-invasive, cost effective, and easy to use [111]. However, results can be unreliable due to differences in the patient's reaction time and ability to fully participate, skin temperature, equipment differences, examiner and participant training, and the anatomical sites of testing used [111, 117, 121–124].

3.9 Bumps Test

Another objective approach for assessing finger and hand tactile threshold (numbness) involves the Bumps Test [125]. The patient runs the index finger over a board that has 12 elevated bumps of different heights. The detection threshold is defined as the height of the smallest bump detected by the patient [125]. When compared to healthy controls ($n = 166$), patients with or at risk for peripheral neuropathy ($n = 103$) had statistically significant ($p < 0.0001$) higher tactile sensation thresholds (could only detect the larger bumps) and took longer to complete the test (mean $= 13.6$ min) [125]. Tactile threshold scores were also positively associated with lower Meissner's corpuscle (mechanoreceptors that detect touch and vibration sensations) density based on skin biopsy findings [125]. Sensitivity and specificity to detect impaired sensation were 71% and 74%, respectively [125]. Further, higher baseline Bumps scores predicted higher CIPN severity ($p = 0.002$) in patients receiving oxaliplatin [105].

Therefore, the Bumps test offers a non-invasive alternative approach for objectively assessing upper extremity sensation in research settings. However, given the requirement for special equipment and the time needed to complete the test, the Bumps test may have limited utility for routine CIPN assessment and monitoring in oncology practice settings.

3.10 Skin Biopsy

Small, unmyelinated epidermal nerve fibers innervate the dermis and epidermis, and die back/disappear when exposed to neurotoxic chemotherapeutic agents [126–130]. Interestingly, some epidermal nerve fibers actually lengthen to compensate for loss in other fibers [126]. Skin biopsy results quantify the magnitude of small nerve fiber loss following treatment with neurotoxic chemotherapy [111, 131]. Patients with CIPN-associated small-fiber neuropathy will have diminished intraepidermal nerve fiber density (IENFD), expressed as the number of fibers per millimeter of epidermal length, and fewer Meissner's corpuscles [127, 131]. Diminished IENFD has been linked with CIPN-associated pain [129, 130], but there is conflicting evidence [108]. In one small pilot study ($N = 12$), some patients with painful CIPN had *improved* IENFD following oxaliplatin or docetaxel treatment when compared to baseline levels [132]. Therefore, clinicians and researchers have not embraced the routine use of skin biopsy as a definitive technique for quantifying CIPN.

The biopsy procedure is invasive and mildly uncomfortable, but generally well tolerated.

However, due to the discomfort associated with the procedure, patients may decline repeated longitudinal testing. Procedures are standardized for the biopsy, laboratory processing, and analysis methods [131]. Tissue collection involves a 3-mm punch biopsy 10 cm above the lateral malleolus [111], but tissue may also be obtained from other sites (e.g., hand, foot, thigh) [126]. Since IENFD is lower in males and older patients, the results are interpreted through comparisons to gender- and age-based norms [131]. Evidence supports the construct validity of IENFD morphologic evaluation based on correlations with sural sensory nerve action potential (SNAP); however, evidence about the associations between IENFD and QST findings is conflicting [131]. Empirical evidence supports excellent intra-rater (stability) and inter-rater (equivalence) reliability (weighted $\kappa \geq 0.90$) when comparing morphologic interpretations provided by the same and different raters [133].

3.11 Biomarkers

Although Chap. 2 outlines numerous biological factors—molecular and genetic— that are associated with an increased risk of developing CIPN, a brief discussion of biomarkers is pertinent here because of their potential to serve as indirect measures of CIPN progression. The caveat is that, despite the potential, no biomarkers are currently available for use in clinical or research settings to monitor CIPN progression over time, or the efficacy of biologically targeted interventions. The main barrier to clinical and research application of biomarkers is directly related to scientific limitations of the current research. Most studies to date were underpowered, used retrospective designs and/or suboptimal phenotype measures, and had inadequate control for confounders [134].

3.12 Nerve Imaging

High-resolution ultrasound (HRUS), magnetic resonance neurography (MRN) and PET/CT imaging are emerging new radiographic methods that may reveal CIPN-associated nerve damage. One recent pilot study revealed that the sural nerve cross-sectional area was smaller in patients receiving paclitaxel ($N = 20$) than in normal controls, and smaller sural nerve diameter was associated with lower IENFD density [135]. Another small pilot study of 20 oxaliplatin-treated patients and matched controls suggests that MRN can reveal dorsal root ganglion hypertrophy in patients with NCS-confirmed CIPN [136]; however, NCS scores did not correlate with dorsal root ganglion size. One interesting case report described NCS and sural nerve biopsy-based evidence of fluorodeoxyglucose (FDG) PET/CT uptake in bilateral peripheral and cranial nerves of a woman who had developed confirmed CIPN after receiving vincristine treatment for lymphoma [137]. Evidence from these early studies is promising, but the data are not conclusive enough to support routine use of imaging studies to quantify CIPN.

3.13   Recommendations for CIPN Assessment in Clinical Practice

The utility of CIPN measurement in clinical practice is limited by the suboptimal effectiveness of available CIPN treatments. For the most part, PROs are the most practical and actionable measures to use in a busy oncology or survivorship clinic because they reveal neurotoxicities that a patient might otherwise forget to mention (and about which a clinician might forget to inquire without routine implementation in the clinical workflow). PROs can be administered as part of a larger symptom assessment (including other important symptoms such as psychosocial distress and pain), and it is generally more feasible and accurate to collect PROs than to require clinician reporting of patient symptoms in routine clinical practice, even with validated scales. Identification of CIPN through biomarkers, nerve conduction studies, and quantitative sensory testing before patients notice (and can report) symptoms is unlikely to be highly useful in a clinical setting given the lack of currently known preventive strategies to employ before symptoms occur or worsen (other than chemotherapy dose reduction).

Improvement of preventive and therapeutic approaches to CIPN may enhance the value of incorporating objective measures of CIPN into clinical practice.

3.14   Conclusion

CIPN measurement can occur via clinician report (using clinical grading scales), patient report, or objective measures of nerve function (which sometimes assess physical function more broadly). Currently, PROs are most clinically useful. Nfl is a promising possible biomarker of CIPN, though more research is needed to further elucidate its utility. High-resolution ultrasound (HRUS), magnetic resonance neurography (MRN), and PET/CT imaging are emerging new radiographic methods that may reveal CIPN-associated nerve damage; however, they are not yet ready for routine clinical use.

References

1. Postma TJ, Heimans JJ, Muller MJ et al (1998) Pitfalls in grading severity of chemotherapy-induced peripheral neuropathy. Ann Oncol 9:739–744. https://doi.org/10.1023/a:1008344507482
2. Hershman DL, Lacchetti C, Dworkin RH et al (2014) Prevention and management of chemotherapy-induced peripheral neuropathy in survivors of adult cancers: American Society of Clinical Oncology Clinical Practice Guideline. J Clin Oncol 32:1941–1967. https://doi.org/10.1200/JCO.2013.54.0914
3. Park SB, Alberti P, Kolb NA et al (2019) Overview and critical revision of clinical assessment tools in chemotherapy-induced peripheral neurotoxicity. J Peripher Nerv Syst 24:S13–S25. https://doi.org/10.1111/jns.12333
4. Cavaletti G, Frigeni B, Lanzani F et al (2010) Chemotherapy-induced peripheral neurotoxicity assessment: a critical revision of the currently available tools. Eur J Cancer 46:479–494. https://doi.org/10.1016/j.ejca.2009.12.008

5. Basch E (2014) The rationale for collecting patient-reported symptoms during routine chemotherapy. Am Soc Clin Oncol Educ B 34:161–165. https://doi.org/10.14694/EdBook_AM.2014.34.161

6. Alberti P, Rossi E, Cornblath DR et al (2014) Physician-assessed and patient-reported outcome measures in chemotherapy-induced sensory peripheral neurotoxicity: two sides of the same coin. Ann Oncol 25:257–264. https://doi.org/10.1093/annonc/mdt409

7. Frigeni B, Piatti M, Lanzani F et al (2011) Chemotherapy-induced peripheral neurotoxicity can be misdiagnosed by the National Cancer Institute Common Toxicity scale. J Peripher Nerv Syst 16:228–236. https://doi.org/10.1111/j.1529-8027.2011.00351.x

8. Lavoie Smith EM, Cohen JA, Pett MA et al (2011) The validity of neuropathy and neuropathic pain measures in patients with cancer receiving taxanes and platinums. Oncol Nurs Forum 38:133–142. https://doi.org/10.1188/11.ONF.133-142

9. U.S. Department of Health and Human Services (2017) NCI-CTCAE ver 5.0. https://ctep.cancer.gov/protocoldevelopment/electronic_applications/docs/CTCAE_v5_Qu ick_Reference_8.5x11.pdf

10. Cavaletti G, Bogliun G, Marzorati L et al (2003) Grading of chemotherapy-induced peripheral neurotoxicity using the Total Neuropathy Scale. Neurology 61:1297–1300. https://doi.org/10.1212/01.wnl.0000092015.03923.19

11. Oken MM, Creech RH, Tormey DC et al (1982) Toxicity and response criteria of the Eastern Cooperative Oncology Group. Am J Clin Oncol Cancer Clin Trials 5:649–655

12. Eastern Cooperative Oncology Group (2007) ECOG common toxicity criteria. http://www.ecog.org/general/ctc.pdf

13. Miller AB, Hoogstraten B, Staquet M, Winkler A (1981) Reporting results of cancer treatment. Cancer 47:207–214. https://doi.org/10.1002/1097-0142(19810101)47:1<207::AID-CNCR2820470134>3.0.CO;2-6

14. Ajani JA, Welch SR, Raber MN et al (1990) Comprehensive criteria for assessing therapy-induced toxicity. Cancer Invest 8:147–159. https://doi.org/10.3109/07357909009017560

15. Levi F, Perpoint B, Garufi C et al (1993) Oxaliplatin activity against metastatic colorectal cancer. A phase II study of 5-day continuous venous infusion at circadian rhythm modulated rate. Eur J Cancer 29:1280–1284. https://doi.org/10.1016/0959-8049(93)90073-O

16. Kautio AL, Haanpää M, Kautiainen H et al (2011) Oxaliplatin scale and National Cancer Institute-common toxicity criteria in the assessment of chemotherapy-induced peripheral neuropathy. Anticancer Res 31:3493–3496

17. Inoue N, Ishida H, Sano M et al (2012) Discrepancy between the NCI-CTCAE and DEB-NTC scales in the evaluation of oxaliplatin-related neurotoxicity in patients with metastatic colorectal cancer. Int J Clin Oncol 17:341–347. https://doi.org/10.1007/s10147-011-0298-z

18. Boku N, Ohtsu A, Hyodo I et al (2007) Phase II study of oxaliplatin in Japanese patients with metastatic colorectal cancer refractory to fluoropyrimidines. Jpn J Clin Oncol 37:440–445. https://doi.org/10.1093/jjco/hym069

19. Cornblath DR, Chaudhry V, Carter K et al (1999) Total neuropathy score: validation and reliability study. Neurology 53:1660–1664

20. Merkies IS, Schmitz PI, van der Meche FG, van Doorn PA (2000) Psychometric evaluation of a new sensory scale in immune-mediated polyneuropathies. Inflammatory Neuropathy Cause and Treatment (INCAT) Group. Neurology 54:943–949

21. Lavoie Smith EM, Li L, Hutchinson RJ et al (2013) Measuring Vincristine-induced peripheral neuropathy in children with acute lymphoblastic leukemia. Cancer Nurs 36:E49–E60. https://doi.org/10.1097/NCC.0b013e318299ad23

22. Smith EML, Cohen JA, Pett MA, Beck SL (2010) The reliability and validity of a modified Total Neuropathy Score-reduced and neuropathic pain severity items when used to measure chemotherapy-induced peripheral neuropathy in patients receiving taxanes and platinums. Cancer Nurs 33:173–183. https://doi.org/10.1097/NCC.0b013e3181c989a3

23. Griffith KA, Merkies IS, Hill EE, Cornblath DR (2010) Measures of chemotherapy-induced peripheral neuropathy: a systematic review of psychometric properties. J Peripher Nerv Syst 15:314–325. https://doi.org/10.1111/j.1529-8027.2010.00292.x
24. Smith EML, Beck SL, Cohen J (2008) The Total Neuropathy Score: a tool for measuring chemotherapy-induced peripheral neuropathy. Oncol Nurs Forum 35:96–102. https://doi.org/10.1188/08.ONF.96-102
25. Smith EML, Kuisell C, Kanzawa-Lee GA et al (2020) Approaches to measure paediatric chemotherapy-induced peripheral neurotoxicity: a systematic review. Lancet Haematol 7: e408–e417. https://doi.org/10.1016/S2352-3026(20)30064-8
26. Chaudhry V, Eisenberger MA, Sinibaldi VJ et al (1996) A prospective study of suramin-induced peripheral neuropathy. Brain 119:2039–2052
27. Cavaletti G, Jann S, Pace A et al (2006) Multi-center assessment of the Total Neuropathy Score for chemotherapy-induced peripheral neurotoxicity. J Peripher Nerv Syst 11:135–141. https://doi.org/10.1111/j.1085-9489.2006.00078.x
28. Cavaletti G, Frigeni B, Lanzani F et al (2007) The Total Neuropathy Score as an assessment tool for grading the course of chemotherapy-induced peripheral neurotoxicity: comparison with the National Cancer Institute-Common Toxicity Scale. J Peripher Nerv Syst 12:210–215. https://doi.org/10.1111/j.1529-8027.2007.00141.x
29. Wampler MA, Topp KS, Miaskowski C et al (2007) Quantitative and clinical description of postural instability in women with breast cancer treated with taxane chemotherapy. Arch Phys Med Rehabil 88:1002–1008. https://doi.org/10.1016/j.apmr.2007.05.007
30. Cavaletti G, Cornblath DR, Merkies IS et al (2013) The chemotherapy-induced peripheral neuropathy outcome measures standardization study: from consensus to the first validity and reliability findings. Ann Oncol 24:454–462. https://doi.org/10.1093/annonc/mds329
31. Chaudhry V, Cornblath D, Corse A et al (2002) Thalidomide-induced neuropathy. Neurology 59:1872–1875
32. Chaudhry V, Rowinsky EK, Sartorius SE et al (1994) Peripheral neuropathy from taxol and cisplatin combination chemotherapy: clinical and electrophysiological studies. Ann Neurol 35:304
33. Binda D, Cavaletti G, Cornblath DR et al (2015) Rasch-Transformed Total Neuropathy Score clinical version (RT-TNSc(©)) in patients with chemotherapy-induced peripheral neuropathy. J Peripher Nerv Syst 20:328–332
34. Lavoie Smith EM, Cohen JA, Pett MA, Beck SL (2011) The validity of neuropathy and neuropathic pain measures in patients with cancer receiving taxanes and platinums. Oncol Nurs Forum 38:133–142
35. Smith EML, Cohen JA, Pett MA, Beck SL (2010) Used to measure chemotherapy-induced receiving taxanes and platinums. Cancer Nurs 33:173–183
36. Panosyan FB, Mountain JM, Reilly MM et al (2016) Rydel-Seiffer fork revisited: beyond a simple case of black and white. Neurology 87:738–740. https://doi.org/10.1212/WNL.0000000000002991
37. Loprinzi CL, Lacchetti C, Bleeker J et al (2020) Prevention and management of chemotherapy-induced peripheral neuropathy in survivors of adult cancers: ASCO guideline update. J Clin Oncol 38:3325–3348. https://doi.org/10.1200/JCO.20.01399
38. Draak THP, Vanhoutte EK, van Nes SI et al (2015) Comparing the NIS vs. MRC and INCAT sensory scale through Rasch analyses. J Peripher Nerv Syst 20:277–288
39. U.S. Food and Drug Administration (2009) Guidance for industry use in medical product development to support labeling claims guidance for industry
40. Tan AC, McCrary JM, Park SB et al (2019) Chemotherapy-induced peripheral neuropathy-patient-reported outcomes compared with NCI-CTCAE grade. Support Care Cancer 27:4771–4777. https://doi.org/10.1007/s00520-019-04781-6
41. Fayers P, Bottomley A (2002) Quality of life research within the EORTC-the EORTC QLQ-C30. Eur J Cancer 38:S125–S133. https://doi.org/10.1007/s10269-006-0412-4

42. European Organisation for Research and Treatment of Cancer (2003) EORTC quality of life chemotherapy-induced peripheral neuropathy. https://qol.eortc.org/questionnaire/qlq-cipn20/

43. Postma TJ, Aaronson NK, Heimans JJ et al (2005) The development of an EORTC quality of life questionnaire to assess chemotherapy-induced peripheral neuropathy: the QLQ-CIPN20. Eur J Cancer 41:1135–1139. https://doi.org/10.1016/j.ejca.2005.02.012

44. Kieffer JM, Postma TJ, van de Poll-Franse L et al (2017) Evaluation of the psychometric properties of the EORTC chemotherapy-induced peripheral neuropathy questionnaire (QLQ-CIPN20). Qual Life Res 26:2999–3010. https://doi.org/10.1007/s11136-017-1626-1

45. Lavoie Smith EM, Barton DL, Qin R et al (2013) Assessing patient-reported peripheral neuropathy: the reliability and validity of the European Organization for Research and Treatment of Cancer QLQ-CIPN20 Questionnaire. Qual Life Res 22:2787–2799. https://doi.org/10.1007/s11136-013-0379-8

46. Smith EML, Banerjee T, Yang JJ et al (2019) Psychometric testing of the European Organisation for Research and Treatment of Cancer Quality of Life Questionnaire-Chemotherapy-Induced Peripheral Neuropathy 20-Item scale using pooled chemotherapy-induced peripheral neuropathy outcome measures standard. Cancer Nurs 42:179–189. https://doi.org/10.1097/NCC.0000000000000596

47. Le-Rademacher J, Kanwar R, Seisler D et al (2017) Patient-reported (EORTC QLQ-CIPN20) versus physician-reported (CTCAE) quantification of oxaliplatin- and paclitaxel/carboplatin-induced peripheral neuropathy in NCCTG/Alliance clinical trials. Support Care Cancer 25:3537–3544. https://doi.org/10.1007/s00520-017-3780-y

48. Smith EML, Knoerl R, Yang JJ et al (2018) In search of a gold standard patient-reported outcome measure for use in chemotherapy-induced peripheral neuropathy clinical trials. Cancer Control 25:1–10. https://doi.org/10.1177/1073274818756608

49. Lavoie Smith EM, Haupt R, Kelly JP et al (2017) The content validity of a chemotherapy-induced peripheral neuropathy patient-reported outcome measure. Oncol Nurs Forum 44:580–588. https://doi.org/10.1188/17.ONF.580-588

50. Wolf SL, Barton DL, Qin R et al (2012) The relationship between numbness, tingling, and shooting/burning pain in patients with chemotherapy-induced peripheral neuropathy (CIPN) as measured by the EORTC QLQ-CIPN20 instrument, N06CA. Support Care Cancer 20:625–632. https://doi.org/10.1007/s00520-011-1141-9

51. Knoerl R, Gray E, Stricker C et al (2017) Electronic versus paper-pencil methods for assessing chemotherapy-induced peripheral neuropathy. Support Care Cancer 25:3437–3446. https://doi.org/10.1007/s00520-017-3764-y

52. Calhoun EA, Welshman EE, Chang CH et al (2003) Psychometric evaluation of the Functional Assessment of Cancer Therapy/Gynecologic Oncology Group - Neurotoxicity (Fact/GOG-Ntx) questionnaire for patients receiving systemic chemotherapy. Int J Gynecol Cancer 13:741–748. https://doi.org/10.1111/j.1525-1438.2003.13603.x

53. Functional Assessment of Chronic Illness Therapy (2007) FACT Questionnaires. https://www.facit.org/FACITOrg/Questionnaires

54. Almadrones L, McGuire DB, Walczak JR et al (2004) Psychometric evaluation of two scales assessing functional status and peripheral neuropathy associated with chemotherapy for ovarian cancer: a gynecologic oncology group study. Oncol Nurs Forum 31:615–623. https://doi.org/10.1188/04.ONF.615-623

55. Huang HQ, Brady MF, Cella D et al (2007) Validation and reduction of FACT/GOG-Ntx subscale for platinum/paclitaxel-induced neurologic symptoms: a gynecologic oncology group study. Int J Gynecol Cancer 17:387–393. https://doi.org/10.1111/j.1525-1438.2007.00794.x

56. Cheng HL, Lopez V, Lam SC et al (2020) Psychometric testing of the Functional Assessment of Cancer Therapy/Gynecologic Oncology Group - Neurotoxicity (FACT/GOG-Ntx) subscale in a longitudinal study of cancer patients treated with chemotherapy. Health Qual Life Outcomes 18:1–9. https://doi.org/10.1186/s12955-020-01493-y

57. Basch EM, Reeve BB, Mitchell SA et al (2011) Electronic toxicity monitoring and patient-reported outcomes. Cancer J 17:231–234. https://doi.org/10.1097/PPO.0b013e31822c28b3

58. Basch E, Reeve BB, Mitchell SA et al (2014) Development of the national cancer institute's patient-reported outcomes version of the common terminology criteria for adverse events (PRO-CTCAE). J Natl Cancer Inst 106. https://doi.org/10.1093/jnci/dju244

59. Dueck AC, Mendoza TR, Mitchell SA et al (2015) Validity and reliability of the US National Cancer Institute's patient-reported outcomes version of the common terminology criteria for adverse events (PRO-CTCAE). JAMA Oncol 1:1051–1059. https://doi.org/10.1001/jamaoncol.2015.2639

60. Basch E, Pugh SL, Dueck AC et al (2017) Feasibility of patient reporting of symptomatic adverse events via the patient-reported outcomes version of the common terminology criteria for adverse events (PRO-CTCAE) in a Chemoradiotherapy Cooperative Group Multicenter Clinical Trial. Int J Radiat Oncol Biol Phys 98:409–418. https://doi.org/10.1016/j.ijrobp.2017.02.002

61. Hausheer FH, Schilsky RL, Bain S et al (2006) Diagnosis, management, and evaluation of chemotherapy-induced peripheral neuropathy. Semin Oncol 33:15–49. https://doi.org/10.1053/j.seminoncol.2005.12.010

62. Kuroi K, Shimozuma K, Ohashi Y et al (2008) A questionnaire survey of physicians' perspectives regarding the assessment of chemotherapy-induced peripheral neuropathy in patients with breast cancer. Jpn J Clin Oncol 38:748–754. https://doi.org/10.1093/jjco/hyn100

63. Kuroi K, Shimozuma K, Ohashi Y et al (2009) Prospective assessment of chemotherapy-induced peripheral neuropathy due to weekly paclitaxel in patients with advanced or metastatic breast cancer (CSP-HOR 02 study). Support Care Cancer 17:1071–1080. https://doi.org/10.1007/s00520-008-0550-x

64. Shimozuma K, Ohashi Y, Takeuchi A et al (2009) Feasibility and validity of the Patient Neurotoxicity Questionnaire during taxane chemotherapy in a phase III randomized trial in patients with breast cancer: N-SAS BC 02. Support Care Cancer 17:1483–1491. https://doi.org/10.1007/s00520-009-0613-7

65. Binda D, Vanhoutte EK, Cavaletti G et al (2013) Rasch-built Overall Disability Scale for patients with chemotherapy-induced peripheral neuropathy (CIPN-R-ODS). Eur J Cancer 49:2910–2918. https://doi.org/10.1016/j.ejca.2013.04.004

66. Driessen CML, De Kleine-Bolt KME, Vingerhoets AJJM et al (2012) Assessing the impact of chemotherapy-induced peripheral neurotoxicity on the quality of life of cancer patients. Support Care Cancer 20:877–881. https://doi.org/10.1007/s00520-011-1336-0

67. Beijers AJM, Mols F, Driessen CML, et al (2014) Chemotherapy-induced peripheral neuropathy and impact on quality of life six months after treatment with taxanes and platinum derivatives. J Community Support Oncol 12:401–406. https://doi.org/10.12788/jcso.0086

68. da Silva Simão DA, Teixeira AL, Souza RS, de Paula Lima EDR (2014) Evaluation of the Semmes–Weinstein filaments and a questionnaire to assess chemotherapy-induced peripheral neuropathy. Support Care Cancer 22:2767–2773. https://doi.org/10.1007/s00520-014-2275-3

69. Tofthagen CS, McMillan SC, Kip KE (2011) Development and psychometric evaluation of the chemotherapy-induced peripheral neuropathy assessment tool. Cancer Nurs 34. https://doi.org/10.1097/NCC.0b013e31820251de

70. Simsek NY, Demir A (2018) Reliability and validity of the Turkish version of Chemotherapy-Induced Peripheral Neuropathy Assessment Tool for breast cancer patients receiving taxane chemotherapy. Asia Pac J Oncol Nurs 5:435–441. https://doi.org/10.4103/apjon.apjon

71. Obaid A, El-Aqoul A, Alafafsheh A et al (2020) Validation of the Arabic version of the chemotherapy-induced peripheral neuropathy assessment tool. Pain Manag Nurs. https://doi.org/10.1016/j.pmn.2020.05.005

72. Kutlutürkan S, Öztürk ES, Arıkan F et al (2017) The psychometric properties of the Turkish version of the Chemotherapy-Induced Peripheral Neuropathy Assessment Tool (CIPNAT). Eur J Oncol Nurs 31:84–89. https://doi.org/10.1016/j.ejon.2017.10.001

73. Mendoza TR, Wang XS, Williams LA et al (2015) Measuring therapy-induced peripheral neuropathy: preliminary development and validation of the Treatment-Induced Neuropathy Assessment Scale. J Pain 16:1032–1043. https://doi.org/10.1016/j.jpain.2015.07.002

74. Williams LA, Garcia-Gonzalez A, Mendoza TR et al (2019) Concept domain validation and item generation for the Treatment-Induced Neuropathy Assessment Scale (TNAS). Support Care Cancer 27:1021–1028. https://doi.org/10.1007/s00520-018-4391-y
75. Mendoza TR, Williams LA, Shi Q et al (2020) The Treatment-induced Neuropathy Assessment Scale (TNAS): a psychometric update following qualitative enrichment. J Patient Rep Outcomes 4. https://doi.org/10.1186/s41687-020-0180-8
76. Oldenburg J, Fosså SD, Dahl AA (2006) Scale for chemotherapy-induced long-term neurotoxicity (SCIN): psychometrics, validation, and findings in a large sample of testicular cancer survivors. Qual Life Res 15:791–800. https://doi.org/10.1007/s11136-005-5370-6
77. Miyoshi Y, Onishi C, Fujie M et al (2015) Validity of the chemotherapy-induced peripheral neuropathy self-check sheet. Intern Med 54:737–742. https://doi.org/10.2169/internalmedicine.54.3318
78. Beijers AJM, Vreugdenhil G, Oerlemans S et al (2016) Chemotherapy-induced neuropathy in multiple myeloma: influence on quality of life and development of a questionnaire to compose common toxicity criteria grading for use in daily clinical practice. Support Care Cancer 24:2411–2420. https://doi.org/10.1007/s00520-015-3032-y
79. Leonard GD, Wright MA, Quinn MG et al (2005) Survey of oxaliplatin-associated neurotoxicity using an interview-based questionnaire in patients with metastatic colorectal cancer. BMC Cancer 5:1–10. https://doi.org/10.1186/1471-2407-5-116
80. Gustafsson E, Litström E, Berterö C, Drott J (2016) Reliability testing of oxaliplatin-associated neurotoxicity questionnaire (OANQ), a pilot study. Support Care Cancer 24:747–754. https://doi.org/10.1007/s00520-015-2838-y
81. Kanzawa-Lee GA, Knoerl R, Donohoe C et al (2019) Mechanisms, predictors, and challenges in assessing and managing painful chemotherapy-induced peripheral neuropathy. Semin Oncol Nurs 35:253–260. https://doi.org/10.1016/j.soncn.2019.04.006
82. Cleeland CS (1989) Measurement of pain by subjective report. In: Chapman CR, Loeser JD (eds) Advances in pain research and therapy. Raven, New York, pp 391–403
83. Cleeland CS (2009) The brief pain inventory user guide. University of Texas MD Anderson Cancer Center. https://www.mdanderson.org/content/dam/mdanderson/documents/Departments-and-Divisions/Symptom-Research/BPI_UserGuide.pdf
84. Smith EML, Pang H, Cirrincione C et al (2013) Effect of duloxetine on pain, function, and quality of life among patients with chemotherapy-induced painful peripheral neuropathy. JAMA 309:1359–1367
85. Farrar JT, Young JP, LaMoreaux L et al (2001) Clinical importance of changes in chronic pain intensity measured on an 11-point numerical pain rating scale. Pain 94:149–158. https://doi.org/10.1016/S0304-3959(01)00349-9
86. Cavaletti G, Cornblath DR, Merkies ISJ et al (2013) The chemotherapy-induced peripheral neuropathy outcome measures standardization study: from consensus to the first validity and reliability findings. Ann Oncol 24:454–462. https://doi.org/10.1093/annonc/mds329
87. Crichton N (2001) Visual analog scale (VAS). J Clin Nurs 10:706. https://doi.org/10.1097/00002060-200110000-00001
88. Maxwell C (1978) Sensitivity and accuracy of the visual analogue scale: a psycho-physical classroom experiment. Br J Clin Pharmacol 6:15–24. https://doi.org/10.1111/j.1365-2125.1978.tb01676.x
89. Takemoto S, Ushijima K, Honda K et al (2012) Precise evaluation of chemotherapy-induced peripheral neuropathy using the visual analogue scale: a quantitative and comparative analysis of neuropathy occurring with paclitaxel-carboplatin and docetaxel-carboplatin therapy. Int J Clin Oncol 17:367–372. https://doi.org/10.1007/s10147-011-0303-6
90. Bouhassira D, Attal N, Fermanian J et al (2004) Development and validation of the Neuropathic Pain Symptom Inventory. Pain 108:248–257. https://doi.org/10.1016/j.pain.2003.12.024

91. Lu LC, Chang SY, Liu CY, Tsay SL (2018) Reliability and validity of the Chinese version neuropathic pain symptom inventory in patients with colorectal cancer. J Formos Med Assoc 117:1019–1026. https://doi.org/10.1016/j.jfma.2017.11.010

92. Askew RL, Cook KF, Keefe FJ et al (2016) A PROMIS measure of neuropathic pain quality. Value Heal 19:623–630. https://doi.org/10.1016/j.jval.2016.02.009

93. Bouhassira D, Attal N, Alchaar H et al (2005) Comparison of pain syndromes associated with nervous or somatic lesions and development of a new neuropathic pain diagnostic questionnaire (DN4). Pain 114:29–36. https://doi.org/10.1016/j.pain.2004.12.010

94. Pérez C, Sánchez-Martínez N, Ballesteros A et al (2015) Prevalence of pain and relative diagnostic performance of screening tools for neuropathic pain in cancer patients: a cross-sectional study. Eur J Pain (UK) 19:752–761. https://doi.org/10.1002/ejp.598

95. Bennett M (2001) The LANSS Pain Scale: the Leeds assessment of neuropathic symptoms and signs. Pain 92:147–157. https://doi.org/10.1016/S0304-3959(00)00482-6

96. Reyes-Gibby C, Morrow PK, Bennett MI et al (2010) Neuropathic pain in breast cancer survivors: using the ID Pain as a screening tool. J Pain Symptom Manage 39:882–889. https://doi.org/10.1016/j.jpainsymman.2009.09.020

97. Winters-Stone KM, Horak F, Jacobs PG et al (2017) Falls, functioning, and disability among women with persistent symptoms of chemotherapy-induced peripheral neuropathy. J Clin Oncol 35:2604–2612. https://doi.org/10.1200/JCO.2016.71.3552

98. Vollmers PL, Mundhenke C, Maass N et al (2018) Evaluation of the effects of sensorimotor exercise on physical and psychological parameters in breast cancer patients undergoing neurotoxic chemotherapy. J Cancer Res Clin Oncol 144:1785–1792. https://doi.org/10.1007/s00432-018-2686-5

99. Gewandter JS, Brell J, Cavaletti G et al (2018) Trial designs for chemotherapy-induced peripheral neuropathy prevention. Neurology 91:403–413. https://doi.org/10.1212/WNL.0000000000006083

100. Podsiadlo D, Richardson S (1991) The timed "Up & Go": a test of basic functional mobility for frail elderly persons. J Am Geriatr Soc 39:142–148

101. Vaz MM, Costa GC, Reis JG et al (2013) Postural control and functional strength in patients with type 2 diabetes mellitus with and without peripheral neuropathy. Arch Phys Med Rehabil 94:2465–2470. https://doi.org/10.1016/j.apmr.2013.06.007

102. Hernandez D, Rose DJ (2008) Predicting which older adults will or will not fall using the Fullerton Advanced Balance scale. Arch Phys Med Rehabil 89:2309–2315. https://doi.org/10.1016/j.apmr.2008.05.020

103. Jeon Y-J, Kim G-M (2017) Comparison of the Berg Balance Scale and Fullerton Advanced Balance Scale to predict falls in community-dwelling adults. J Phys Ther Sci 29:232–234. https://doi.org/10.1589/jpts.29.232

104. Vasquez S, Guidon M, McHugh E et al (2014) Chemotherapy induced peripheral neuropathy: the modified total neuropathy score in clinical practice. Ir J Med Sci 183:53–58. https://doi.org/10.1007/s11845-013-0971-5

105. Schlenstedt C, Brombacher S, Hartwigsen G et al (2016) Comparison of the Fullerton Advanced Balance Scale, Mini-BESTest, and Berg Balance Scale to predict falls in Parkinson disease. Phys Ther 96:494–501. https://doi.org/10.2522/ptj.20150249

106. McCrary JM, Goldstein D, Wyld D et al (2019) Mobility in survivors with chemotherapy-induced peripheral neuropathy and utility of the 6-min walk test. J Cancer Surviv 13:495–502. https://doi.org/10.1007/s11764-019-00769-7

107. Niu H-X, Wang R-H, Xu H-L et al (2017) Nine-hole Peg Test and Ten-meter Walk Test for evaluating functional loss in Chinese Charcot-Marie-Tooth disease. Chin Med J (Engl) 130:1773–1778. https://doi.org/10.4103/0366-6999.211550

108. Argyriou AA, Park SB, Islam B et al (2019) Neurophysiological, nerve imaging and other techniques to assess chemotherapy-induced peripheral neurotoxicity in the clinical and research settings. J Neurol Neurosurg Psychiatry 90:1361–1369. https://doi.org/10.1136/jnnp-2019-320969

109. England JD, Gronseth GS, Franklin G et al (2005) Distal symmetric polyneuropathy: a definition for clinical research—report of the American Academy of Neurology, the American Association of Electrodiagnostic Medicine, and the American Academy of Physical Medicine and Rehabilitation. Neurology 64:199–207. https://doi.org/10.1212/01.WNL.0000149522. 32823.EA

110. Perkins BA, Bril V (2003) Diabetic neuropathy: a review emphasizing diagnostic methods. Clin Neurophysiol 114:1167–1175. https://doi.org/10.1016/S1388-2457(03)00025-7

111. Timmins HC, Li T, Kiernan MC et al (2019) Quantification of small fiber neuropathy in chemotherapy-treated patients. J Pain 21:44–58. https://doi.org/10.1016/j.jpain.2019.06.011

112. Park SB, Goldstein D, Krishnan AV et al (2013) Chemotherapy-induced peripheral neurotoxicity: a critical analysis. CA Cancer J Clin 63:419–437. https://doi.org/10.1002/caac.21204

113. Alberti P, Rossi E, Argyriou AA et al (2018) Risk stratification of oxaliplatin induced peripheral neurotoxicity applying electrophysiological testing of dorsal sural nerve. Support Care Cancer 26:3143–3151. https://doi.org/10.1007/s00520-018-4170-9

114. Griffith KA, Dorsey SG, Renn CL et al (2014) Correspondence between neurophysiological and clinical measurements of chemotherapy-induced peripheral neuropathy: secondary analysis of data from the CI-PeriNomS study. J Peripher Nerv Syst 19:127–135. https://doi.org/10. 1111/jns5.12064

115. Dorsey SG, Kleckner IR, Barton D et al (2019) The National Cancer Institute Clinical Trials Planning Meeting for prevention and treatment of chemotherapy-induced peripheral neuropathy. J Natl Cancer Inst 111:531–537. https://doi.org/10.1093/jnci/djz011

116. England JD, Gronseth GS, Franklin G et al (2009) Practice parameter: evaluation of distal symmetric polyneuropathy: role of autonomic testing, nerve biopsy, and skin biopsy (an evidence-based review): report of the American Academy of Neurology, American Association of Neuromuscular and Electrodiagnostic Medicine, and American Academy of Physical Medicine and Rehabilitation. Neurology 72:177–184. https://doi.org/10.1212/01. wnl.0000336345.70511.0f

117. Rolke R, Andrews K, Magerl W, Treede R-D (2010) German Research Network on Neuropathic Pain. Quantitative sensory testing investigator's manual, version 2.1

118. Mücke M, Cuhls H, Radbruch L et al (2016) Quantitative sensory testing (QST). English version. Schmerz. https://doi.org/10.1007/s00482-015-0093-2

119. Maier C, Baron R, Tölle TR et al (2010) Quantitative sensory testing in the German Research Network on Neuropathic Pain (DFNS): somatosensory abnormalities in 1236 patients with different neuropathic pain syndromes. Pain 150:439–450. https://doi.org/10.1016/j.pain.2010. 05.002

120. Rolke R, Magerl W, Campbell KA et al (2006) Quantitative sensory testing: a comprehensive protocol for clinical trials. Eur J Pain 10:77–88. https://doi.org/10.1016/j.ejpain.2005.02.003

121. Cazzato D, Lauria G (2017) Small fibre neuropathy. Curr Opin Neurol 30:490–499. https:// doi.org/10.1097/WCO.0000000000000472

122. Andriamamonjy M, Delmotte J-B, Savinelli F et al (2017) Quantification of chronic oxaliplatin-induced hypesthesia in two areas of the hand. J Clin Neurophysiol 34:126–131. https://doi.org/10.1097/WNP.0000000000000347

123. Velasco R, Videla S, Villoria J et al (2015) Reliability and accuracy of quantitative sensory testing for oxaliplatin-induced neurotoxicity. Acta Neurol Scand 131:282–289. https://doi.org/ 10.1111/ane.12331

124. Backonja M-M, Walk D, Edwards RR et al (2009) Quantitative sensory testing in measurement of neuropathic pain phenomena and other sensory abnormalities. Clin J Pain 25:641–647. https://doi.org/10.1097/AJP.0b013e3181a68c7e

125. Kennedy WR, Selim MM, Brink TS et al (2011) A new device to quantify tactile sensation in neuropathy. Neurology 76:1642–1649

126. Waller LA, Särkkä A, Olsbo V et al (2011) Second-order spatial analysis of epidermal nerve fibers. Stat Med 30:2827–2841. https://doi.org/10.1002/sim.4315

127. Mangus LM, Rao DB, Ebenezer GJ (2020) Intraepidermal nerve fiber analysis in human patients and animal models of peripheral neuropathy: a comparative review. Toxicol Pathol 48:59–70. https://doi.org/10.1177/0192623319855969

128. Burakgazi AZ, Messersmith W, Vaidya D et al (2011) Longitudinal assessment of oxaliplatin-induced neuropathy. Neurology 77:980–986. https://doi.org/10.1212/WNL. 0b013e31822cfc59

129. Boyette-Davis JA, Cata JP, Zhang H et al (2011) Follow-up psychophysical studies in bortezomib-related chemoneuropathy patients. J Pain 12:1017–1024. https://doi.org/10. 1016/j.jpain.2011.04.008

130. Boyette-Davis JA, Cata JP, Driver LC et al (2013) Persistent chemoneuropathy in patients receiving the plant alkaloids paclitaxel and vincristine. Cancer Chemother Pharmacol 71:619–626. https://doi.org/10.1007/s00280-012-2047-z

131. Lauria G, Hsieh ST, Johansson O et al (2010) European Federation of Neurological Societies/ Peripheral Nerve Society Guideline on the use of skin biopsy in the diagnosis of small fiber neuropathy. Report of a joint task force of the European Federation of Neurological Societies and the Peripheral Nerve Society. Eur J Neurol 17:903–912 e44–49. https://doi.org/10.1111/j. 1468-1331.2010.03023.x

132. Koskinen MJ, Kautio A-L, Haanpää ML et al (2011) Intraepidermal nerve fibre density in cancer patients receiving adjuvant chemotherapy. Anticancer Res 31:4413–4416

133. Bakkers M, Merkies ISJ, Lauria G et al (2009) Intraepidermal nerve fiber density and its application in sarcoidosis. Neurology 73:1142–1148. https://doi.org/10.1212/WNL. 0b013e3181bacf05

134. Chan A, Hertz DL, Morales M et al (2019) Biological predictors of chemotherapy-induced peripheral neuropathy (CIPN): MASCC Neurological Complications Working Group overview. Support Care Cancer 27:3729–3737. https://doi.org/10.1007/s00520-019-04987-8

135. Lycan TW, Hsu F-C, Ahn CS et al (2020) Neuromuscular ultrasound for taxane peripheral neuropathy in breast cancer. Muscle Nerve 61:587–594. https://doi.org/10.1002/mus.26833

136. Apostolidis L, Schwarz D, Xia A et al (2017) Dorsal root ganglia hypertrophy as in vivo correlate of oxaliplatin-induced polyneuropathy. PLoS One 12:e0183845. https://doi.org/10. 1371/journal.pone.0183845

137. Ng KS, Leung JMS (2019) Vincristine-induced polyneuropathy at FDG PET/CT. Radiology 293:36. https://doi.org/10.1148/radiol.2019190861

Preventive Strategies for Chemotherapy-Induced Peripheral Neuropathy

4

Basic Science and Models for Drug Development

Sebastian Werngreen Nielsen and Jørn Herrstedt

Abstract

There are no clinically relevant, evidence-based preventive strategies for chemotherapy-induced peripheral neuropathy (CIPN). In this chapter we discuss how limitations in current animal models lead to insufficient understanding of CIPN pathophysiology and how drug development for neurodegenerative diseases in general suffers because of this. We draw on previous studies of CIPN prevention to reflect upon what can be learned, but this chapter is not a historical account of past CIPN strategies nor it is an exhaustive list of CIPN mechanisms in rodents and mice. There are several succinctly well-written and recent reviews that cover these topics.

We look towards the horizon of CIPN drug development and provide an overview of the strategies that are emerging. We argue that some of these strategies herald early signs of methodological change for CIPN research, where basic science researchers begin to employ a systems biology approach to model neurological diseases such as CIPN in greater pathophysiological detail. Here diseases are caused by disruption to biological networks such as the neuron/neuroglia homeostasis rather than singular mechanisms within individual cell types. In this new perspective, we suggest three "core mechanisms" of CIPN that could be modeled within a systems biology methodology. We present studies that show how new methods, such as single cell multi-omics and bioengineering of

S. W. Nielsen · J. Herrstedt (✉)
Department of Clinical Oncology and Palliative Care, Zealand University Hospital, Roskilde, Denmark

Institute of Clinical Medicine, Faculty of Health and Medical Sciences, University of Copenhagen, Copenhagen, Denmark
e-mail: sewn@regionsjaelland.dk; jherr@regionsjaelland.dk

© The Author(s), under exclusive license to Springer Nature Switzerland AG 2021
M. Lustberg, C. Loprinzi (eds.), *Diagnosis, Management and Emerging Strategies for Chemotherapy-Induced Neuropathy*,
https://doi.org/10.1007/978-3-030-78663-2_4

human 3D organoids, can be analyzed with machine learning algorithms to aid CIPN drug development.

Keywords

Chemotherapy-induced peripheral neuropathy · Pathophysiological mechanisms · Drug development · Explanatory models · Animal studies · Systems biology · Machine learning · Multi-omics

4.1   Introduction

Normal science, the activity in which most scientists inevitably spend almost all their time, is predicated on the assumption that the scientific community knows what the world is like.
– Thomas S. Kuhn, Philosopher of Science, Physicist

A patient recently described his side-effects, as he was filling out a neuropathy questionnaire. *"Everything has changed. When I tighten metal bolts it feel like someone is tearing off my fingernails. I can only do a few, before I have to stop."* As a single provider working as a certified electrician, he had become unsure whether adjuvant chemotherapy had been a good idea given the severity and impact of his side-effects. He was now a cancer survivor. Surgery and oxaliplatin-based chemotherapy had *cured* him, yet his peripheral neuropathy would not allow him to (ever) *heal.*

Chemotherapy-induced peripheral neuropathy (CIPN) is one among many late-effects of cancer treatments that keep many patients in a complex state of survivorship as CIPN symptoms can continue years after treatment has ended. The term "survivorship" itself implies a transition from patient to survivor and not normality [1]. This is why prevention of long-term neurotoxicity is of paramount importance. Prevention is not about treating CIPN, but thwarting its emergence in the first place and, by extension, offering the possibility of alleviating the impact of survivorship.

The history of CIPN prevention is also the story of drug development and technological advancement within basic sciences. The paradigm of drug development is well known; a new method is applied within biology and it produces new insights into cellular and/or protein based mechanisms. Targetable proteins/lipids are reviewed and targeted drugs are tested. First, in cells, then animals and, lastly, in humans. It is a powerful machine that churns out numerous hopeful mechanisms and accompanied drugs. Yet, the failure rate of pre-clinical drug development is a staggering 96% and comes with a lofty price tag for successful drugs [2]. This is true for CIPN drug development, as well, except, here, the failure rate is 100%, so far, as there are no effective preventive drugs.

The mechanisms of CIPN that inform drug development have been reviewed in some 60 review articles (see [3, 4] for 2020 updates of commonly used drugs such as taxanes and platinum). Many candidate drugs are successful in the early phases of drug development, but fail the transition from animal to human studies. A

Fig. 4.1 Knowledge domains that can be useful for CIPN preventive strategy development

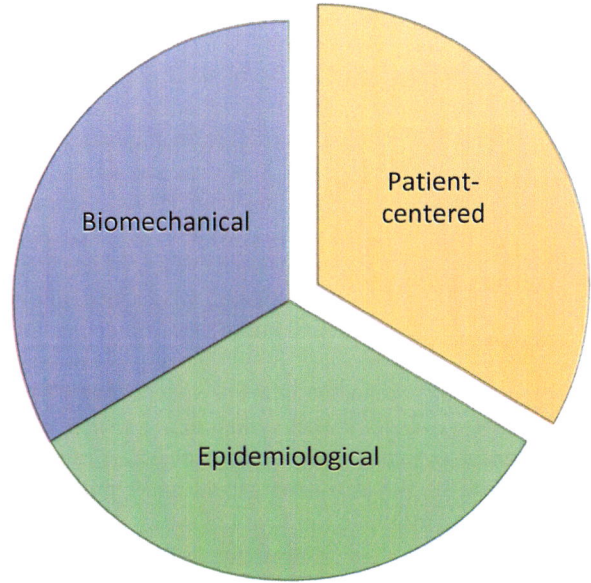

phenomenon aptly called the "valley of death" in drug development as this is where most promising drugs meet their end [5]. After more than 50 clinical trials of CIPN treatment and prevention, we do not seem to be much closer to a solution. The answer to these failures seems to be that it was the drug that failed, not the methodology used to derive or test it. Hence we should try again, but only better [2]. The entire field of CIPN pathobiology seems so caught up in molecular biology that we forgot to ask the patients if we are researching the right thing. Qualitative studies of patient perspectives on CIPN show that patients will accept transient CIPN symptoms in exchange for certainty of treatment efficacy. They only become concerned about long-term chronic symptoms which become apparent in the weeks and months after treatment has started or ended [6–8]. Since the mechanisms involved in acute and chronic CIPN may not always be the same [9], it is concerning that almost all animal research investigating CIPN pathophysiology applies to the hours and days following administration. Most emerging strategies will be founded on mechanistic insight from animal models; so, we are in fact trying to prevent a form of CIPN (the acute and transient) that patients are not really concerned about, when we should try to prevent long-term chronic and painful CIPN [10]. In the span of almost 40 years of CIPN research we argue that no real epistemological change has occurred in drug development. The first step is almost solely based in a non-iterative biomechanical reductionism [11], yet this is changing rapidly as new ideas are implemented in drug development methodology [12].

On the horizon, a novel type of analyses is anticipated to yield potential drugs. Artificial intelligence (AI) can effectively incorporate the different types of data that we already have into new models of CIPN [12] (Fig. 4.1). By greatly expanding the data output on CIPN with the use of "multi-omics" we can continue to use biology to derive new data on the mechanisms of CIPN [13]. Associations from epidemiological studies of patient populations could also provide useful data and promising drugs such as the repurposing of the oral antidiabetic drug, metformin [14]. High quality epidemiological studies based on validated registries could also help us determine whether cancer patients receiving the antihypertensive drug, carvedilol are at less risk of developing CIPN [15]. Lastly, as care becomes patients-centered, treatments become personalized and funding agencies demand patient-involvement in research, we might want to start taking patient experiences and values seriously. This represents a shift from the traditional evidence-based model where expert and patient values and experience are considered a poor level of evidence [16]. However, these "local" forms of evidence and knowledge from clinicians and their patients may hold many insights and nuances of worth which warrant scientific investigation. An example of this is cannabinoids: For years patients have been using various cannabinoid compounds for cancer symptom relief such as neuropathic pain [17] and the endocannabinoidome is now a major object of research interest within the field of neurology [18]. We also need patient and clinician insights and classifications to adequately supervise the formation of future AI based models in order to adequately predict the outcomes of interest within the field of CIPN [19].

This chapter describes CIPN prevention by tracing the traditional path of mechanistic insights from molecular biology into an emergent reframing of CIPN as networked mechanisms unfolding within biological systems. We delve into the limitations of animal models to understand the reasons why every successful animal study should not be heralded as a promising new emerging strategy. We provide a comprehensive list of promising drugs and strategies emerging within the last 10 years and unfold the most promising strategies that point towards a new approach to CIPN pathophysiology based on systems biology. But first, we reflect upon several overarching problems in CIPN research which we should keep in mind as we go forward.

In all science, *"conceptualization precedes operationalization"* [11]. In order to effectively prevent CIPN, we must first be able to classify and measure CIPN. The classical categories of central and peripheral neuropathy further detailed by motor, sensory, and autonomic neuropathy has not led to an accepted gold standard of CIPN ascertainment or grading [20]. The mechanisms that induce acute and chronic neuropathy can be distinct from each other and a novel drug that targets acute neurotoxic mechanisms may not protect against chronic neuropathy [21]. We do not have a specific biomarker, test, or tool, which ultimately leaves inadequate methods of measurement. In other words, we lack a good understanding of what CIPN is. This is emphasized in the discussions and conclusions of almost all CIPN review articles. But is there a valid overarching concept called CIPN in patients? Or does the concept of CIPN simply fulfill a need for a common general term, while we may in fact be dealing with heterogeneous drug-specific neuropathies? At first glance, this question may seem overly philosophical. However, the implications regarding it are dire. If the conceptualization of CIPN is vague, classification suffers equally and subsequent scientific investigations become unfocused. If we are, in fact, dealing with distinct types of neuropathy, then searching for *ONE* central mechanism of CIPN or *ONE* preventive measure is futile, for it will simply not exist. This also entails that if we found one for paclitaxel, it might not work for vincristine and vice versa [22, 23]. This problem of classification and measurement also has implications for prevention trials. Meta-analysis of studies and the development of evidence-based treatment options are hampered by the lack of measurement consensus as well as the incommensurability of differing measurement methods [20, 24]. As nerve conduction studies are not easily applicable to clinical practice [20], researchers have turned to the apparent objectivity of psychometrics in the form of patient-reported outcomes (PROs) as a measure of CIPN categories. However, the construct validity and reliability of CIPN psychometric scores are far from perfect and their validity may be confounded by unmet assumptions of the item response theory such as unidimensionality [25]. Paired with the continued crises of reproducibility within psychology [26], psychometrics begins to lose its appeal as a scientific method of CIPN diagnostics. For instance, the psychometrically validated 20-item questionnaire from the European Organisation for Research and Treatment of Cancer (EORTC-CIPN20) was designed to distinguish between motor, sensory, and autonomic chemotherapy-induced neuropathy. This would enable researchers to adequately assess specific interventions targeted at specific types of chemotherapy-induced neuropathy. Yet, the psychometric properties of the EORTC-CIPN20 subscales did not perform as expected when modeled mathematically, which led researchers to suggest rejecting the CIPN construct and adopt a simple summation of items representing symptom burden [25, 27].

In this context, CIPN begins to look like an illness that degrades patients' quality of life in a myriad of ways which may not always be captured accurately by state-of-the-art instruments and measures. PROs do not adequately measure underlying CIPN pathophysiology, and measurements based on technology such as nerve conduction cannot capture CIPN impact on patient life. Instead, it may be that we have to do both, when we design CIPN prevention and intervention trials, in order to capture the complex interrelatedness of the CIPN *disease* and the CIPN *illness* in a comparative and reproducible manner [23, 28].

4.2 Limitations of Extrapolating from Animal Models

Laboratory experiments are faster, cheaper, safer, and easier to do than are clinical trials in humans. While novel laboratory models based on human biology are being developed [29, 30], we must contend with the fact that almost all insights on CIPN mechanisms are from single cell type studies or animal models and that almost 80% of animals were administered paclitaxel or oxaliplatin [31]. We believe that this poses significant limitations which make extrapolation from the laboratory difficult. Below we consider three of these limitations.

4.2.1   Lab Animals Are Not Human Beings

While this statement seems trivial, we think that the implication goes beyond an important discussion of different mammalian pharmacokinetic and pharmacodynamic profiles [2]. In 2014, the Lancet commissioned a work on culture and health, Napier and colleagues appraised the impact of culture on health and biomedicine. In short, human beings are situated in complex political, economic, and cultural relationships, which shape human action upon—and experience of the diseased body and self [32]. As we have briefly touched upon, there if often a very low correlation between objective and subjective measures of CIPN [28], it may be hard to predict how human beings will value the effect of a drug by evaluating measurements from animal studies. Another example of the influence of dynamic relationships in human worlds is the phenomenon of placebo/nocebo effects, which are unpredictable and may change over the course of time, causing RCT methodology to weaken [33]. In some ways, the controlled environment of the lab is a world apart from the environment of patients, as patients *"will often have their own interpretation of what is going on in these trials, and this interpretation may influence their responses over and above the behavior intended by the experimenters"* [34].

4.2.2   Administration and Measurement Methods in Lab Animals Are Not Suited for Human Trial Designs

In a 2014 article, Höke and Ray [35] succinctly describe numerous problems with current animal models, in regard to selection of animals, mode of administration, and outcome measurements. In short, intrathecal or intraperitoneal administration of chemotherapy is one of the many disparate problems in animal models. Simply factoring in the pharmacodynamic difference between administration methods is not enough, as other synergies and effects may be underestimated or neglected. These may be chemotherapy-induced dysregulation of gut microbiome, which mediate microglia maturation [36], microbiome specific metabolism and regulation [37, 38], and the subsequent impact and emergence of different metabolomic profiles [39]. These field specific methods also apply to measurement of CIPN outcomes. However, we would press the issue further than Höke and Ray's criticism and argue that extrapolating from hind paw retraction to the experience of pain in human being requires a certain leap of faith, not only due to the implications of limitation 1, but also because the CIPN classifications contain more than pain signal transduction.

4.2.3   Using Monotherapy Models in a World of Combination Chemotherapy and Complex Patient Trajectories

Only monotherapy has been investigated in animal models. As clinicians, we very rarely use drugs such as paclitaxel and oxaliplatin as monotherapies, but in combination with other (sometimes neurotoxic) chemotherapies and numerous supportive care drugs. For instance, a retrospective cohort study showed that concurrent use of bevacizumab is associated with increased risk of CIPN [40] and the neurotoxic effect of cisplatin combined with paclitaxel is different (and greater) than the neurotoxicity induced by single agent administration [41, 42]. Perhaps the greatest impact of limitation 3 is imparted by the fact that 55% of cancer patients presented with at least one comorbidity and 35% were subjected to polypharmacy at diagnosis [43]. New models of CIPN, such as neuroinflammation and mitochondrial mitostasis, indicate that the road from neurotoxic damage to CIPN is influenced by many factors, other than the neurotoxic drug itself. Recent evidence even suggests that there may be a significant interaction between the physiological consequences of cancer and the development of CIPN [44].

In addition to these three limitations, several more exist; their impacts have been investigated in a large meta-analysis of 337 animal studies of CIPN [31]. Some have argued that the limitations of using animal models to derive new causative pathways are not fundamental [45, 46] and simply need to be improved [47]. However, in other multifactorial neurological diseases, such as Alzheimer's disease, drug development has also been disappointing [48]. In a 2009 systematic meta-analysis of 100 experiments, Perel and colleagues concluded that the lack of translatability between non-human animals and humans, across many different diseases, may entail that current animal studies do not accurately represent human diseases [49]. Based on this, we are concerned that mechanistic insight from an animal model is overemphasized in CIPN drug development. Despite these fundamental limitations and failures, animal research continue to receive more funding than clinical research [50].

4.3 Research on Preventive Models of CIPN

Potential preventive drugs for CIPN started appearing in the mid-1980s to 1990s. Hydration therapy for cisplatin-induced nephrotoxicity had been implemented, treatments for cisplatin-induced nausea and vomiting had emerged and cisplatin-induced neurotoxicity was becoming a dose limiting factor for patients with ovarian, bladder, and testicular cancers [51]. Two candidate drugs (Org2766; ACTH analogue [52] and Ethiofos; or amifostine [53]) were found to elicit neuroregenerative properties in cellular and animal models. Initial trials were promising, but each drug failed replication in larger RCTs [54, 55] and Org2776 even seemed to increase CIPN severity.

Throughout the 1990s, 2000s, and 2010s several potential drugs have come and gone in this way, while the list of anti-cancer therapies that induce neurotoxicity has grown ever longer [56]. A quick search reveals that around 60 review articles hold the sum of our mechanistic knowledge about CIPN. Eight of these have been published so far in 2020 [3, 4, 21–23, 57–59]! Upon reading these reviews, it seems we have accumulated vast amounts of knowledge about CIPN pathophysiology. However, piecing it together is not an easy task and roaming through the information leaves a person with pieces of a puzzle that do not offer an explanation. When confronted by an evident change from normality brought on by neurotoxic chemotherapy, one might ask: Are we looking at the beginning, the middle, or the end of a cascade of molecular and cellular events and are these changes representative of what happens in humans? Take the case of paclitaxel-induced neuropathy, the dorsal root ganglion (DRG) is invaded by inflammatory cells [60], why? The authors point to upregulation of proteins such as monocyte chemoattractant protein-1 (MCP-1) via activation of toll-like receptor 4 (TLR4), they show that animals develop allodynia and that this phenomenon could be reversed by administration of anti-MCP-1 antibodies. At first it seems elegant, maybe we should initiate a trial of anti-MCP-1 antibodies in humans? But then you remember the limitations of animal studies. Paclitaxel was administered as a single agent via intrathecal injection in male-only rats. Animals were followed for 21 days, in total. Combined with the fact that TLR4 receptor sequence, expression, and function in humans are very different from what has been observed in animals [61], how should we begin to interpret the implications of this study? Is it useful? Hypothesis-generating? Is it just a laboratory phenomenon with no clinical implication?

There must more be consistency between the proposed drug target, its role in confirmed CIPN pathways and it must be able to predict alleviation of pathology confirmed in patients, such as swelling of the DRG or loss of IENF [62]. State-of-the-art neuroscientific insights tell a story of a complex corporation between neurons, glia, and immune cells that CIPN animal studies do not encompass. They expose the fact that CIPN animal models suffer from reductionist thinking that frames CIPN as a mechanical problem when we might, in fact, be dealing with a relational one.

Next, we will review some of the most important major sites of neurotoxic damage. Several recent and well written reviews covering new preventive measures and mechanisms exist (Pellacani and Eleftheriou [57] for drug-specific neurotoxic mechanisms, Hu et al. [63] for an overview of CIPN mechanistic pathways in relation to emerging drugs and Argyriou et al. [23] for field expert opinion). With so many recent reviews we will approach the subject from an untraditional angle of basic neuroscience, in order to reach a different perspective on CIPN. We will review central neuronal structures and try to relate these to confirmed pathological lesions in humans. Singular mechanisms are elicited in animal models, but appear to be less important within molecular and cellular systems that are networked in the human body. So we begin each section by providing the most recent insights of the complex networks that CIPN mechanisms change and unfold within. We do this in order to provide a relevant backdrop for the numerous animal models of CIPN. We hope this will leave the reader with a representative impression of core CIPN pathophysiology in relation to neuronal basic science context which animal models try to solicit. The denotation of "core" applies to mechanisms that are sufficiently established by translational research and that manifest in major classes of neurotoxic drugs: platinum, taxanes, vinca alkaloids, and bortezomib. It is also important to remember that every type of chemotherapy has more than one drug-specific

Table 4.1 Mechanisms associated with development of CIPN

Mechanism	Associated with	Implicated components	References
Neuroinflammation	Platinum, taxanes, bortezomib, vinca alkaloids	TNF-α↑, IL-1β↑, IL-6↑, IL-10↓, IL-4 ↓, MCP-1↑, TLR4, α7 nAChR, CAMKII, CB2↑,	[25, 60, 64–66]
Mitochondrial dysfunction	Platinum, taxanes, bortezomib, vinca alkaloids	mPTP, VDAC, mDNA, mETC-Ps, ROS↑, β-Tubulin, Bcl-2	[4, 23, 67, 68]
Axon degeneration	Taxanes, platinum, vinca alkaloids	DLK, MAPKs, Sarm-1, IP3R1	[23, 69, 70]
Lipid membrane dysregulation	Platinum, taxanes, bortezomib	S1P↑, GM1, ROS-LP	[23, 71, 72]
Ion channel dysregulation	Platinum, taxanes, bortezomib	Nav1.7↑, Kv7 ↓, Cav3.2 T↓↑, Cav2.3↑, Cav2.2↑, NMDAR, TRPV1↑, TRPA1↑, TRPM8↑, Na+/K +-ATPase↓, K2P2.1↓, K2P 4.1↓ (more in [73])	[3, 4, 21, 67, 74, 75]
DNA modification	Platinum	pt-DNA, APE1 (NER)	[4, 23]
Intracellular signalling transduction	Platinum, taxanes, vinca alkaloids	Ca2+, HSF-1; PKC, NFκB, AT1R↑, p53,p38,p75, MAPKs, ATF3, JNK	[3, 4, 23, 57, 76]
Extracellular matrix dysregulation	Paclitaxel	MMP2,9&13	[3, 9]
Selective organic ion transporters (SLCs)	Oxaliplatin, paclitaxel	OCT-N1(2), OATP-1B1, OATP-1B3	[23]
Intercellular signal mediation	Paclitaxel, vincristine	SGC and astrocyte gap-junctions↑	[3, 77]

TNF-α, tumor necrosis factor Alpha; IL-, interleukin-; TLR4, toll-like receptor 4; MCP-1, monocyte chemoattractant protein-1; CAMKII, calcium/calmodulin-dependent protein kinase II; mPTP, mitochondrial permeability transition pore; VDAC, voltage-dependent anion channel; mDNA, mitochondrial DNA; ROS, reactive oxygen species; mETC-Ps, mitochrondrial electron transport chain proteins; Bcl-2, B-cell lymphoma 2; DLK, dual leucine zipper kinase; MAPK, mitogen-activated protein kinases; IP3R1, inositol 1,4,5-triphosphate receptor 1; Sarm-1, sterile alpha and TIR motif containing 1; CB2, cannabinoid receptor 2; S1P, sphingosine 1-phosphate; GM1, ganglioside-monosialic acid 1; ROS-LP, ROS-mediated lipid peroxidation; α7 nAChR, alpha 7 nicotinic acetylcholine receptors; KV7, voltage-gated potassium channel 7; Cav-, T-type calcium channel; NMDAR, N-methyl-D-aspartate receptor; TRPV1, transient receptor potential vanilloid-1; TRPA1, transient receptor potential ankyrin 1; TRPM8, transient receptor potential melastatin 8; K2P-, potassium channel subfamily K member 2.1 and 4.1; APE-1, apyrimidinic endonuclease/redox effector factor-1; NER, nuclear excision repair; pt-DNA, platinum DNA adducts; Ca2+, calcium; HSF-1, heat shock transcription factor 1; PKC, protein kinase C; NFκB, nuclear translocation of nuclear factor-κB; AT1R, angiotensin II receptor type 1; ATF3, activating transcription factor 3; JNK, c-Jun N-terminal kinase; MMP-, matrix metalloproteinases; SLC, solute carriers transporter superfamily; OCTN1, organic cation transporters novel type 1 (and 2); OATP, organic anion transporting polypeptides; SGC, satellite glial cell mechanism of inducing neuropathy (see Table 4.1 for an overview of mechanisms associated with CIPN development).

4.4   The Core Mechanisms

4.4.1   Neuroinflammation

Neuroinflammation can be found in many neuroanatomical structures in humans and animal models of CIPN. However, the Dorsal Root Ganglion (DRG) takes center stage in CIPN research. It is a well-defined area located in the neuroforaminae of the spinal cord, on the borderline between the peripheral nervous system and the central nervous system. It can be claimed to be neither or both, displaying its own unique signature of cell types and structure of relevance to CIPN [78]. It has been claimed to be the site of the central sensory cell body in the peripheral nervous system. Here sensory neurons relay information, from the peripheral sensory bodies, to the spinal cord. The body of the neuron (the soma) is surrounded by, and encased in, satellite glial cells, which regulate, nourish, and support it. In fact, there may be eight times more glial cells, than neurons, in the DRG [78]. In recent years, basic science has shown that a complex and codependent relationship exists between neuroglia and neurons [79]. Even though sparse post-mortem material has been collected from heterogeneous patient populations, conclusions have been conflicting [80] and the human DRG remains a black box.

Since we cannot get a biopsy from the DRG in humans we cannot achieve any iterative transnationality in research where target lesion response provides further guidance for intervention development. However, new nerve imaging technology can allow us to peek inside the DRG and other nerve structures, as CIPN manifests [81]. In an elegant study, Apostolidis et al. used in vivo NMR scans of patients undergoing treatment with oxaliplatin to show that the size of the DRG increases significantly, compared with controls [82]. This kind of evidence can be used to validate mechanisms and link findings from the laboratory to the clinic. An increase in size of the DRG may be a sign of inflammation and drugs that successfully target neuroinflammation in animals may translate better into human trials when evidence of neuronal inflammation in CIPN patients already exist. Indeed, a neuroinflammatory model of CIPN is emerging [64, 83] and several drugs modulating neuroinflammation are under investigation for prevention of CIPN [63], Table 4.1. Most of the promising drugs which emerge from a neuroinflammatory model of CIPN involve modulation of neuroglia and their response to neurotoxic chemotherapy. There has been some disagreement about which types of neuroglia are involved in CIPN neuroinflammation [76, 84, 85] and whether neuroinflammation is located in the PNS, CNS, or both [64, 86]. The case of minocycline for CIPN prevention gives us some important hints in our attempt to understand neuroinflammation. Minocycline was found to exhibit preventive effects on CIPN in several animal models via a proposed modulatory effect on microglia inflammatory pathways [87, 88]. Unfortunately, a randomized trial in 47 patients recently assessing the anti-neuroinflammatory effect of minocycline in paclitaxel-induced peripheral neuropathy did not show any effect in humans measured on the EORTC-CIPN20 [89]. There was a small and significant effect on paclitaxel-acute-pain-syndrome (P-APS) and fatigue; however, the effect on fatigue was not replicated in another RCT trial of 66 patients [90]. This study supports that neuroinflammation may still be relevant, but also that animal models will probably not yield an exact answer to how we should approach neuroinflammation in regard to CIPN in humans. Differences between mouse and human DRG injury transcriptomes show that response to injury is species-specific [91] and basic science of neurodegenerative diseases shows that glial cells are involved in a complex interrelationship with the microbiome, sleep, and exercise, through inflammatory and metabolic processes that are human specific [92]. In the face of such complex species-specificity, only clinical trials in patients will be able to delineate between effective and ineffective neuromodulatory strategies.

Basic science and animal research show that neuroinflammation is a broad homeostatic process consisting of a network of intertwined mechanistic pathways spread across multiple physiological systems connected to CIPN development [92, 93]. We have unfolded how injury to this system in humans is set apart from animals on many levels. Inhibition or activation of one path in the system and others may cause up- or downregulation. This has led some to suggest that a successful prevention of CIPN requires multimodel drug approaches that mitigate neurotoxic damage in more than one way [23, 94].

Cannabidiol (CBD) is a new player in this field that may fulfill the need for a multimodal drug approach, as it displays an interesting plethora of activity in receptors and systems of interest to CIPN [95]. Recent advances in cannabinoid pharmacology and CIPN animal models have pushed cannabinoids into the focus of CIPN research interest [96, 97]. Investigations into the newly coined term *endocannabinoidome* have already yielded new treatments for other complex neurological diseases such as refractory childhood epilepsy and multiple sclerosis [18]. These discoveries build on established research that show cannabinoids have neuroprotective and anti-neuroinflammatory properties [98–100] mediated, in large part, by effects on neuronal support cells [101–103]. In animal models of paclitaxel-induced pain and neuropathy, cannabinoids have been shown to prevent the development of CIPN in animals [104–106] without compromising chemotherapy efficacy [107]. With concern to CIPN, CBD is a more potent regulator of neuroinflammation, cellular stress, and redox homeostasis than is tetrahydrocannabinol (THC) [108]. CBD may exert its anti-inflammatory effect on microglia via competitive inhibition of adenosine transport, leading to increased signaling through adenosine-receptors [109, 110] which, in itself, is interesting, since agonism of the adenosine-3A-receptor has been shown to inhibit development of CIPN via spinal astrocytes [111]. Evidence also suggest that CBD may be able to inhibit the upregulation of connexin-43 via its inhibitory effect of the FAAH enzyme [110], which hydrolyze neuronal anandamide (AEA), which is involved in neuroglial/neuron hemi-channel regulation [112, 113]. This is promising, since recent evidence shows that these channels can spread neurotoxic damage among glia cells in models of CIPN, leading to chronic pain conditioning [77, 85, 114, 115]. As CBD does not elicit the severe psychotropic side effects of THC [116], it represents an interesting option for a clinical trial of CIPN prevention.

4.4.2 Neuronal Mitochondrial Dysfunction

Neurons are some of the most energy demanding cells in existence. A large sensory neuron may have a diameter of 50μm, but an axonal length of up to 1,000,000μm. That means that many newly synthesized proteins from the soma may need to travel 20,000 times the length of the soma to get to the distal part of the extremities [117]. The neuron (and glia cells) accomplishes this feat with the help of mitochondria and a specific form of homeostasis called mitostasis [118]. Mitostasis is the combined effects of mitochondrial genesis, maintenance, transport, fusion/ fission, and eventual clearance from the cell. Based on these findings, mitostasis is now being implicated in all complex neurodegenerative diseases and it is difficult to assess whether mitochondrial dysfunction is a driver of disease or collateral damage [118]. In animal models, bortezomib, oxaliplatin, paclitaxel, vincristine, and cisplatin have all been associated with neuronal mitochondrial dysfunction leading to an imbalance in production of reactive oxygen species (ROS) and ATP [67]. The "mitotoxicity hypothesis" of CIPN states that each neurotoxic drug damages mitochondria in a drug-specific way. Drugs targeting oxidative stress in models of CIPN were successful in a few smaller clinical trials but failed in later and larger clinical trials [119, 120]. Studies of antioxidant treatments for CIPN taught us that oxidative stress, in and of itself, is an important CIPN mechanism in humans. However the drugs tested so far have been ineffective, produced side-effects, or

Table 4.2 Emerging strategies for CIPN associated neuroinflammation

Drug class	Specific drug(s)	Primary target	References
APE-1	APX3330	DNA repair of APE1	[121–123]
Tetracycline	Minocycline	Microglia	[87, 88]
Cannabinoids	Win 55,212-2, MDA-7, Cannabidiol	NAM-CB1, TRPV1, PPAR-γ, A1-2A-agonism	[104, 105, 124–127]
α7 nAChR	R-47	Microglia	[65]
S1PR agonist/ antagonists	Fingolimod, Ponesimod, CYM-5478	SP11, glia cells	[71, 72, 128, 129]
Adenosine A3 receptor agonists	MRS5698	Spinal astrocytes	[111]
MMP inhibitors	N/a	MMP 2 and 9	[130]

α7 nAChR, alpha 7 nicotinic acetylcholine receptors; TRPV1, transient receptor potential vanilloid-1; APE-1, apyrimidinic endonuclease/redox effector factor-1; MMP-, matrix metalloproteinases; NAM-CB1, negative allosteric modulator of cannabinoid receptor 1; PPAR-γ, peroxisome proliferator-activated receptor gamma even showed signs of worsening CIPN [3]. As an alternative, targeting mitostasis might provide an adequate avenue for new CIPN discoveries (Table 4.2).

Analysis of human breast cancer survivors with persistent paclitaxel-induced peripheral neuropathy and patients receiving vincristine or bortezomib shows gene deficiencies in many functions of mitostasis, such as fission, clearance, and maintenance [131, 132]. The molecular genetics of CIPN has recently been reviewed by Cliff et al. [133]. Mitochondrial transfer between mesenchymal stem cells, astrocytes, and neurons has been observed in recent studies of cisplatin-induced neurotoxicity [134, 135], leading to normal cell function via restoration of mitostasis. This suggests that, just as cellular crosstalk is fundamental for injury and restoration in the neuroinflammatory model of CIPN, this is also the case in CIPN-induced mitochondrial dysfunction. While we cannot (yet) manipulate the Miro-1 pathway in a way that allows mitochondrial transfer from healthy glia to damaged neurons, other emerging strategies may be able to restore mitostasis by improving mitochondrial function in CIPN. Evidence from animal studies of pifithrin-μ shows us that restoring mitostasis may prevent CIPN. Pifithrin-μ inhibits the accumulation of p53 in mitochondria in response to chemotherapy in a recent model of cisplatin-induced neuropathy [136]. Besides its well-known involvement in cancer where p53 is downregulated, p53 upregulation has been shown to be involved in neurodegenerative diseases [137] and CIPN [136]. Basic science shows that a complex redistribution and translocation to mitochondria of p53 occur in glia and neurons following injury, leading to bioenergetic failure, neuroinflammation, and neurodegeneration [138, 139]. This evidence show that pifithrin-μ prevents CIPN by mitigating damage to mitostasis; however, this effect may also lead to downregulation of resulting neuroinflammation, as evidenced in a model of spinal cord injury [140]. Thus, mitochondrial dysfunction does not develop in isolation, but bridges to neuroinflammation [141] and programmed axon degeneration [142], illustrating the need for an explanatory model of CIPN based in pathophysiological systems and not just mechanisms.

4.4.3 Wallerian-Like Axon Degeneration

While the axons of motor neurons can be involved in CIPN [143], CIPN is more often associated with sensory symptoms such as allodynia, hyperalgesia, or dysesthesia manifesting initially in feet and hands. The pseudo-unipolar axon of the sensory neuron carries and relays tactile information between the peripheral and central nervous system involving multiple types of axons and associated skin mechanoreceptors [144] that undergo constant remodeling in the epidermis [145]. The complexity and amount of sensory input from the hand equals that of the eye [144]. From the perspective of the neuron, this is an enormous feat, requiring substantial energy and resources, making it vulnerable to change [145]. CIPN research shows that axonal transport of these resources (proteins, mitochondria, mRNA, etc.) is impaired by microtubule-stabilizing agents such as vincristine and paclitaxel. Based on this, it was hypothesized that the cross-linking of microtubules

Table 4.3 Emerging strategies for CIPN associated mitochondrial dysfunction

Drug class	Specific drug(s)	Primary target	References
Biguanides	Metformin	mTOR, AMPK	[14, 153, 154]
MnSOD mimec	Calmangafodipir	ROS generation	[155, 156]
Small-molecule inhibitors of p53	Pifithrin-μ	Mitochondrial p53	[136, 157]
HDAC6-inhibitors	Ricolinostat, ACY-1215	α-Tubulin	[158, 159]
Sigma-1-receptor antagonists	MR309	Sigma-1-receptor	[160, 161]
Beta/alpha-blockers	Carvedilol	MnSOD	[15]
mPTP stabilizer	Olesoxime	VDAC, TSPO	[162, 163]

mPTP, mitochondrial permeability transition pore; VDAC, voltage-dependent anion channel; ROS, reactive oxygen species; MnSOD, manganese superoxide dismutase; HDAC6, histone deacetylase 6; mTOR, mammalian target of rapamycin; AMPK, 5' adenosine monophosphate-activated protein kinase; TSPO, translocator protein

caused transport impairment, eventually leading to the "dying-back"-axonopathy of CIPN. New evidence show that axonopathy can happen independently of axonal transport [145] and that microtubules are maintained independently in axonal segments, a concept called local axon homeostasis [146]. This may explain how axonal damage is seen beyond terminal arbors of the epidermis, but not in the proximal subdermal segment of the axon next to it, as witnessed in a CIPN model associated with a low dose of paclitaxel [147]. It is not that axonal transport is without importance in CIPN [148], it is more that it is becoming evident that microtubules, mitochondria, and the endoplasmic reticulum mutually regulate each other in normal axon biology and injury [146, 149]. This complex interaction of organelles and proteins at a local axonal level may also explain how neurotoxic drugs, that do not target microtubules directly, eventually lead to the same outcome as those that do, namely axon degeneration [150]. Although there is some diversity in the types of axons that are lost in chemotherapy-induced axonopathy, biopsy studies in patients verify that all major classes of neurotoxic chemotherapy can induce a loss of intra- epidermal-nerve-fibers (IENFs) [151, 152]. Preventing axon-degeneration or enhancing axon-regeneration could provide the means of combating CIPN development. Strategies that target axon degeneration are summarized in Table 4.3.

Wallerian axon degeneration describes the process of programmed axon degeneration that exist in all species [164]. Wallerian axon degeneration originally referred to the process induced by physical cutting of the axon, yet recent evidence show that the molecular pathway is triggered in CIPN development, leading to activation of Sterile Alpha And TIR Motif Containing 1 (Sarm-1) [165–167]. Sarm-1 specific CIPN research is still in its infancy. Genetic knockout of Sarm-1 prevents CIPN axon degeneration, by blocking biodegradation of nicotinamide adenine dinucleotide (NAD+) leading to sustained axon integrity despite injury [165, 168]; however, much is still unknown and there is still no drug that specifically inhibits Sarm-1.

Table 4.4 Emerging strategies for CIPN associated axonal degeneration

Drug class	Specific drug(s)	Primary target	References
Sarm-1 inhibitors	None exists yet	SARM-1	[165, 166]
DLK-inhibitors	GNE-3511	DLK	[171, 176, 177]
Cryotherapy	n/a	Vasoconstriction	[178–180]
NAMPT inhibitor	FK866	NAMPT	[167]
HSP protein modulation	Ethoxyquin	HSP 27 and 90	[170, 172, 181–183]
MMP-13 inhibitor	DB04760, CL-82198	MMP-13	[184]
GM1[a]	GM1	Trks, NGF, Na+/K+-ATPase	[185–188]

DLK, dual leucine zipper kinase; Sarm-1, sterile alpha and TIR motif containing 1; GM1, ganglioside-monosialic acid 1; MMP-, matrix metalloproteinases; NAMPT, nicotinamide phosphoribosyltransferase; HSP, heat shock protein; Trk, tyrosine receptor kinase; NGF, nerve growth factor
[a]GM1 has three positive clinical studies [185–187], but results from [188] ($n = 196$) show no benefit

Targeting associated pathways upstream from Sarm-1, with known small-molecules, is an alternative option for CIPN prevention, but may only work for specific types of neurotoxic chemotherapy. For example, activating the expression of heat shock protein (HSP) 27 restores caspase 3 and RhoA levels in neuronal cells bodies, a process involved in development of bortezomib-induced axon degeneration [169, 170]. Furthermore, vincristine induced axonal degeneration can be prevented by inhibiting MAPK or HSP90 dependent dual leucine zipper kinase (DLK) activation [169, 171, 172] or lastly by generation of surplus NAD[+] through inhibition of nicotinamide phosphoribosyltransferase (NAMPT) [167]. It seems that many pathways can lead to axon degeneration [69] and, as with CIPN pathophysiology, there are examples of complex relationships with other pathways and organelles. For instance, research show that mitochondrial dysfunction may precede Sarm-1 activation [142, 173]. This is also what makes Sarm-1 a very promising candidate since Sarm-1 seems to present an obligatory passage to axon degeneration. [169]. Since the molecular structure of Sarm-1 has recently been described in detail [174] we may see a Sarm-1 inhibitor in the near future. Sarm-1 inhibition does, however, also raise some concerns, as inhibition of it may lead to long-term cognitive side effects stemming from Sarm-1 effects on axon modulation and homeostasis [175] (Table 4.4).

While we wait for research on programmed axon degeneration to become clinically applicable, we may be able to protect axons by physically lowering the temperature of patient extremities. Cryotherapy has been used for chemotherapy-induced alopecia and onycholysis [189]. The first evidence of using cryotherapy for CIPN emerged from a retrospective study showing lower occurrence of docetaxel-induced neuropathy among patient using cryotherapy for onycholysis [190]. Subsequent studies reported mixed results [21, 178]. The success or failure of cryotherapy may potentially teach us something important about the relationship between peripheral and central damage in the development of CIPN. Given that CIPN has been described as a "dying-back"-axonopathy, logic might dictate that protecting the distal sites of initial damage will translate into alleviation of CIPN. But will this also result in durable prevention, since cryotherapy will hardly effect central nerve structures involved in CIPN? Non-mammalian models of paclitaxel-induced neuropathy have shown that several epidermal changes occur, such as keratinocyte-derived formation of hydrogen peroxide and upregulation of matrix metalloproteinase 13 (MMP-13) [191, 192]. This leads to axon degeneration which can be prevented by inhibiting MMP-13 in a mammalian model of paclitaxel and glucose-induced neuropathy [184]. We could only find one non-human animal study that used cryotherapy to investigate the pathophysiology of CIPN outcomes. Cooling of the lower back of mice receiving paclitaxel showed a markedly reduction of CIPN surrogate outcomes and significantly reduced invasion of inflammatory cells into the DRG [193]. One might be concerned that cryotherapy only protects against the acute transient symptoms of neurotoxic chemotherapy, but should the effect extend to long term prevention, it would teach us something essential about our understanding of CIPN and axon homeostasis in general.

4.5   Towards a Future of Better CIPN Models

We have discussed three core explanatory models of CIPN development. Our discussion that neuroinflammation, mitochondrial dysfunction, and axon degeneration lie at the heart of CIPN development is hardly new. Previous review articles showcase these and other related CIPN pathophysiological mechanisms in independent paragraphs with far more detail. We wanted to show that this textual separation is somewhat artificial, as CIPN mechanisms interact across different scales (tissue, cells, and proteins) and systems (immunological, gastrointestinal, and neurological), a point, which has not received much attention in previous reviews. Neuroinflammation is maintained through interaction between different cell types; mitochondrial dysfunction arises from injury to mitostasis which is just one of the ways in which neurotoxic chemotherapy can activate programmed axon degeneration.

Basic science shows that the intersection and interaction between these scales and systems are important in the pathophysiology of disease [198]. For complex neurological diseases such as CIPN, we may learn something new if we consider remodeling it, as change within a system of mechanisms unfolding simultaneously in tissue, cells, and proteins. In order to do this, we need new models of CIPN that adequately captures CIPN as a non-linear change to a system and is not the result of just one linear mechanism. To rectify this, we must begin to use models based on human biology; new in vitro models may provide a different modeling fit for future CIPN drug development [199]. Using bioengineering of multi-cellular 3D organoids it has been possible to create a fully functional human peripheral nerve in vitro [200]. This technology can also be used to create a human model of neuron and non-neuron cells that is able to represent neuroimmune or neuroendocrine interactions [201]. These in vitro structures can be inhabited by induced pluripotent stem cells derived (iPSC) human neuronal and endothelial cell types [202] and will be more representative than previous animal models when extrapolating from the laboratory to the clinic [199]. We can even use cells derived from patients, opening the possibility of evaluating patient specific genetics in vitro and comparing these to their real world outcomes [203]. Combined with new methods of evaluating injurious change such as "multi-omics" (metabolomics, (epi)-genomics, transcriptomics, and proteinomics), we will gather new types of data and more data on CIPN development than ever before [13]. In this regard, we already have a lot of CIPN data; more than 300 animal studies spanning almost 15,000 animals, more than 50 RCTs, multiple genetic and epidemiological studies. We, in this chapter, are arguing whether we are using these data as efficiently as can be done.

So far, we have relied heavily on human cognition and parametric statistics to analyze CIPN data. Yet in recent years, we have begun to see artificial intelligence applied in the field of drug discovery [204, 205]. Machine learning has already been used to predict the neurotoxicity of new anti-cancer drugs as well as suggest new potential preventive drugs for CIPN based on a combined dataset regarding neurotoxic drugs, their molecular descriptors and the neuropathy incidence they cause in patients [206]. Recent reviews stress that a multi-disciplinary approach may be key to success in CIPN [9, 207, 208]. Involvement of data scientists may be quite helpful. An artificial intelligence model such as supervised deep learning could incorporate the data we already have with the future so-called multi-omics datasets, which are emerging from new in vitro models. With the combination of multiple types of data, we may be able to predict the effects (and side-effects) of emerging drugs in patients with more accuracy [209].

4.6 The Potential of Tyrosine Kinase Inhibitors

There is one other promising strategy that deserve attention, but does not accurately fit into the categories of neuroinflammation, mitochondrial dysfunction or axon degeneration. Studies have shown that inhibiting uptake of paclitaxel and oxaliplatin in neuronal cells, by inhibiting specific organic anion transporting polypeptides (OATPs) and organic cation transporters (OCTs), can prevent development of CIPN in rats and mice [194, 195]. Using known and already approved tyrosine kinase inhibitors to block influx channels is an elegant solution if it does not interfere with anti-tumor efficacy of chemotherapy. Dasatinib specifically blocks the channel OCT2, preventing influx into neuronal cells in a model of oxaliplatin-induced peripheral neuropathy [196]. Transcriptional profiling, animal modeling and cell cultures of tumor cells incubated with dasatinib have demonstrated that chemotherapy uptake into tumor cells is not effected by inhibition of OCT2 [196]. Clinical trials have been initiated for dasatinib and nilotinib (for paclitaxel-induced and oxaliplatin-induced peripheral neuropathy), and it will be important to remember that transport channels such as OCT2 has broad selective drug transport capabilities, raising the possibility of drug-drug interactions with commonly used drugs such as metformin [197]. Given that 35% of newly diagnosed cancer patients present with polypharmacy [43], blocking drug transport channels may potentially present a myriad of complications in many patients.

4.7 Conclusion

In 1962 Thomas Kuhn described the process of scientific progression in his highly influential work *The Structure of Scientific Revolutions*. Scientific progression, Kuhn says, happens in leaps when we are faced with an anomaly that is not explained by the current model of research [210]. Kuhnian philosophy opens the possibility of admitting that we are wrong about the most basic assumptions of CIPN. Over the years, we have collected vast amounts of knowledge about the mechanisms of CIPN, yet every review article states that we do not know enough about the pathophysiology of CIPN. Something is amiss. The striking failure of CIPN drug development over the span of 40 years demands some reflection on underlying methodological reasons. We have questioned the knowledge that animal models produce in the context of complex neurodegenerative diseases such as CIPN. The limitations of animal models and the species-specificity of the human nervous system may render this knowledge misleading. In general, drug development has been in a state of crises as the rate of new drugs has declined and value-for-money has become low. Animal models may still be useful in dose-finding and in estimating toxicity, but biology is having less success with reducing complex biological diseases—within neurology and immunology—to singular mechanisms [11, 211]. In brief, when biological mechanisms are connected in a system, new biological phenomena can emerge that are not predicted by examining its parts [212]. We believe this is also the case for CIPN and so we may need to shift our focus to understanding the relationship between systems and scales involved in CIPN within the framework of systems biology [11]. This reconceptualization enables a new perspective on diseases such as CIPN. Here disease manifests as a disturbance in a network and not an alteration of a molecular structure; therefore, intervention is no longer about targeting something specific, but restoration of the network homeostasis [11]. For instance, Romoe-Guitar et al. recently used machine learning algorithms based on preclinical nerve avulsion proteomic data. The resulting model suggested a multimodal drug-based therapy that boosted neuro-regenerative mechanisms, not necessarily associated with the disease condition [204, 213]. In subsequent studies they showed that their therapy (NeuroHeal) was effective in motor and sensory neuropathy derived from nerve root avulsion [214].

These early studies show how epistemological reframing of a specific disease can reveal novel therapy options that are not predicted by examining disease specific mechanisms. This may help us answer a recent call within the field of CIPN and find a way to target multiple mechanisms of CIPN simultaneously [23]. Advances in data generation and modeling can potentially improve the success rate of drugs selected for clinical trials, optimizing CIPN preventive efforts for the benefit of patients and society.

References

1. Mayer DK, Nasso SF, Earp JA (2017) Defining cancer survivors, their needs, and perspectives on survivorship health care in the USA. Lancet Oncol 18:e11–e18
2. Hingorani AD, Kuan V, Finan C et al (2019) Improving the odds of drug development success through human genomics: modelling study. Sci Rep. https://doi.org/10.1038/s41598-019-54849-w
3. Staff NP, Fehrenbacher JC, Caillaud M, Damaj MI, Segal RA, Rieger S (2020) Pathogenesis of paclitaxel-induced peripheral neuropathy: a current review of in vitro and in vivo findings using rodent and human model systems. Exp Neurol. https://doi.org/10.1016/j.expneurol.2019.113121
4. Lazic A, Popović J, Paunesku T, Woloschak GE, Stevanović M (2020) Insights into platinum-induced peripheral neuropathy-current perspective. Neural Regen Res 15:1623–1630
5. Butler D (2008) Translational research: crossing the valley of death. Nature. https://doi.org/10.1038/453840a
6. Hertz D, Krumbach E, Nobles B, Erickson S, Farris K (2018) Abstract P4-11-06: The role of patient perceptions in under reporting chemotherapy induced peripheral neuropathy (CIPN). In: San Antonio breast cancer symposium. https://doi.org/10.1158/1538-7445.sabcs17-p4-11-06
7. Tanay MA, Armes J (2019) Lived experiences and support needs of women who developed chemotherapy-induced peripheral neuropathy following treatment for breast and ovarian cancer. Eur J Cancer Care (Engl). https://doi.org/10.1111/ecc.13011
8. Drott J, Starkhammar H, Kjellgren K, Berterö C (2016) The trajectory of neurotoxic side effects' impact on daily life: a qualitative study. Support Care Cancer 24:3455–3461
9. Colvin LA (2020) Europe PMC Funders Group. Chemotherapy-induced peripheral neuropathy (CIPN): where are we now? Pain 160:1–22
10. Bonhof CS, Trompetter HR, Vreugdenhil G, van de Poll-Franse LV, Mols F (2020) Painful and non-painful chemotherapy-induced peripheral neuropathy and quality of life in colorectal cancer survivors: results from the population-based PROFILES registry. Support Care Cancer. https://doi.org/10.1007/s00520-020-05438-5
11. Margineanu DG (2016) Neuropharmacology beyond reductionism—a likely prospect. BioSystems 141:1–9
12. Freedman DH (2019) Hunting for new drugs with AI. Nature. https://doi.org/10.1038/d41586-019-03846-0
13. (2020) Method of the year 2019: single-cell multimodal omics. Nat Methods. https://doi.org/10.1038/s41592-019-0703-5
14. Martinez NW, Sánchez A, Diaz P et al (2020) Metformin protects from oxaliplatin induced peripheral neuropathy in rats. Neurobiol Pain 8:100048
15. Areti A, Komirishetty P, Kumar A (2017) Carvedilol prevents functional deficits in peripheral nerve mitochondria of rats with oxaliplatin-evoked painful peripheral neuropathy. Toxicol Appl Pharmacol 322:97–103
16. Fernandez A, Sturmberg J, Lukersmith S, Madden R, Torkfar G, Colagiuri R, Salvador-Carulla L (2015) Evidence-based medicine: is it a bridge too far? Heal Res Policy Syst 13:1–9
17. Martell K, Fairchild A, LeGerrier B, Sinha R, Baker S, Liu H, Ghose A, Olivotto IA, Kerba M (2018) Rates of cannabis use in patients with cancer. Curr Oncol 25:219–225
18. Cristino L, Bisogno T, Di Marzo V (2019) Cannabinoids and the expanded endocannabinoid system in neurological disorders. Nat Rev Neurol. https://doi.org/10.1038/s41582-019-0284-z
19. Deo RC (2015) Machine learning in medicine. Circulation. https://doi.org/10.1161/CIRCULATIONAHA.115.001593
20. McCrary JM, Goldstein D, Boyle F et al (2017) Optimal clinical assessment strategies for chemotherapy-induced peripheral neuropathy (CIPN): a systematic review and Delphi survey. Support Care Cancer 25:3485–3493

21. Gordon-Williams R, Farquhar-Smith P (2020) Recent advances in understanding chemotherapy-induced peripheral neuropathy [version 1; peer review: 2 approved]. F1000Research 9:1–13
22. Eldridge S, Guo L, Hamre J (2020) A comparative review of chemotherapy-induced peripheral neuropathy in in vivo and in vitro models. Toxicol Pathol 48:190–201
23. Argyriou AA, Bruna J, Park SB, Cavaletti G (2020) Emerging pharmacological strategies for the management of chemotherapy-induced peripheral neurotoxicity (CIPN), based on novel CIPN mechanisms. Expert Rev Neurother 00:1–12
24. Farquhar-Smith P (2011) Chemotherapy-induced neuropathic pain. Curr Opin Support Palliat Care 5:1–7
25. Kieffer JM, Postma TJ, van de Poll-Franse L, Mols F, Heimans JJ, Cavaletti G, Aaronson NK (2017) Evaluation of the psychometric properties of the EORTC chemotherapy-induced peripheral neuropathy questionnaire (QLQ-CIPN20). Qual Life Res 26:2999–3010
26. Baker M (2015) Over half of psychology studies fail reproducibility test. Nature. https://doi.org/10.1038/nature.2015.18248
27. Smith EML, Knoerl R, Yang JJ, Kanzawa-Lee G, Lee D, Bridges CM (2018) In search of a gold standard patient-reported outcome measure for use in chemotherapy-induced peripheral neuropathy clinical trials. Cancer Control 25:1073274818756608
28. Saito T, Makiura D, Inoue J et al (2020) Comparison between quantitative and subjective assessments of chemotherapy-induced peripheral neuropathy in cancer patients: a prospective cohort study. Phys Ther Res 7–10
29. Vojnits K, Mahammad S, Collins TJ, Bhatia M (2019) Chemotherapy-induced neuropathy and drug discovery platform using human sensory neurons converted directly from adult peripheral blood. Stem Cells Transl Med. https://doi.org/10.1002/sctm.19-0054
30. Mostajo-Radji MA, Schmitz MT, Montoya ST, Pollen AA (2020) Reverse engineering human brain evolution using organoid models. Brain Res. https://doi.org/10.1016/j.brainres.2019.146582
31. Currie GL, Angel-Scott HN, Colvin L et al (2019) Animal models of chemotherapy-induced peripheral neuropathy: a machine-assisted systematic review and meta-analysis. PLoS Biol. https://doi.org/10.1371/journal.pbio.3000243
32. Napier AD, Ancarno C, Butler B et al (2014) Culture and health. Lancet 384:1607–1639
33. Tuttle AH, Tohyama S, Ramsay T, Kimmelman J, Schweinhardt P, Bennett GJ, Mogil JS (2015) Increasing placebo responses over time in U.S. clinical trials of neuropathic pain. Pain. https://doi.org/10.1097/j.pain.0000000000000333
34. Radder H (2009) The philosophy of scientific experimentation: a review. Autom Exp 1:2
35. Höke A, Ray M (2014) Rodent models of chemotherapy-induced peripheral neuropathy. ILAR J 54:273–281
36. Bajic JE (2019) From the bottom up: chemotherapy-induced gut toxicity, glial reactivity and cognitive impairment
37. Montassier E, Gastinne T, Vangay P et al (2015) Chemotherapy-driven dysbiosis in the intestinal microbiome. Aliment Pharmacol Ther. https://doi.org/10.1111/apt.13302
38. Alexander JL, Wilson ID, Teare J, Marchesi JR, Nicholson JK, Kinross JM (2017) Gut microbiota modulation of chemotherapy efficacy and toxicity. Nat Rev Gastroenterol Hepatol. https://doi.org/10.1038/nrgastro.2017.20
39. Louise Pouncey A, James Scott A, Leslie Alexander J, Marchesi J, Kinross J (2018) Gut microbiota, chemotherapy and the host: the influence of the gut microbiota on cancer treatment. Ecancermedicalscience. https://doi.org/10.3332/ecancer.2018.868
40. Matsuoka A, Maeda O, Mizutani T, Nakano Y, Tsunoda N, Kikumori T, Goto H, Ando Y (2016) Bevacizumab exacerbates paclitaxel-induced neuropathy: a retrospective cohort study. PLoS One. https://doi.org/10.1371/journal.pone.0168707
41. Carozzi V, Chiorazzi A, Canta A et al (2009) Effect of the chronic combined administration of cisplatin and paclitaxel in a rat model of peripheral neurotoxicity. Eur J Cancer. https://doi.org/10.1016/j.ejca.2008.10.038

42. Zhang J, Tuckett RP (2008) Comparison of paclitaxel and cisplatin effects on the slowly adapting type I mechanoreceptor. Brain Res 1214:50–57

43. Loeppenthin K, Dalton SO, Johansen C et al (2020) Total burden of disease in cancer patients at diagnosis—a Danish nationwide study of multimorbidity and redeemed medication. Br J Cancer. https://doi.org/10.1038/s41416-020-0950-3

44. Housley SN, Nardelli P, Carrasco D, Rotterman TM, Pfahl E, Matyunina LV, McDonald JF, Cope TC (2020) Cancer exacerbates chemotherapy-induced sensory neuropathy. Cancer Res. https://doi.org/10.1158/0008-5472.can-19-2331

45. Sikandar S, Dickenson AH (2013) II. No need for translation when the same language is spoken. Br J Anaesth 111:3–6

46. Bruna J, Alberti P, Calls-Cobos A, Caillaud M, Damaj MI, Navarro X (2020) Methods for in vivo studies in rodents of chemotherapy induced peripheral neuropathy. Exp Neurol. https://doi.org/10.1016/j.expneurol.2019.113154

47. Gadgil S, Ergün M, van den Heuvel SA, van der Wal SE, Scheffer GJ, Hooijmans CR (2019) A systematic summary and comparison of animal models for chemotherapy induced (peripheral) neuropathy (CIPN). PLoS One 14:1–17

48. Reardon S (2018) Frustrated Alzheimer's researchers seek better lab mice. Nature. https://doi.org/10.1038/d41586-018-07484-w

49. Perel P, Roberts I, Sena E, Wheble P, Briscoe C, Sandercock P, Macleod M, Mignini LE, Jayaram P, Khan KS (2007) Comparison of treatment effects between animal experiments and clinical trials: systematic review. BMJ 334:197

50. Pound P, Ebrahim S, Sandercock P, Bracken MB, Roberts I (2004) Where is the evidence that animal research benefits humans? Br Med J. https://doi.org/10.1136/bmj.328.7438.514

51. Thompson SW, Davis LE, Kornfeld M, Hilgers RD, Standefer JC (1984) Cisplatin neuropathy. Clinical, electrophysiologic, morphologic, and toxicologic studies. Cancer. https://doi.org/10.1002/1097-0142(19841001)54:7<1269::AID-CNCR2820540707>3.0.CO;2-9

52. Gary PH, JBM, SCT et al (1990) The New England Journal of Medicine Downloaded from nejm.org on April 1, 2015. For personal use only. No other uses without permission. Copyright © 1990 Massachusetts Medical Society. All rights reserved. N Engl J Med 323:1120–1123

53. Mollman JE, Glover DJ, Hogan WM, Furman RE (1988) Cisplatin neuropathy. Risk factors, prognosis, and protection by WR-2721. Cancer. https://doi.org/10.1002/1097-0142(19880601)61:11<2192::AID-CNCR2820611110>3.0.CO;2-A

54. Roberts JA, Jenison EL, Kim K, Ph D, Clarke-pearson D, Langleben A (1998) A randomized, multicenter, double-blind, placebo-controlled, dose-finding study of org 2766 in the prevention or delay of cisplatin-induced neuropathies in women with ovarian cancer. Int J Gynecol Obstet 61:95–95

55. Albers JW, Chaudhry V, Cavaletti G, Donehower RC (2014) Interventions for preventing neuropathy caused by cisplatin and related compounds. Cochrane Database Syst Rev. https://doi.org/10.1002/14651858.CD005228.pub4

56. Stone JB, DeAngelis LM (2016) Cancer-treatment-induced neurotoxicity-focus on newer treatments. Nat Rev Clin Oncol. https://doi.org/10.1038/nrclinonc.2015.152

57. Pellacani C, Eleftheriou G (2020) Neurotoxicity of antineoplastic drugs: mechanisms, susceptibility, and neuroprotective strategies. Adv Med Sci 65:265–285

58. Cirrincione AM, Rieger S (2020) Analyzing chemotherapy-induced peripheral neuropathy in vivo using non-mammalian animal models. Exp Neurol 323:113090

59. Ibrahim EY, Ehrlich BE (2020) Prevention of chemotherapy-induced peripheral neuropathy: a review of recent findings. Crit Rev Oncol Hematol. https://doi.org/10.1016/j.critrevonc.2019.102831

60. Zhang H, Li Y, De Carvalho-Barbosa M, Kavelaars A, Heijnen CJ, Albrecht PJ, Dougherty PM (2016) Dorsal root ganglion infiltration by macrophages contributes to paclitaxel chemotherapy-induced peripheral neuropathy. J Pain. https://doi.org/10.1016/j.jpain.2016.02.011

61. Vaure C, Liu Y (2014) A comparative review of toll-like receptor 4 expression and functionality in different animal species. Front Immunol. https://doi.org/10.3389/fimmu.2014.00316
62. Mangus LM, Rao DB, Ebenezer GJ (2020) Intraepidermal nerve fiber analysis in human patients and animal models of peripheral neuropathy: a comparative review. Toxicol Pathol. https://doi.org/10.1177/0192623319855969
63. Hu S, Huang KM, Adams EJ, Loprinzi CL, Lustberg MB (2019) Recent developments of novel pharmacologic therapeutics for prevention of chemotherapy-induced peripheral neuropathy. Clin Cancer Res 25:6295–6301
64. Lees JG, Makker PGS, Tonkin RS, Abdulla M, Park SB, Goldstein D, Moalem-Taylor G (2017) Immune-mediated processes implicated in chemotherapy-induced peripheral neuropathy. Eur J Cancer. https://doi.org/10.1016/j.ejca.2016.12.006
65. Toma W, Kyte SL, Bagdas D et al (2019) The α7 nicotinic receptor silent agonist R-47 prevents and reverses paclitaxel-induced peripheral neuropathy in mice without tolerance or altering nicotine reward and withdrawal. Exp Neurol 320:113010
66. Brandolini L, D'Angelo M, Antonosante A, Allegretti M, Cimini A (2019) Chemokine signaling in chemotherapy-induced neuropathic pain. Int J Mol Sci 20:1–13
67. Canta A, Pozzi E, Carozzi V (2015) Mitochondrial dysfunction in chemotherapy-induced peripheral neuropathy (CIPN). Toxics. https://doi.org/10.3390/toxics3020198
68. Griffiths LA, Flatters SJL (2015) Pharmacological modulation of the mitochondrial electron transport chain in paclitaxel-induced painful peripheral neuropathy. J Pain. https://doi.org/10.1016/j.jpain.2015.06.008
69. Fukuda Y, Li Y, Segal RA (2017) A mechanistic understanding of axon degeneration in chemotherapy-induced peripheral neuropathy. Front Neurosci 11:481
70. McLeary F, Davis A, Rudrawar S, Perkins A, Anoopkumar-Dukie S (2019) Mechanisms underlying select chemotherapeutic-agent-induced neuroinflammation and subsequent neurodegeneration. Eur J Pharmacol 842:49–56
71. Janes K, Little JW, Li C et al (2014) The development and maintenance of paclitaxel-induced neuropathic pain require activation of the sphingosine 1-phosphate receptor subtype 1. J Biol Chem 289:21082–21097
72. Stockstill K, Doyle TM, Yan X et al (2018) Dysregulation of sphingolipid metabolism contributes to bortezomib-induced neuropathic pain. J Exp Med. https://doi.org/10.1084/jem.20170584
73. Aromolaran KA, Goldstein PA (2017) Ion channels and neuronal hyperexcitability in chemotherapy-induced peripheral neuropathy; cause and effect? Mol Pain 13:1744806917714693
74. Poupon L, Lamoine S, Pereira V et al (2018) Targeting the TREK-1 potassium channel via riluzole to eliminate the neuropathic and depressive-like effects of oxaliplatin. Neuropharmacology 140:43–61
75. Li Y, North RY, Rhines LD et al (2018) DRG voltage-gated sodium channel 1.7 is upregulated in paclitaxel-induced neuropathy in rats and in humans with neuropathic pain. J Neurosci. https://doi.org/10.1523/jneurosci.0899-17.2017
76. Imai S, Koyanagi M, Azimi Z et al (2017) Taxanes and platinum derivatives impair Schwann cells via distinct mechanisms. Sci Rep 7:1–14
77. Zhou L, Ao L, Yan Y, Li C, Li W, Ye A, Liu J, Hu Y, Fang W, Li Y (2020) Levo-corydalmine attenuates vincristine-induced neuropathic pain in mice by upregulating the Nrf2/HO-1/CO pathway to inhibit connexin 43 expression. Neurotherapeutics 17:340–355
78. Esposito MF, Malayil R, Hanes M, Deer T (2019) Unique characteristics of the dorsal root ganglion as a target for neuromodulation. Pain Med (United States) 20:S23–S30
79. Verkhratsky A, Ho MS, Zorec R, Parpura V (2019) The concept of neuroglia. Adv Exp Med Biol. https://doi.org/10.1007/978-981-13-9913-8_1
80. Krarup-Hansen A, Rietz B, Krarup C, Heydorn K, Rørth M, Schmalbruch H (1999) Histology and platinum content of sensory ganglia and sural nerves in patients treated with cisplatin and carboplatin: an autopsy study. Neuropathol Appl Neurobiol 25:28–39

81. Argyriou AA, Park SB, Islam B, Tamburin S, Velasco R, Alberti P, Bruna J, Psimaras D, Cavaletti G, Cornblath DR (2019) Neurophysiological, nerve imaging and other techniques to assess chemotherapy-induced peripheral neurotoxicity in the clinical and research settings. J Neurol Neurosurg Psychiatry 90(12):1361–1369

82. Apostolidis L, Schwarz D, Xia A et al (2017) Dorsal root ganglia hypertrophy as in vivo correlate of oxaliplatin-induced polyneuropathy. PLoS One 12:1–15

83. Makker PGS, Duffy SS, Lees JG, Perera CJ, Tonkin RS, Butovsky O, Park SB, Goldstein D, Moalem-Taylor G (2017) Characterisation of immune and neuroinflammatory changes associated with chemotherapy-induced peripheral neuropathy. PLoS One 12:e0170814

84. Robinson CR, Zhang H, Dougherty PM (2014) Astrocytes, but not microglia, are activated in oxaliplatin and bortezomib-induced peripheral neuropathy in the rat. Neuroscience. https://doi.org/10.1016/j.neuroscience.2014.05.051

85. Warwick RA, Hanani M (2013) The contribution of satellite glial cells to chemotherapy-induced neuropathic pain. Eur J Pain (United Kingdom). https://doi.org/10.1002/j.1532-2149.2012.00219.x

86. Peters CM, Jimenez-Andrade JM, Kuskowski MA, Ghilardi JR, Mantyh PW (2007) An evolving cellular pathology occurs in dorsal root ganglia, peripheral nerve and spinal cord following intravenous administration of paclitaxel in the rat. Brain Res 1168:46–59

87. Boyette-Davis J, Xin W, Zhang H, Dougherty PM (2011) Intraepidermal nerve fiber loss corresponds to the development of taxol-induced hyperalgesia and can be prevented by treatment with minocycline. Pain 152:308–313

88. Boyette-Davis J, Dougherty PM (2011) Protection against oxaliplatin-induced mechanical hyperalgesia and intraepidermal nerve fiber loss by minocycline. Exp Neurol 229:353–357

89. Pachman DR, Dockter T, Zekan PJ et al (2017) A pilot study of minocycline for the prevention of paclitaxel-associated neuropathy: ACCRU study RU221408I. Support Care Cancer. https://doi.org/10.1007/s00520-017-3760-2

90. Wang XS, Shi Q, Bhadkamkar NA, Cleeland CS, Garcia-Gonzalez A, Aguilar JR, Heijnen C, Eng C (2019) Minocycline for symptom reduction during oxaliplatin-based chemotherapy for colorectal cancer: a phase II randomized clinical trial. J Pain Symptom Manage. https://doi.org/10.1016/j.jpainsymman.2019.06.018

91. Wangzhou A, McIlvried LA, Paige C et al (2020) Pharmacological target-focused transcriptomic analysis of native versus cultured human and mouse dorsal root ganglia. Pain. https://doi.org/10.1097/j.pain.0000000000001866

92. Madore C, Yin Z, Leibowitz J, Butovsky O (2020) Microglia, lifestyle stress, and neurodegeneration. Immunity. https://doi.org/10.1016/j.immuni.2019.12.003

93. Davies AJ, Rinaldi S, Costigan M, Oh SB (2020) Cytotoxic immunity in peripheral nerve injury and pain. Front Neurosci. https://doi.org/10.3389/fnins.2020.00142

94. Duggett NA, Griffiths LA, Flatters SJL (2017) Paclitaxel-induced painful neuropathy is associated with changes in mitochondrial bioenergetics, glycolysis, and an energy deficit in dorsal root ganglia neurons. Pain. https://doi.org/10.1097/j.pain.0000000000000939

95. Di Marzo V (2018) New approaches and challenges to targeting the endocannabinoid system. Nat Rev Drug Discov 17:623–639

96. Blanton HL, Brelsford J, DeTurk N, Pruitt K, Narasimhan M, Morgan DJ, Guindon J (2019) Cannabinoids: current and future options to treat chronic and chemotherapy-induced neuropathic pain. Drugs 79:969–995

97. O'Hearn S, Diaz P, Wan BA, DeAngelis C, Lao N, Malek L, Chow E, Blake A (2017) Modulating the endocannabinoid pathway as treatment for peripheral neuropathic pain: a selected review of preclinical studies. Ann Palliat Med 6:S209–S214

98. Maccarone M, Bab I, Bíró T et al (2015) Endocannabinoid signaling at the periphery: 50 years after THC. Trends Pharmacol Sci 36:277–296

99. Crippa JA, Guimarães FS, Campos AC, Zuardi AW (2018) Translational investigation of the therapeutic potential of cannabidiol (CBD): toward a new age. Front Immunol 9:2009

100. Turcotte C, Chouinard F, Lefebvre JS, Flamand N (2015) Regulation of inflammation by cannabinoids, the endocannabinoids 2-arachidonoyl-glycerol and arachidonoyl-ethanolamide, and their metabolites. J Leukoc Biol 97:1049–1070
101. Kozela E, Juknat A, Vogel Z (2017) Modulation of astrocyte activity by cannabidiol, a nonpsychoactive cannabinoid. Int J Mol Sci 18:1669
102. Perez M, Benitez SU, Cartarozzi LP, Del Bel E, Guimarães FS, Oliveira ALR (2013) Neuroprotection and reduction of glial reaction by cannabidiol treatment after sciatic nerve transection in neonatal rats. Eur J Neurosci 38:3424–3434
103. dos-Santos-Pereira M, Guimarães FS, Del-Bel E, Raisman-Vozari R, Michel PP (2020) Cannabidiol prevents LPS-induced microglial inflammation by inhibiting ROS/NF-κB-dependent signaling and glucose consumption. Glia 68:561–573
104. Rahn EJ, Deng L, Thakur GA, Vemuri K, Zvonok AM, Lai YY, Makriyannis A, Hohmann AG (2014) Prophylactic cannabinoid administration blocks the development of paclitaxel-induced neuropathic nociception during analgesic treatment and following cessation of drug delivery. Mol Pain 10:27
105. Naguib M, Xu JJ, Diaz P, Brown DL, Cogdell D, Bie B, Hu J, Craig S, Hittelman WN (2012) Prevention of paclitaxel-induced neuropathy through activation of the central cannabinoid type 2 receptor system. Anesth Analg 114:1104–1120
106. King KM, Myers AM, Soroka-Monzo AJ, Tuma RF, Tallarida RJ, Walker EA, Ward SJ (2017) Single and combined effects of Δ(9)-tetrahydrocannabinol and cannabidiol in a mouse model of chemotherapy-induced neuropathic pain. Br J Pharmacol 174:2832–2841
107. Ward SJ, Ramirez MD, Neelakantan H, Walker EA (2011) Cannabidiol prevents the development of cold and mechanical allodynia in paclitaxel-treated female C57Bl6 mice. Anesth Analg 113:947–950
108. Juknat A, Gao F, Coppola G, Vogel Z, Kozela E (2019) MiRNA expression profiles and molecular networks in resting and LPS-activated BV-2 microglia-Effect of cannabinoids. PLoS One. https://doi.org/10.1371/journal.pone.021203
109. Carrier EJ, Auchampach JA, Hillard CJ (2006) Inhibition of an equilibrative nucleoside transporter by cannabidiol: a mechanism of cannabinoid immunosuppression. Proc Natl Acad Sci U S A. https://doi.org/10.1073/pnas.0511232103
110. Turner SE, Williams CM, Iversen L, Whalley BJ (2017) Molecular pharmacology of phytocannabinoids. Prog Chem Org Nat Prod. https://doi.org/10.1007/978-3-319-45541-9_3
111. Janes K, Wahlman C, Little JW, Doyle T, Tosh DK, Jacobson KA, Salvemini D (2015) Spinal neuroimmune activation is independent of T-cell infiltration and attenuated by A3 adenosine receptor agonists in a model of oxaliplatin-induced peripheral neuropathy. Brain Behav Immun 44:91–99
112. Labra VC, Santibáñez CA, Gajardo-Gómez R, Díaz EF, Gómez GI, Orellana JA (2018) The neuroglial dialog between cannabinoids and hemichannels. Front Mol Neurosci 11:1–17
113. Vázquez C, Tolón RM, Pazos MR, Moreno M, Koester EC, Cravatt BF, Hillard CJ, Romero J (2015) Endocannabinoids regulate the activity of astrocytic hemichannels and the microglial response against an injury: in vivo studies. Neurobiol Dis 79:41–50
114. Retamal MA, Riquelme MA, Stehberg J, Alcayaga J (2017) Connexin43 hemichannels in satellite glial cells, can they influence sensory neuron activity? Front Mol Neurosci 10:1–9
115. Yoon S-Y, Robinson CR, Zhang H, Dougherty PM (2013) Spinal astrocyte gap junctions contribute to oxaliplatin-induced mechanical hypersensitivity. J Pain 14:205–214
116. Chesney E, Oliver D, Green A, Sovi S, Wilson J, Englund A, Freeman TP, McGuire P (2020) Adverse effects of cannabidiol: a systematic review and meta-analysis of randomized clinical trials. Neuropsychopharmacology. https://doi.org/10.1038/s41386-020-0667-2
117. Haberberger RV, Barry C, Dominguez N, Matusica D (2019) Human dorsal root ganglia. Front Cell Neurosci. https://doi.org/10.3389/fncel.2019.00271
118. Misgeld T, Schwarz TL (2017) Mitostasis in neurons: maintaining mitochondria in an extended cellular architecture. Neuron. https://doi.org/10.1016/j.neuron.2017.09.055

119. Schloss JM, Colosimo M, Airey C, Masci PP, Linnane AW, Vitetta L (2013) Nutraceuticals and chemotherapy induced peripheral neuropathy (CIPN): a systematic review. Clin Nutr 32:888–893

120. Hershman DL, Lacchetti C, Dworkin RH et al (2014) Prevention and management of chemotherapy-induced peripheral neuropathy in survivors of adult cancers: American Society of Clinical Oncology Clinical Practice Guideline. J Clin Oncol. https://doi.org/10.1200/JCO.2013.54.0914

121. Fehrenbacher JC, Guo C, Kelley MR, Vasko MR (2017) DNA damage mediates changes in neuronal sensitivity induced by the inflammatory mediators, MCP-1 and LPS, and can be reversed by enhancing the DNA repair function of APE1. Neuroscience. https://doi.org/10.1016/j.neuroscience.2017.09.039

122. Baek H, Lim CS, Byun HS et al (2016) The anti-inflammatory role of extranuclear apurinic/apyrimidinic endonuclease 1/redox effector factor-1 in reactive astrocytes. Mol Brain. https://doi.org/10.1186/s13041-016-0280-9

123. Kelley MR, Messmann RA, Fehrenbacher J (2018) Novel first-in-class small molecule targeting APE1/Ref-1 to prevent and treat chemotherapy-induced peripheral neuropathy (CIPN). J Clin Oncol 36:229

124. Burgos E, Gómez-Nicola D, Pascual D, Martín MI, Nieto-Sampedro M, Goicoechea C (2012) Cannabinoid agonist WIN 55,212-2 prevents the development of paclitaxel-induced peripheral neuropathy in rats. Possible involvement of spinal glial cells. Eur J Pharmacol 682:62–72

125. Xu JJ, Diaz P, Bie B, Astruc-Diaz F, Wu J, Yang H, Brown DL, Naguib M (2014) Spinal gene expression profiling and pathways analysis of a CB2 agonist (MDA7)-targeted prevention of paclitaxel-induced neuropathy. Neuroscience 260:185–194

126. Ward SJ, McAllister SD, Kawamura R, Murase R, Neelakantan H, Walker EA (2014) Cannabidiol inhibits paclitaxel-induced neuropathic pain through 5-HT(1A) receptors without diminishing nervous system function or chemotherapy efficacy. Br J Pharmacol 171:636–645

127. Wu J, Hocevar M, Bie B, Foss JF, Naguib M (2019) Cannabinoid type 2 receptor system modulates paclitaxel-induced microglial dysregulation and central sensitization in rats. J Pain. https://doi.org/10.1016/j.jpain.2018.10.007

128. Wang W, Xiang P, Chew WS et al (2020) Activation of sphingosine 1-phosphate receptor 2 attenuates chemotherapy-induced neuropathy. J Biol Chem 295:1143–1152

129. Chua KC, Xiong C, Ho C et al (2020) Genome-wide meta-analysis validates a roles for s1pr1 in microtubule targeting agent-induced sensory peripheral neuropathy. Clin Pharmacol Ther 1–10

130. Tonello R, Lee SH, Berta T (2019) Monoclonal antibody targeting the matrix metalloproteinase 9 prevents and reverses paclitaxel-induced peripheral neuropathy in mice. J Pain. https://doi.org/10.1016/j.jpain.2018.11.003

131. Kober KM, Olshen A, Conley YP et al (2018) Expression of mitochondrial dysfunction-related genes and pathways in paclitaxel-induced peripheral neuropathy in breast cancer survivors. Mol Pain. https://doi.org/10.1177/1744806918816462

132. Broyl A, Corthals SL, Jongen JLM et al (2010) Mechanisms of peripheral neuropathy associated with bortezomib and vincristine in patients with newly diagnosed multiple myeloma: a prospective analysis of data from the HOVON-65/GMMG-HD4 trial. Lancet Oncol. https://doi.org/10.1016/S1470-2045(10)70206-0

133. Cliff J, Jorgensen AL, Lord R, Azam F, Cossar L, Carr DF, Pirmohamed M (2017) The molecular genetics of chemotherapy-induced peripheral neuropathy: a systematic review and meta-analysis. Crit Rev Oncol Hematol. https://doi.org/10.1016/j.critrevonc.2017.09.009

134. English K, Shepherd A, Uzor NE, Trinh R, Kavelaars A, Heijnen CJ (2020) Astrocytes rescue neuronal health after cisplatin treatment through mitochondrial transfer. Acta Neuropathol Commun 8:1–14

135. Boukelmoune N, Chiu GS, Kavelaars A, Heijnen CJ (2018) Mitochondrial transfer from mesenchymal stem cells to neural stem cells protects against the neurotoxic effects of cisplatin. Acta Neuropathol Commun. https://doi.org/10.1186/s40478-018-0644-8

136. Maj MA, Ma J, Krukowski KN, Kavelaars A, Heijnen CJ (2017) Inhibition of mitochondrial p53 accumulation by PFT-μ prevents cisplatin-induced peripheral neuropathy. Front Mol Neurosci. https://doi.org/10.3389/fnmol.2017.00108

137. Houck AL, Seddighi S, Driver JA (2018) At the crossroads between neurodegeneration and cancer: a review of overlapping biology and its implications. Curr Aging Sci. https://doi.org/10.2174/1874609811666180223154436

138. Rodkin S, Khaitin A, Pitinova M, Dzreyan V, Guzenko V, Rudkovskii M, Sharifulina S, Uzdensky A (2020) The localization of p53 in the crayfish mechanoreceptor neurons and its role in axotomy-induced death of satellite glial cells remote from the axon transection site. J Mol Neurosci. https://doi.org/10.1007/s12031-019-01453-2

139. Turnquist C, Horikawa I, Foran E, Major EO, Vojtesek B, Lane DP, Lu X, Harris BT, Harris CC (2016) P53 isoforms regulate astrocyte-mediated neuroprotection and neurodegeneration. Cell Death Differ. https://doi.org/10.1038/cdd.2016.37

140. Caponegro MD, Torres LF, Rastegar C, Rath N, Anderson ME, Robinson JK, Tsirka SE (2019) Pifithrin-μ modulates microglial activation and promotes histological recovery following spinal cord injury. CNS Neurosci Ther. https://doi.org/10.1111/cns.13000

141. Aloi MS, Su W, Garden GA (2015) The p53 transcriptional network influences microglia behavior and neuroinflammation. Crit Rev Immunol 35:401–415

142. Loreto A, Hill CS, Hewitt VL et al (2020) Mitochondrial impairment activates the Wallerian pathway through depletion of NMNAT2 leading to SARM1-dependent axon degeneration. Neurobiol Dis 134:104678

143. Molassiotis A, Cheng HL, Lopez V et al (2019) Are we mis-estimating chemotherapy-induced peripheral neuropathy? Analysis of assessment methodologies from a prospective, multinational, longitudinal cohort study of patients receiving neurotoxic chemotherapy. BMC Cancer 19:1–19

144. Abraira VE, Ginty DD (2013) The sensory neurons of touch. Neuron 79:618–639

145. Gornstein EL, Schwarz TL (2017) Neurotoxic mechanisms of paclitaxel are local to the distal axon and independent of transport defects. Exp Neurol 288:153–166

146. Hahn I, Voelzmann A, Liew YT, Costa-Gomes B, Prokop A (2019) The model of local axon homeostasis—explaining the role and regulation of microtubule bundles in axon maintenance and pathology. Neural Dev 14:1–28

147. Bennett GJ, Liu GK, Xiao WH, Jin HW, Siau C (2011) Terminal arbor degeneration—a novel lesion produced by the antineoplastic agent paclitaxel. Eur J Neurosci 33:1667–1676

148. Tasnim A, Rammelkamp Z, Slusher AB, Wozniak K, Slusher BS, Farah MH (2016) Paclitaxel causes degeneration of both central and peripheral axon branches of dorsal root ganglia in mice. BMC Neurosci 17:1–8

149. Öztürk Z, O'Kane CJ, Pérez-Moreno JJ (2020) Axonal endoplasmic reticulum dynamics and its roles in neurodegeneration. Front Neurosci 14:1–33

150. Malacrida A, Meregalli C, Rodriguez-Menendez V, Nicolini G (2019) Chemotherapy-induced peripheral neuropathy and changes in cytoskeleton. Int J Mol Sci. https://doi.org/10.3390/ijms20092287

151. Boehmerle W, Huehnchen P, Peruzzaro S, Balkaya M, Endres M (2014) Electrophysiological, behavioral and histological characterization of paclitaxel, cisplatin, vincristine and bortezomib-induced neuropathy in C57Bl/6 mice. Sci Rep. https://doi.org/10.1038/srep06370

152. Burakgazi AZ, Messersmith W, Vaidya D, Hauer P, Hoke A, Polydefkis M (2011) Longitudinal assessment of oxaliplatin-induced neuropathy. Neurology 77:980–986

153. El-Fatatry BM, Ibrahim OM, Hussien FZ, Mostafa TM (2018) Role of metformin in oxaliplatin-induced peripheral neuropathy in patients with stage III colorectal cancer: randomized, controlled study. Int J Colorectal Dis. https://doi.org/10.1007/s00384-018-3104-9

154. Inyang KE, McDougal TA, Ramirez ED, Williams M, Laumet G, Kavelaars A, Heijnen CJ, Burton M, Dussor G, Price TJ (2019) Alleviation of paclitaxel-induced mechanical

hypersensitivity and hyperalgesic priming with AMPK activators in male and female mice. Neurobiol Pain. https://doi.org/10.1016/j.ynpai.2019.100037

155. Karlsson JOG, Ignarro LJ, Lundström I, Jynge P, Almén T (2015) Calmangafodipir [Ca4Mn (DPDP)5], mangafodipir (MnDPDP) and MnPLED with special reference to their SOD mimetic and therapeutic properties. Drug Discov Today. https://doi.org/10.1016/j.drudis.2014.11.008

156. Glimelius B, Manojlovic N, Pfeiffer P et al (2018) Persistent prevention of oxaliplatin-induced peripheral neuropathy using calmangafodipir (PledOx®): a placebo-controlled randomised phase II study (PLIANT). Acta Oncol (Madr). https://doi.org/10.1080/0284186X.2017.1398836

157. Krukowski K, Nijboer CH, Huo X, Kavelaars A, Heijnen CJ (2015) Prevention of chemotherapy-induced peripheral neuropathy by the small-molecule inhibitor pifithrin-µ. Pain. https://doi.org/10.1097/j.pain.0000000000000290

158. Krukowski K, Ma J, Golonzhka O, Laumet GO, Gutti T, Van Duzer JH, Mazitschek R, Jarpe MB, Heijnen CJ, Kavelaars A (2017) HDAC6 inhibition effectively reverses chemotherapy-induced peripheral neuropathy. Pain. https://doi.org/10.1097/j.pain.0000000000000893

159. Ma J, Trinh RT, Mahant ID, Peng B, Matthias P, Heijnen CJ, Kavelaars A (2019) Cell-specific role of histone deacetylase 6 in chemotherapy-induced mechanical allodynia and loss of intraepidermal nerve fibers. Pain 160:2877–2890

160. Nieto FR, Cendán CM, Cañizares FJ, Cubero MA, Vela JM, Fernández-Segura E, Baeyens JM (2014) Genetic inactivation and pharmacological blockade of sigma-1 receptors prevent paclitaxel-induced sensory-nerve mitochondrial abnormalities and neuropathic pain in mice. Mol Pain. https://doi.org/10.1186/1744-8069-10-11

161. Bruna J, Videla S, Argyriou AA et al (2018) Efficacy of a novel sigma-1 receptor antagonist for oxaliplatin-induced neuropathy: a randomized, double-blind, placebo-controlled phase IIa clinical trial. Neurotherapeutics. https://doi.org/10.1007/s13311-017-0572-5

162. Xiao WH, Zheng FY, Bennett GJ, Bordet T, Pruss RM (2009) Olesoxime (cholest-4-en-3-one, oxime): analgesic and neuroprotective effects in a rat model of painful peripheral neuropathy produced by the chemotherapeutic agent, paclitaxel. Pain 147:202–209

163. Rovini A (2019) Tubulin-VDAC interaction: molecular basis for mitochondrial dysfunction in chemotherapy-induced peripheral neuropathy. Front Physiol. https://doi.org/10.3389/fphys.2019.00671

164. Rosell AL, Neukomm LJ (2019) Axon death signalling in Wallerian degeneration among species and in disease. Open Biol. https://doi.org/10.1098/rsob.190118

165. Tian W, Czopka T, López-Schier H (2020) Systemic loss of Sarm1 protects Schwann cells from chemotoxicity by delaying axon degeneration. Commun Biol 3:49

166. Geisler S, Doan RA, Strickland A, Huang X, Milbrandt J, DiAntonio A (2016) Prevention of vincristine-induced peripheral neuropathy by genetic deletion of SARM1 in mice. Brain. https://doi.org/10.1093/brain/aww251

167. Liu HW, Smith CB, Schmidt MS, Cambronne XA, Cohen MS, Migaud ME, Brenner C, Goodman RH (2018) Pharmacological bypass of NAD+ salvage pathway protects neurons from chemotherapy-induced degeneration. Proc Natl Acad Sci U S A. https://doi.org/10.1073/pnas.1809392115

168. DiAntonio A (2019) Axon degeneration. Pain 160:S17–S22

169. Krauss R, Bosanac T, Devraj R, Engber T, Hughes RO (2020) Axons matter: the promise of treating neurodegenerative disorders by targeting SARM1-mediated axonal degeneration. Trends Pharmacol Sci 41:281–293

170. Chine VB, Au NPB, Kumar G, Ma CHE (2019) Targeting axon integrity to prevent chemotherapy-induced peripheral neuropathy. Mol Neurobiol 56:3244–3259

171. Geisler S, Doan RA, Cheng GC, Cetinkaya-Fisgin A, Huang SX, Höke A, Milbrandt J, DiAntonio A (2019) Vincristine and bortezomib use distinct upstream mechanisms to activate a common SARM1-dependent axon degeneration program. JCI Insight 4:1–17

172. Karney-Grobe S, Russo A, Frey E, Milbrandt J, DiAntonio A (2018) HSP90 is a chaperone for DLK and is required for axon injury signaling. Proc Natl Acad Sci U S A 115:E9899–E9908
173. Summers DW, DiAntonio A, Milbrandt J (2014) Mitochondrial dysfunction induces Sarm1-dependent cell death in sensory neurons. J Neurosci 34:9338–9350
174. Sporny M, Guez-Haddad J, Lebendiker M, Ulisse V, Volf A, Mim C, Isupov MN, Opatowsky Y (2019) Structural evidence for an octameric ring arrangement of SARM1. J Mol Biol. https://doi.org/10.1016/j.jmb.2019.06.030
175. Loring HS, Thompson PR (2020) Emergence of SARM1 as a potential therapeutic target for Wallerian-type diseases. Cell Chem Biol 27:1–13
176. Miller BR, Press C, Daniels RW, Sasaki Y, Milbrandt J, Diantonio A (2009) A dual leucine kinase-dependent axon self-destruction program promotes Wallerian degeneration. Nat Neurosci. https://doi.org/10.1038/nn.2290
177. Patel S, Cohen F, Dean BJ et al (2015) Discovery of dual leucine zipper kinase (DLK, MAP3K12) inhibitors with activity in neurodegeneration models. J Med Chem. https://doi.org/10.1021/jm5013984
178. Schaper T, Rezai M, Petruschke G, Gross B, Franzmann L, Darsow M (2019) Efficiency of controlled cryotherapy in prevention of chemotherapy induced peripheral neuropathy (CIPN). Ann Oncol 30:718–746
179. Sundar R, Bandla A, Tan SSH et al (2017) Limb hypothermia for preventing paclitaxel-induced peripheral neuropathy in breast cancer patients: a pilot study. Front Oncol 6:1–10
180. Hanai A, Ishiguro H, Sozu T et al (2018) Effects of cryotherapy on objective and subjective symptoms of paclitaxel-induced neuropathy: prospective self-controlled trial. J Natl Cancer Inst 110:141–148
181. Chine VB, Au NPB, Ma CHE (2019) Therapeutic benefits of maintaining mitochondrial integrity and calcium homeostasis by forced expression of Hsp27 in chemotherapy-induced peripheral neuropathy. Neurobiol Dis. https://doi.org/10.1016/j.nbd.2019.104492
182. Zhu J, Chen W, Zhou C, Reed N, Höke A (2013) Ethoxyquin provides neuroprotection via HSP90 to ameliorate chemotherapy-induced peripheral neuropathy. J Peripher Nerv Syst. https://doi.org/10.1111/jns5.12025
183. Zhu J, Carozzi VA, Reed N, Mi R, Marmiroli P, Cavaletti G, Hoke A (2016) Ethoxyquin provides neuroprotection against cisplatin-induced neurotoxicity. Sci Rep. https://doi.org/10.1038/srep28861
184. Cirrincione AM, Pellegrini AD, Dominy JR, Benjamin ME, Utkina-Sosunova I, Lotti F, Jergova S, Sagen J, Rieger S (2020) Paclitaxel-induced peripheral neuropathy is caused by epidermal ROS and mitochondrial damage through conserved MMP-13 activation. Sci Rep 10:1–12
185. Zhu Y, Yang J, Jiao S, Ji T (2013) Ganglioside-monosialic acid (GM1) prevents oxaliplatin-induced peripheral neurotoxicity in patients with gastrointestinal tumors. World J Surg Oncol 11:1–7
186. Chen XF, Wang R, Yin YM, Røe OD, Li J, Zhu LJ, Guo RH, Wu T, Shu YQ (2012) The effect of monosialotetrahexosylganglioside (GM1) in prevention of oxaliplatin induced neurotoxicity: a retrospective study. Biomed Pharmacother. https://doi.org/10.1016/j.biopha.2012.01.002
187. Su Y, Huang J, Wang S et al (2020) The effects of ganglioside-monosialic acid in taxane-induced peripheral neurotoxicity in patients with breast cancer: a randomized trial. J Natl Cancer Inst. https://doi.org/10.1093/jnci/djz086
188. Wang DS, Wang ZQ, Chen G et al (2020) Phase III randomized, placebo-controlled, double-blind study of monosialotetrahexosylganglioside for the prevention of oxaliplatin-induced peripheral neurotoxicity in stage II/III colorectal cancer. Cancer Med 9:151–159
189. Kadakia KC, Rozell SA, Butala AA, Loprinzi CL (2014) Supportive cryotherapy: a review from head to toe. J Pain Symptom Manage 47:1100–1115

190. Eckhoff L, Knoop AS, Jensen MB, Ejlertsen B, Ewertz M (2013) Risk of docetaxel-induced peripheral neuropathy among 1,725 Danish patients with early stage breast cancer. Breast Cancer Res Treat 142:109–118

191. Lissea TS, Middletona LJ, Pellegrinia AD, Martina PB, Spauldinga EL, Lopesa O, Brochua EA, Cartera EV, Waldrona A, Riegera S (2016) Paclitaxel-induced epithelial damage and ectopic MMP-13 expression promotes neurotoxicity in zebrafish. Proc Natl Acad Sci U S A. https://doi.org/10.1073/pnas.1525096113

192. Waldron AL, Schroder PA, Bourgon KL, Bolduc JK, Miller JL, Pellegrini AD, Dubois AL, Blaszkiewicz M, Townsend KL, Rieger S (2018) Oxidative stress-dependent MMP-13 activity underlies glucose neurotoxicity. J Diabetes Complications. https://doi.org/10.1016/j.jdiacomp.2017.11.012

193. Beh ST, Kuo YM, Chang WSW, Wilder-Smith E, Tsao CH, Tsai CH, Chen LT, De Liao L (2019) Preventive hypothermia as a neuroprotective strategy for paclitaxel-induced peripheral neuropathy. Pain 160:1505–1521

194. Leblanc AF, Sprowl JA, Alberti P et al (2018) OATP1B2 deficiency protects against paclitaxel-induced neurotoxicity. J Clin Invest 128:816–825

195. Sprowl JA, Ong SS, Gibson AA et al (2016) A phosphotyrosine switch regulates organic cation transporters. Nat Commun 7:1–11

196. Huang KM, Leblanc AF, Uddin ME et al (2020) Neuronal uptake transporters contribute to oxaliplatin neurotoxicity in mice. J Clin Invest. https://doi.org/10.1172/jci136796

197. Wright SH (2019) Molecular and cellular physiology of organic cation transporter 2. Am J Physiol Ren Physiol 317:F1669–F1679

198. Loscalzo J (2011) Systems biology and personalized medicine: a network approach to human disease. Proc Am Thorac Soc 8:196–198

199. Costamagna G, Andreoli L, Corti S, Faravelli I (2019) iPSCs-based neural 3D systems: a multidimensional approach for disease modeling and drug discovery. Cells. https://doi.org/10.3390/cells8111438

200. Sharma AD, McCoy L, Jacobs E, Willey H, Behn JQ, Nguyen H, Bolon B, Curley JL, Moore MJ (2019) Engineering a 3D functional human peripheral nerve in vitro using the Nerve-on-a-Chip platform. Sci Rep. https://doi.org/10.1038/s41598-019-45407-5

201. Chukwurah E, Osmundsen A, Davis SW, Lizarraga SB (2019) All together now: modeling the interaction of neural with non-neural systems using organoid models. Front Neurosci. https://doi.org/10.3389/fnins.2019.00582

202. Nzou G, Wicks RT, VanOstrand NR et al (2020) Multicellular 3D neurovascular unit model for assessing hypoxia and neuroinflammation induced blood-brain barrier dysfunction. Sci Rep. https://doi.org/10.1038/s41598-020-66487-8

203. Papapetrou EP (2016) Patient-derived induced pluripotent stem cells in cancer research and precision oncology. Nat Med. https://doi.org/10.1038/nm.4238

204. Romeo-Guitart D, Forés J, Herrando-Grabulosa M et al (2018) Neuroprotective drug for nerve trauma revealed using artificial intelligence. Sci Rep 8:1–15

205. Ekins S, Puhl AC, Zorn KM, Lane TR, Russo DP, Klein JJ, Hickey AJ, Clark AM (2019) Exploiting machine learning for end-to-end drug discovery and development. Nat Mater 18:435–441

206. Bloomingdale P, Mager DE (2019) Machine learning models for the prediction of chemotherapy-induced peripheral neuropathy. Pharm Res. https://doi.org/10.1007/s11095-018-2562-7

207. Chan A, Hertz DL, Morales M et al (2019) Biological predictors of chemotherapy-induced peripheral neuropathy (CIPN): MASCC neurological complications working group overview. Support Care Cancer 27:3729–3737

208. Crevenna R, Keilani M (2019) Chemotherapy-induced peripheral neuropathy—more high-quality research is needed. Support Care Cancer 27:5–6

209. Camacho DM, Collins KM, Powers RK, Costello JC, Collins JJ (2018) Next-generation machine learning for biological networks. Cell 173:1581–1592

210. Pajares F (2003) The structure of scientific revolutions by Thomas S. Kuhn outline and study guide. Emory University
211. Rivas AL, Leitner G, Jankowski MD et al (2017) Nature and consequences of biological reductionism for the immunological study of infectious diseases. Front Immunol 8:1–8
212. Turnbull L, Hütt MT, Ioannides AA et al (2018) Connectivity and complex systems: learning from a multi-disciplinary perspective. Appl Netw Sci. https://doi.org/10.1007/s41109-018-0067-2
213. Romeo-Guitart D, Casas C (2019) Network-centric medicine for peripheral nerve injury: treating the whole to boost endogenous mechanisms of neuroprotection and regeneration. Neural Regen Res 14:1122
214. Romeo-Guitart D, Casas C (2020) NeuroHeal treatment alleviates neuropathic pain and enhances sensory axon regeneration. Cells. https://doi.org/10.3390/cells9040808

Prevention of Chemotherapy-Induced Peripheral Neuropathy (CIPN): Current Clinical Data and Future Directions

5

Paola Alberti and Christopher B. Steer

Abstract

Chemotherapy-induced peripheral neurotoxicity (CIPN) is a treatment related toxicity that burdens the quality of life of cancer survivors. Unfortunately, no efficacious treatment (symptomatic or preventive) is available for this condition for many reasons. First, a still incomplete pathogenetic knowledge hampers the recognition of a strong biological rationale for clinical trials. Second, there are some methodological issues in clinical trial design that still need to be addressed. In this chapter we will present an overview of strategies that were undertaken in the past and some that are now undergoing clinical investigation for prevention of CIPN. This is a complex challenge that will require multidisciplinary collaborative research between basic scientists, health care professionals, and patient representatives.

Keywords

Chemotherapy-induced peripheral neurotoxicity · Clinical trial · Prevention · Treatment · Neuroprotection

P. Alberti (✉)
Experimental Neurology Unit, School of Medicine and Surgery, University of Milano-Bicocca, Monza (MB), Italy

NeuroMI (Milan Center for Neuroscience), Milan, Italy
e-mail: paola.alberti@unimib.it

C. B. Steer
Border Medical Oncology, Albury Wodonga Regional Cancer Centre, Albury, NSW, Australia

UNSW Rural Clinical School, Albury Campus, Albury, NSW, Australia
e-mail: Christopher.Steer@bordermedonc.com.au

5.1 Introduction

Chemotherapy-induced peripheral neurotoxicity (CIPN) is one of the most common non-haematological toxicities of several cornerstone anticancer drugs—platinum compounds (cisplatin, carboplatin, oxaliplatin), taxanes (paclitaxel, docetaxel), vinca alkaloids (vinorelbine, vincristine, vinblastine), proteasome inhibitors (bortezomib), epothilones (ixabepilone), eribulin, thalidomide. This toxicity can lead to dose reductions and discontinuations which may impact cancer related outcomes and quality of life in cancer survivors. The prevention and treatment of CIPN is still a major challenge for physicians who treat patients with cancer with neurotoxic agents as there are still several issues to be addressed to adequately manage this toxicity [1]. Whilst, in a some cases (mainly oxaliplatin-related), CIPN can be acute and occur early during treatment, it is the tendency for this toxicity to be insidious, occur late, and be permanent and irreversible and this leads to its negative impact on patient quality of life (QOL). If not adequately prevented, CIPN can be the cause of significant symptom burden and reduced health-related QOL in cancer survivors and lead to increased healthcare costs [2, 3].

There are currently numerous pitfalls in research aiming at discovering new treatment and prevention strategies for CIPN. Among these, the most relevant ones are the absence of a fully pathogenetic knowledge of CIPN—which goes beyond the scope of this chapter (for more information, see Chap. 4)—as well as the need to develop a gold standard outcome measure to accurately evaluate CIPN, allowing a precise definition of its incidence, risk factors, and clinical picture. Notably, an accurate risk stratification for the development of severe CIPN would be a crucial aspect in the testing of potential neuroprotectant agents [4]. So far, no blood/serum biomarkers have been identified as gold standard to stratify patients at higher risk of CIPN development [5]. Therefore, efforts from the scientific community are required to push the search of potential biomarkers for risk stratification.

In this chapter, we will present at first some methodological considerations which should be considered in CIPN clinical trials and then we will present past and future perspectives.

5.2 Methodological Considerations

In this section we provide an overview of some methodological considerations that should be carefully weighted when evaluating/designing a CIPN clinical trial.

5.2.1 Issues in CIPN Clinical Trials

It has been reported that patients experience dose-limiting and - often persistent - CIPN during, but also after administration of the aforementioned drugs. Incidences have been reported up to 80% of exposed patients, with prevalence estimates of 68.1% (57.7–78.4%) in the 1st month after treatment completion, 60.0%

(36.4–81.6%) after 3 months, and 30.0% (6.4–53.5%) after at least 6 months [6]. However, an accurate estimation of the true prevalence of CIPN in patients with cancer is lacking. This is due to a number of factors including: the fact that epidemiological data may greatly vary from study to study, due to the use of different methodological approaches (e.g. the outcome measures used and timing of assessments), and different study design (e.g. prospective vs retrospective study) [7]. The estimation of the incidence of CIPN is made more challenging by the fact that each neurotoxic drug exhibits a slightly different neurotoxicity pattern. Many different outcome measures, among which different scales, have been proposed. They can be divided into toxicity scales (e.g. NCI-CTC scale), physician-based (e.g. TNS scale), and Patient Reported Outcome (PRO) measures. For more details on these instruments, see Chap. 3. The relevance of clinimetric issues in CIPN assessment and clinical trial design can be understood considering international initiatives aiming at dissecting this methodological issue. In 2017 the National Cancer Institute Symptom Management and Health-Related Quality of Life Steering Committee Clinical Trials Planning Meeting (CTPM) was created to shed light on possible solutions. This group noted the lack of a validated gold standard to measure the impact of CIPN and developed a mechanism for the formulation of consensus expert opinion. This involved the creation of specific working groups to unravel the issue [8]. In parallel, another international initiative—the Analgesic, Anaesthetic, and Addiction Clinical Trial Translations, Innovations, Opportunities and Networks (ACTTION)–Consortium on Clinical Endpoints and Procedures for Peripheral Neuropathy Trials (CONCEPPT)—met to develop guidelines to drive future trial design and despite the absence of a gold standard in CIPN assessment made suggestions on eligibility criteria, outcome measures, endpoints, and sample size estimation. They suggested that a robust clinical trial must include both physician-based tools and PROs [9].

5.2.2 Surrogate Biomarkers: Some Promising Neurophysiological Options for CIPN Risk Stratification

The assessment of CIPN involves both PROs and objective assessments such as nerve conduction studies (NCS). NCS are central in the diagnosis—for clinical purposes—of peripheral neuropathies (Fuglsang—[10]). There is a suggestion that testing for neurotoxicity prior to the use of neurotoxic therapies may predict of the development of CIPN. In particular, abnormal sensory NCS have been reported in patients with CIPN prior to symptoms onset, suggesting NCS might be an important early surrogate biomarker for axonal damage [11]. Given that CIPN is a length-dependent process, that is expected to ensue first at limbs extremities, dorsal sural nerve (DSN), a more distal branch of the sural nerve [12] that is included in polyneuropathy assessment protocol in any EMG lab, was tested longitudinally in a cohort of 200 patient with colorectal cancer (CRC) treated with oxaliplatin [13]. DSN neurophysiological alterations (drop in amplitude in particular), in fact, were yet proved more sensitive than sural nerve in detecting early dysfunction due to chemotherapy [14] and polyneuropathies due to other causes (e.g. diabetes, vitamin B12 deficiency) [15–17]. In this trial [13], NCS of the DSN were performed at mid-treatment and after oxaliplatin chemotherapy completion (FOLFOX-4 or XELOX regimens). The authors were able to develop an algorithm and demonstrated that the mid-treatment DSN NCS could assign each patient to a 'risk class' predictive of neurological outcome at end of treatment with high correlation value [13]. Therefore, it was the first and promising tentative to stratify patients (high vs low risk for CIPN development) that could be taken into account in future clinical trials, to be combined with physician and patient based tools, even if of course—it should be further tested in a larger population and in patients treated with other regimens than oxaliplatin-based.

5.3 Prior Completed Studies

Unfortunately, as stated above, there is no efficacious pharmacological approach for the prevention of CIPN. The American Society of Clinical Oncology (ASCO) produced guidelines who extensively addressed, with a systematic review, the prevention and management of CIPN; both in 2014 [18] and in the updated version published in 2020 [19]. Authors did not find any strong evidence for all the molecules tested over the years in CIPN patients, apart from the use of duloxetine as symptomatic treatment (moderate recommendation) in patients with painful peripheral neuropathy. In the next couple of sections, we provide a brief overview of the molecules that were tested but were not proved to be efficacious.

5.3.1 Neuroprotectant Agents

To prevent CIPN, the most sensible option is to target the underlying mechanism. The problem is that neurotoxic mechanisms differ among different drug classes and the exact cascade for neuronal damage is not fully elucidated [20–22]. For a detailed description of mechanisms involved in CIPN development, see Chap. 2. Despite decades of translational research and drug repurposing studies involving a wide variety of molecules, no agent has been shown to successfully prevent CIPN. The list of potential candidates, many tested in rigorous randomized control trials, includes: recombinant human leukaemia inhibitory factor (rhuLIF, Emfilermin, AM424), amifostine (WR-2721), pregabalin, acetyl-l-carnitine, glutathione, minocycline, omega-3 fatty acids, retinoic acid, lafutidine, vitamin E and vitamin B complexes. As a consequence, the aforementioned ASCO guidelines, unfortunately, concluded that no neuroprotective agent can be addressed with confidence as a neuroprotective strategy against CIPN in the general setting [18, 19]. The only options still available in daily practice therefore remain treatment modification (e.g. dose reduction, prolongation of infusion time or dose fractionation) and treatment withdrawal [19].

5.4   Potential Options for Future Clinical Trials

Given the significant symptom burden in patients with CIPN, there is an urgent need for effective prevention strategies. Research in the underlying mechanisms continues to drive drug discovery. We provide a brief overview of ongoing trials whose results might be of potential interest in the next few years for patients at risk of developing CIPN.

First, we will present data of yet completed studies targeting ganglioside monosialic pathway, oxidative stress, and sigma-1 receptors. Then, we will address some options that are currently being tested targeting neuronal uptake transporters, glutamatergic neurotransmission, serotoninergic receptors, and sphingolipids.

5.4.1   Ganglioside Monosialic Acid Pathway

Targeting the ganglioside monosialic acid pathway with delivery of ganglioside monosialic acid (GM-1) may have neurotrophic and neuroprotective effects [23]. Therefore, GM1 was tested in a randomized, double-blind, placebo-controlled trial: 206 breast cancer patients scheduled to receive taxane-based adjuvant chemotherapy were randomized to intravenous infusion of GM1 (80 mg, Day -1 to Day 2) or placebo. GM1-treated patients had a lower incidence of chemotherapy dose reductions/delays and taxane-induced neuropathic pain was significantly reduced [24]. GM1 was then also tested in another randomized, double-blind, multicenter, placebo-controlled, phase III trial, in a population of 196 patients with stage II/III colorectal cancer: unfortunately, GM1 did not prevent chronic CIPN manifestations due to oxaliplatin, even though acute oxaliplatin neurotoxicity syndrome was partially contained [25]. Therefore, other independent conducted RCTs are warranted to give a final judgement, even though, as stated above, different anticancer drugs have different neurotoxic mechanisms and neuroprotective agents might be efficacious with some drug classes and not with others.

5.4.2   Oxidative Stress

Oxidative stress has been described as a factor in the development and progression of CIPN [26] and molecules involved in oxidative stress modulation are now being tested in clinical trials.

Calmangafodipir is intended to target oxidative stress and neuronal mitochondrial injury and was tested in a placebo-controlled, double-blinded randomized phase II study in patients with metastatic colorectal cancer: in the PLIANT trial, 173 patients were randomized to calmangafodipir 2 mmol/kg ($n = 57$), calmangafodipir 5 mmol/kg ($n = 45$; initially 10 mmol/kg, $n = 11$), or placebo ($n = 60$). The 5 mmol/kg dose was reported to be associated with reduction of both acute and chronic neurotoxicity symptoms due to oxaliplatin [27]. However, as highlighted by Karlsson and Jynge [28] the study raised several methodological pitfalls making impossible to draw any strong conclusion from this trial (in particular: frequency of adverse events; change of primary endpoint once the study was yet started; and discrepancies of endpoints in different reports leading to questionable data handling and interpretation). Calmangafodipir is currently being tested in the management of oxaliplatin neurotoxicity in 2 phase III double-blind placebo-controlled trials, POLAR-M and POLAR-A: POLAR-M is evaluating the use of 5μmol/kg calmangafodipir, 2μmol/kg calmangafodipir, or placebo (NCT03654729) in patients with metastatic colorectal cancer. In a similar trial, POLAR-A is evaluating the use of 5μmol/kg calmangafodipir in patients with stage III or high-risk stage II colorectal cancer being treated with oxaliplatin-based regimens in the adjuvant setting (NCT04034355) [29]. Data collection was completed on 14th October 2020 and results, as reported in a recent press release (MFN.se > PledPharma > Results from the prematurely closed PledOx POLAR program), were disappointing.

5.4.3 Sigma-1 Receptor Antagonist

The sigma-1 receptor is a transmembrane protein found in the endoplasmic reticulum, specifically at the mitochondria associated endoplasmic reticulum membrane, and has a modulatory role in nociception, attenuating intracellular signal transduction cascades related to noxious stimuli and sensitization phenomena [30]. A Phase IIa, double-blind, RCT of the sigma-1 receptor antagonist MR309 (previously developed as E-52862) enrolled 124 colorectal cancer patients to active treatment (400 mg/day, 5 days per cycle) or placebo ($n = 62$). Intermittent treatment with MR309 reduced the incidence and severity of acute oxaliplatin-induced neurotoxicity and allowed higher oxaliplatin exposure [31].

5.4.4 Emerging Options: Ongoing Translational Research

In this section we present some ongoing clinical trials based on sound biological rationales that might offer solutions to the significant unmet clinical and scientific needs.

5.4.4.1 Neuronal Uptake Transporters
Organic-anion-transporting polypeptides (OATP) and organic cation transporters (OCT) were suggested in preclinical studies as a possible neurotoxicity prevention strategy as they transport anticancer drug inside dorsal root ganglia (DRG) neurons triggering neurotoxicity development [32, 33].

Among the option to modulate this pathway, there is dasatinib, an orally active targeted therapy used to treat haematological malignancies, part of SRC-family protein tyrosine kinase inhibitor; in fact, Sprowl and collaborators [34] demonstrated that it can inhibit organic cation transporter 2 (OCT2) inhibitor. OCT2 receptors are of particular interest in CIPN research since they are widely expressed in dorsal root ganglia and the peripheral nervous system [35, 36]. A phase Ib open-label clinical trial is evaluating the role of dasatinib in the prevention of oxaliplatin-related neurotoxicity in patients with metastatic colorectal cancer (NCT04164069).

Another drug currently under consideration is nilotinib, a Bcr-Abl tyrosine kinase inhibitor used to treat haematological malignancies. Nilotinib can inhibit an organic-anion-transporting polypeptide B (OATP1B) uptake transporter inhibitor [37], which is expressed in peripheral nervous system [38]. Thus, drug is undergoing a phase Ib/II randomized parallel double-blind study; the aim is evaluating safety and addressing its use against paclitaxel neurotoxicity in breast cancer patients (NCT04205903).

5.4.4.2 Glutamatergic Neurotransmission

The neurotransmitter glutamate is known leading to toxicity when present in an excessive amount (the so-called excitotoxicity) in a wide range of neurological conditions [39]. Preclinical data showed glutamate signalling was altered as a consequence of cisplatin, paclitaxel, and bortezomib administration and inhibition of glutamate decarboxypeptidase enzyme (resulting in decreased production of glutamate) may ameliorate neurotoxicity [29]. Notably, rodents exposed to a polyamine-deficient diet did not show neuropathic pain behaviour after oxaliplatin exposure, thanks to the fact that polyamines positively modulate the NR2B subunit of N-methyl-D-aspartate receptors (NMDAR) on which glutamate acts [40].

On the basis of this evidence, some molecules are being tested to modulate glutamate excitotoxicity.

A drug able to modulate glutamate transmission is riluzole, which is currently used in patients with amyotrophic lateral sclerosis [41], is under investigation in CIPN. The mechanism of the effect in ALS is not yet known but the blockade of glutamate transmission has been hypothesized as a pivotal event [42]. In the setting of CIPN, riluzole is being tested in the clinical trial RILUZOX-01 which is a phase II randomized double-blind trial (vs placebo) aimed at preventing oxaliplatin neuro-toxicity (NCT03722680).

5.4.4.3 Serotoninergic Receptors

Serotonin or 5-hydroxytryptamin (5-HT) is a neurotransmitter involved in pain modulation and its effects are modulated by various receptors, among which the most numerous are part of the GPCR family. The 5-HT2C receptor (5-HT2CR) subtype has been involved in the modulation of neuropathic pain in various animal models [43]; moreover, preclinical data showed that oxaliplatin administration increases 5-HT2CRmRNA expression in spinal cord and in midbrain [44]. Among compounds modulating this axis, there is a lorcaserin, a 5-HT2CR activator, which is prescribed for weight loss. Lorcaserin was under evaluation in a phase I trial assessing its effects on acute neurotoxicity manifestations of oxaliplatin and taxanes (NCT04205071), and in a phase II study comparing lorcaserin and duloxetine in the treatment of oxaliplatin chronic CIPN (NCT03812523). However, lorcaserin trials were recently halted due to increased risk cancer as emerged by FDA revision; therefore, FDA requested that the manufacturer to voluntarily withdraw the drug from the market [45].

Duloxetine is also being tested as neuroprotectant in a phase II/III trial to prevent CIPN in patients undergoing treatment with oxaliplatin for colorectal cancer (NCT04137107).

5.4.4.4 Sphingolipids

Sphingolipids are a family of membrane lipids which control cellular processes (cell division and differentiation, and cell death). Their metabolism alterations has been related to neurodegenerative diseases [46], as well as to neuropathic pain development [47]. In clinical practice, fingolimod, a sphingosine-1-phosphate (S1P) receptor 1 (S1PR1) antagonist, is one of the treatment options for multiple sclerosis [48]. This drug is a sphingosine analogue which is phosphorylated via cellular sphingosine kinase. S1P receptors are widely expressed in the central and peripheral nervous system and involved in process of regeneration [49, 50]. The drug is currently being tested in an early phase I trial to test its efficacy in CIPN patients. The drug is being tested to obtain preliminary data to support whether it prevents CIPN in patients receiving weekly adjuvant/neoadjuvant paclitaxel treatment (NCT03941743).

5.4.5 Non-pharmacological Treatments

Non-pharmacological treatments have been investigated in recent years in CIPN patients. Physical therapy can be useful, in particular, for accelerating recovery after CIPN onset or to ameliorate physical performance once stable CIPN had ensued. This is crucial in particular when patients develop sensory ataxia, a condition quite frequent especially after the administration of platinum compounds [51]. This condition is due to impairment of spinal cord dorsal columns sensory modalities: proprioception. The patient, even if s/he has no motor impairment, has difficulty in manipulation, and develops gait unsteadiness/unbalance [4, 52]. At the state of the art, phase III clinical trials are still lacking to draw a definite conclusion of the efficacy of physical treatments. For a detailed description of these, see Chaps. 8 and 9.

5.5 Conclusion

Even if at the present there is no efficacious approach for CIPN prevention, many research groups are actively working on finding a solution for this detrimental toxicity of anticancer drugs. The lesson learnt from past studies is that a multidisciplinary effort is warranted. A sound biological rationale is needed to be addressed at bench-side to provide a strong background for future translational clinical research. At bedside, a careful study design is needed to increase the chance to accurately test efficacy of the tested molecule. The best environment where such initiatives can grow can be provided by working groups such as the CIPN working group of the Multinational Association of Supportive Care in Cancer (MASCC) or the Toxic Neuropathy Consortium (part of the Peripheral Nerve Society), in which basic scientists, health care professionals, and patients' representative can cooperate in novel CIPN research projects.

References

1. Cavaletti G, Alberti P, Argyriou AA, Lustberg M, Staff NP, Tamburin S, Toxic Neuropathy Consortium of the Peripheral Nerve Society (2019) Chemotherapy-induced peripheral neurotoxicity: A multifaceted, still unsolved issue. J Peripher Nerv Syst 24(Suppl 2):S6–S12
2. Pike CT, Birnbaum HG, Muehlenbein CE, Pohl GM, Natale RB (2012) Healthcare costs and workloss burden of patients with chemotherapy-associated peripheral neuropathy in breast, ovarian, head and neck, and nonsmall cell lung cancer. Chemother Res Pract 2012:913848
3. Song X, Wilson KL, Kagan J, Panjabi S (2019) Cost of peripheral neuropathy in patients receiving treatment for multiple myeloma: a US administrative claims analysis. Ther Adv Hematol 10:2040620719839025
4. Cavaletti G, Cornblath DR, ISJ M, Postma TJ, Rossi E, Alberti P, Bruna J, Argyriou AA, Briani C, Velasco R, Kalofonos HP, Psimaras D, Ricard D, Pace A, Faber CG, Lalisang RI, Brandsma D, Koeppen S, Kerrigan S, Schenone A, Grisold W, Mazzeo A, Padua L, Dorsey SG, Penas-Prado M, Valsecchi MG, CI-PeriNomS Group (2019) Patients' and physicians' interpretation of chemotherapy-induced peripheral neurotoxicity. J Peripher Nerv Syst. 24(1):111–119
5. Argyriou AA, Bruna J, Genazzani AA, Cavaletti G (2017) Chemotherapy-induced peripheral neurotoxicity: management informed by pharmacogenetics. Nat Rev Neurol 13:492–504
6. Seretny M, Currie GL, Sena ES, Ramnarine S, Grant R, MacLeod MR, Colvin LA, Fallon M (2014) Incidence, prevalence, and predictors of chemotherapy-induced peripheral neuropathy: A systematic review and meta-analysis. Pain 155:2461–2470
7. Argyriou AA, Park SB, Islam B, Tamburin S, Velasco R, Alberti P, Bruna J, Psimaras D, Cavaletti G, Cornblath DR, TNC (2019) Neurophysiological, nerve imaging and other techniques to assess chemotherapy-induced peripheral neurotoxicity in the clinical and research settings. J Neurol Neurosurg Psychiatry 90(12):1361–1369
8. Dorsey SG, Kleckner IR, Barton D, Mustian K, O'Mara A, St Germain D, Cavaletti G, Danhauer SC, Hershman D, Hohmann AG, Hoke A, Hopkins JO, Kelly KP, Loprinzi CL, McLeod HL, Mohile S, Paice J, Rowland JH, Salvemini D, Segal RA, Lavoie Smith E, McCaskill Stevens W, Janelsins MC (2019) NCI Clinical Trials Planning Meeting for prevention and treatment of chemotherapy-induced peripheral neuropathy. J Natl Cancer Inst 111 (6):531–537
9. Gewandter JS, Brell J, Cavaletti G, Dougherty PM, Evans S, Howie L, McDermott MP, O'Mara A, Smith AG, Dastros-Pitei D, Gauthier LR, Haroutounian S, Jarpe M, Katz NP, Loprinzi C, Richardson P, Lavoie-Smith EM, Wen PY, Turk DC, Dworkin RH, Freeman R (2018) Trial designs for chemotherapy-induced peripheral neuropathy prevention: ACTTION recommendations. Neurology 91:403–413
10. Fuglsang-Frederiksen A, Pugdahl K (2011) Current status on electrodiagnostic standards and guidelines in neuromuscular disorders. Clin Neurophysiol 122:440–455
11. Mileshkin L, Stark R, Day B, Seymour JF, Zeldis JB, Prince HM (2006) Development of neuropathy in patients with myeloma treated with thalidomide: patterns of occurrence and the role of electrophysiologic monitoring. J Clin Oncol 24:4507–4514
12. Frigeni B, Cacciavillani M, Ermani M, Briani C, Alberti P, Ferrarese C, Cavaletti G (2012) Neurophysiological examination of dorsal sural nerve. Muscle Nerve 46:895–898
13. Alberti P, Rossi E, Argyriou AA, Kalofonos HP, Briani C, Cacciavillani M, Campagnolo M, Bruna J, Velasco R, Cazzaniga ME, Cortinovis D, Valsecchi MG, Cavaletti G (2018) Risk stratification of oxaliplatin induced peripheral neurotoxicity applying electrophysiological testing of dorsal sural nerve. Support Care Cancer 26:3143–3151

14. Dalla Torre C, Zambello R, Cacciavillani M, Campagnolo M, Berno T, Salvalaggio A, De March E, Barilà G, Lico A, Lucchetta M, Ermani M, Briani C (2016) Lenalidomide long-term neurotoxicity: Clinical and neurophysiologic prospective study. Neurology 87:1161–1166

15. Koçer A, Domaç FM, Boylu E, Us O, Tanridağ T (2007) A comparison of sural nerve conduction studies in patients with impaired oral glucose tolerance test. Acta Neurol Scand 116:399–405

16. Turgut B, Turgut N, Akpinar S, Balci K, Pamuk GE, Tekgündüz E, Demir M (2006) Dorsal sural nerve conduction study in vitamin B(12) deficiency with megaloblastic anemia. J Peripher Nerv Syst 11:247–252

17. Uluc K, Isak B, Borucu D, Temucin CM, Cetinkaya Y, Koytak PK, Tanridag T, Us O (2008) Medial plantar and dorsal sural nerve conduction studies increase the sensitivity in the detection of neuropathy in diabetic patients. Clin Neurophysiol 119:880–885

18. Hershman DL, Lacchetti C, Dworkin RH, Lavoie Smith EM, Bleeker J, Cavaletti G, Chauhan C, Gavin P, Lavino A, Lustberg MB, Paice J, Schneider B, Smith ML, Smith T, Terstriep S, Wagner-Johnston N, Bak K, Loprinzi CL, American Society of Clinical Oncology (2014) Prevention and management of chemotherapy-induced peripheral neuropathy in survivors of adult cancers: American Society of Clinical Oncology clinical practice guideline. J Clin Oncol 32:1941–1967

19. Loprinzi CL, Lacchetti C, Bleeker J, Cavaletti G, Chauhan C, Hertz DL, Kelley MR, Lavino A, Lustberg MB, Paice JA, Schneider BP, Lavoie Smith EM, Smith ML, Smith TJ, Wagner-Johnston N, Hershman DL (2020) Prevention and management of chemotherapy-induced peripheral neuropathy in survivors of adult cancers: ASCO guideline update. J Clin Oncol 38 (28):3325–3348

20. Calls A, Carozzi V, Navarro X, Monza L, Bruna J (2020) Pathogenesis of platinum-induced peripheral neurotoxicity: Insights from preclinical studies. Exp Neurol 325:113141

21. Geisler S (2021) Vincristine- and bortezomib-induced neuropathies—from bedside to bench and back. Exp Neurol 336:113519

22. Staff NP, Fehrenbacher JC, Caillaud M, Damaj MI, Segal RA, Rieger S (2020) Pathogenesis of paclitaxel-induced peripheral neuropathy: A current review of in vitro and in vivo findings using rodent and human model systems. Exp Neurol 324:113121

23. Aureli M, Mauri L, Ciampa MG, Prinetti A, Toffano G, Secchieri C, Sonnino S (2016) GM1 ganglioside: past studies and future potential. Mol Neurobiol 53:1824–1842

24. Su Y, Huang J, Wang S, Unger JM, Arias-Fuenzalida J, Shi Y, Li J, Gao Y, Shi W, Wang X, Peng R, Xu F, An X, Xue C, Xia W, Hong R, Zhong Y, Lin Y, Huang H, Zhang A, Zhang L, Cai L, Zhang J, Yuan Z (2020) The effects of ganglioside-monosialic acid in taxane-induced peripheral neurotoxicity in patients with breast cancer: a randomized trial. J Natl Cancer Inst 112:55–62

25. Wang DS, Wang ZQ, Chen G, Peng JW, Wang W, Deng YH, Wang FH, Zhang JW, Liang HL, Feng F, Xie CB, Ren C, Jin Y, Shi SM, Fan WH, Lu ZH, Ding PR, Wang F, Xu RH, Li YH (2020) Phase III randomized, placebo-controlled, double-blind study of monosialotetrahexosylganglioside for the prevention of oxaliplatin-induced peripheral neuro-toxicity in stage II/III colorectal cancer. Cancer Med 9:151–159

26. Carozzi VA, Canta A, Chiorazzi A (2015) Chemotherapy-induced peripheral neuropathy: What do we know about mechanisms? Neurosci Lett 596:90–107

27. Glimelius B, Manojlovic N, Pfeiffer P, Mosidze B, Kurteva G, Karlberg M, Mahalingam D, Buhl Jensen P, Kowalski J, Bengtson M, Nittve M, Näsström J (2018) Persistent prevention of oxaliplatin-induced peripheral neuropathy using calmangafodipir (PledOx). Acta Oncol 57:393–402

28. Karlsson JOG, Jynge P (2018) Is it possible to draw firm conclusions from the PLIANT trial? Acta Oncol 57:862–864

29. Bouchenaki H, Danigo A, Sturtz F, Hajj R, Magy L, Demiot C (2020) An overview of ongoing clinical trials assessing pharmacological therapeutic strategies to manage

chemotherapy-induced peripheral neuropathy, based on preclinical studies in rodent models. Fundam Clin Pharmacol. https://doi.org/10.1111/fcp.12617

30. Vela JM, Merlos M, Almansa C (2015) Investigational sigma-1 receptor antagonists for the treatment of pain. Expert Opin Investig Drugs 24:883–896

31. Bruna J, Videla S, Argyriou AA, Velasco R, Villoria J, Santos C, Nadal C, Cavaletti G, Alberti P, Briani C, Kalofonos HP, Cortinovis D, Sust M, Vaqué A, Klein T, Plata-Salamán C (2018) Efficacy of a novel sigma-1 receptor antagonist for oxaliplatin-induced neuropathy: a randomized, double-blind, placebo-controlled phase IIa clinical trial. Neurotherapeutics 15 (1):178–189

32. Fujita S, Hirota T, Sakiyama R, Baba M, Ieiri I (2019) Identification of drug transporters contributing to oxaliplatin-induced peripheral neuropathy. J Neurochem 148:373–385

33. Liu JJ, Kim Y, Yan F, Ding Q, Ip V, Jong NN, Mercer JF, McKeage MJ (2013) Contributions of rat Ctr1 to the uptake and toxicity of copper and platinum anticancer drugs in dorsal root ganglion neurons. Biochem Pharmacol 85:207–215

34. Sprowl JA, Ong SS, Gibson AA, Hu S, Du G, Lin W, Li L, Bharill S, Ness RA, Stecula A, Offer SM, Diasio RB, Nies AT, Schwab M, Cavaletti G, Schlatter E, Ciarimboli G, Schellens JHM, Isacoff EY, Sali A, Chen T, Baker SD, Sparreboom A, Pabla N (2016) A phosphotyrosine switch regulates organic cation transporters. Nat Commun 7:10880

35. Roth M, Obaidat A, Hagenbuch B (2012) OATPs, OATs and OCTs: the organic anion and cation transporters of the SLCO and SLC22A gene superfamilies. Br J Pharmacol 165:1260–1287

36. Sprowl JA, Ciarimboli G, Lancaster CS, Giovinazzo H, Gibson AA, Du G, Janke LJ, Cavaletti G, Shields AF, Sparreboom A (2013) Oxaliplatin-induced neurotoxicity is dependent on the organic cation transporter OCT2. Proc Natl Acad Sci U S A 110:11199–11204

37. Hu S, Mathijssen RH, de Bruijn P, Baker SD, Sparreboom A (2014) Inhibition of OATP1B1 by tyrosine kinase inhibitors: in vitro-in vivo correlations. Br J Cancer 110:894–898

38. Leblanc AF, Sprowl JA, Alberti P, Chiorazzi A, Arnold WD, Gibson AA, Hong KW, Pioso MS, Chen M, Huang KM, Chodisetty V, Costa O, Florea T, de Bruijn P, Mathijssen RH, Reinbolt RE, Lustberg MB, Sucheston-Campbell LE, Cavaletti G, Sparreboom A, Hu S (2018) OATP1B2 deficiency protects against paclitaxel-induced neurotoxicity. J Clin Invest 128:816–825

39. Beretta S, Begni B, Ferrarese C (2003) Pharmacological manipulation of glutamate transport. Drug News Perspect 16:435–445

40. Ferrier J, Bayet-Robert M, Pereira B, Daulhac L, Eschalier A, Pezet D, Moulinoux JP, Balayssac D (2013) A polyamine-deficient diet prevents oxaliplatin-induced acute cold and mechanical hypersensitivity in rats. PLoS One 8:e77828

41. Miller RG, Mitchell JD, Moore DH (2012). Riluzole for amyotrophic lateral sclerosis (ALS)/ motor neuron disease (MND). Cochrane Database Syst Rev (2):CD001447.

42. Kretschmer BD, Kratzer U, Schmidt WJ (1998) Riluzole, a glutamate release inhibitor, and motor behavior. Naunyn Schmiedebergs Arch Pharmacol 358:181–190

43. Nakae A, Nakai K, Tanaka T, Hosokawa K, Mashimo T (2013) Serotonin 2C receptor alternative splicing in a spinal cord injury model. Neurosci Lett 532:49–54

44. Baptista-de-Souza D, Di Cesare ML, Zanardelli M, Micheli L, Nunes-de-Souza RL, Canto-de-Souza A, Ghelardini C (2014) Serotonergic modulation in neuropathy induced by oxaliplatin: effect on the 5HT2C receptor. Eur J Pharmacol 735:141–149

45. Sharretts J, Galescu O, Gomatam S, Andraca-Carrera E, Hampp C, Yanoff L (2020) Cancer risk associated with lorcaserin—the FDA's review of the CAMELLIA-TIMI 61 trial. N Engl J Med 383:1000–1002

46. Pralhada Rao R, Vaidyanathan N, Rengasamy M, Mammen Oommen A, Somaiya N, Jagannath MR (2013) Sphingolipid metabolic pathway: an overview of major roles played in human diseases. J Lipids 2013:178910

47. Patti GJ, Yanes O, Shriver LP, Courade JP, Tautenhahn R, Manchester M, Siuzdak G (2012) Metabolomics implicates altered sphingolipids in chronic pain of neuropathic origin. Nat Chem Biol 8:232–234

48. Alroughani R, Inshasi J, Al-Asmi A, Alkhabouri J, Alsaadi T, Alsalti A, Boshra A, Canibano B, Ahmed SF, Shatila A (2020) Disease-modifying drugs and family planning in people with multiple sclerosis: a consensus narrative review from the Gulf region. Neurol Ther 9:265–280

49. Heinen A, Beyer F, Tzekova N, Hartung HP, Küry P (2015) Fingolimod induces the transition to a nerve regeneration promoting Schwann cell phenotype. Exp Neurol 271:25–35

50. Szepanowski F, Derksen A, Steiner I, Meyer Zu Hörste G, Daldrup T, Hartung HP, Kieseier BC (2016) Fingolimod promotes peripheral nerve regeneration via modulation of lysophospholipid signaling. J Neuroinflammation 13:143

51. Alberti P (2019) Platinum-drugs induced peripheral neurotoxicity: clinical course and preclinical evidence. Expert Opin Drug Metab Toxicol 15:487–497

52. Windebank AJ, Grisold W (2008) Chemotherapy-induced neuropathy. J Peripher Nerv Syst 13:27–46

Treatment of Established Chemotherapy-Induced Peripheral Neuropathy: Basic Science and Animal Models

6

Manuel Morales and Nathan P. Staff

Abstract

Advancement of effective therapies to treat established CIPN will require a deeper understanding of CIPN pathomechanisms. Simplified models of CIPN have been developed using whole-animal systems, primary cultures, and immortalized cell lines to allow for detailed mechanistic studies. Recently, human stem-cell derived neuronal cultures have also allowed new opportunities to study CIPN. In this chapter, we provide an overview of studies that used model systems to investigate the treatment of established CIPN. We have divided the chapter into two main areas. First, there are studies that investigate CIPN-related nerve damage through the lens of neurogenesis, Schwann cells, and axonal regrowth. Next, we review model approaches to treat CIPN-related pain that have focused on voltage-gated ion channels, neuroinflammation, sphingosine metabolism, and endocannabinoids. The broad approaches that are being employed to study the treatment of established CIPN in model systems provide hope for future beneficial therapeutics.

M. Morales
Hospital Universitario Nuestra Señora de Candelaria, Santa Cruz de Tenerife, Spain
e-mail: mmoraleg@ull.edu.es

N. P. Staff (✉)
Mayo Clinic, Rochester, MN, USA
e-mail: staff.nathan@mayo.edu

6.1 Introduction

Despite a growing understanding of the pathophysiology of CIPN few therapies have shown success in humans. Only the antidepressant medication duloxetine has shown moderate efficacy to treat established pain due to CIPN [1]. Animal models appear to be important for identifying appropriate therapies for treating established CIPN. Experimental models of CIPN can be induced in different strains of rats or mice through intraperitoneal (ip), subcutaneous (sc), or intravenous (iv) administration of the desired drug [2]. "In vitro" studies are also important to further study the effects of the different drugs at the cellular level and for the search of potential therapy targets against CIPN. These studies can be performed with cultures of dorsal root ganglion (DRG)-neurons obtained from rats or mice [3] or with immortalized and commercially available murine sensory neurons cell lines [4, 5]. Nonetheless "in vitro" studies have limitations due to the biologic differences between humans versus mice or rats. To overcome this problem, sensory neurons can be induced from human skin fibroblasts or multipotential CD34$^+$ hematopoietic stem cells obtained from peripheral blood [6, 7].

6.2 Models of CIPN

6.2.1 In Vivo Animal Models of CIPN

About 70% of in vivo animal studies are conducted with rats and 30% with mice, the drugs commonly used to induce CIPN are oxaliplatin, paclitaxel, vincristine, cisplatin, and bortezomib [8]. The doses and schedules of the different chemotherapy agents for the induction of CIPN in rodents are listed in Table 6.1.

After the administration of the drug in the required dosage, behavioral tests are performed to assess the establishment of neuropathy. These tests are directed to test motor coordination, mechanical allodynia, and thermal sensitivity. Neuromuscular coordination is assessed with the rotarod test, which consists of a circular rod turning at different speeds. The amount of time in which an animal stays on the rotating rod is related to its motor coordination. Mechanical allodynia is measured with the electronic von Frey hair test, placing the mouse or rat in an inverted plastic cage with a wire-mesh floor. Semiflexible filaments are then applied to the center of the hind paws, gradually increasing the pressure for 5 s, in order to establish a pain threshold [18]. Cold hyperalgesia and alterations in thermal sensibility are tested with the acetone test and the hot plate test, respectively. The acetone test consists of touching the plantar skin of a hind paw with a 100 µl droplet of acetone from a syringe, while the hot plate test is performed by placing animals on an aluminum plate which is uniformly heated. For the hot plate a cut-off time of 30 s is used, to prevent damage [19].

Table 6.1 Doses and schedules for experimental models of CIPN in mice and rats

Drug	Animal	Dose	Route	Schedule	References
Oxaliplatin	Rat	4 mg/kg	Ip	Twice a week × 4	[9]
	Rat	5 mg/kg	Ip	Days 0, 3, 6, and 9	[10]
	Mouse	4 mg/kg	Ip	Days 0, 2, 4, and 6	[10]
Paclitaxel	Rat	2 m/kg	Ip	Days 0, 2,4, and 6	[11]
	Mouse	4 mg/kg	Ip	Days 0, 2, 4, and 6	[12]
Vincristine	Rat	200 μg/kg	Iv	Single dose	[13]
	Mouse	200 μg/kg	Ip	Single dose	
Cisplatin	Rat	2 mg/kg	Ip	4 consecutive days	[14]
	Mouse	2.3 mg/kg	Ip	2 cycles of 5 consecutive days with 5 days rest in between.	[15]
Bortezomib	Rat	0.1–0.2 mg/kg	Ip	Days 0, 3, 7, and 10	[16]
	Mouse	400 μg/kg	Ip	3 days /week × 4 weeks	[17]

6.2.2   In Vitro Models of CIPN

The difficulties in obtaining human neurons for study make cell culture models an important tool for CIPN pathophysiological and pharmacological research. The commercially available rat PC12 pheochromocytoma cell line differentiates to neurons in the presence of forskolin, stimulating neurite outgrowth [20]. Forskolin is a diterpenoid obtained from the plant *Coleus forskohlii* that penetrates cell membranes and increases the levels of adenylyl cyclase (cAMP), which is involved in many transduction pathways [21]. The 50B11 neuronal cell line is another commercially available cell line derived from rat DRG [4].

Primary cell cultures can be performed with DRG neurons obtained from embryonic or early-postnatal rats after surgical removal, cultivation with collagenase I, centrifugation and seeding in neurobasal medium [3]. Schwann cells derived from the sciatic nerves of neonatal rats are also used for primary culture [22].

The biologic differences between mice or rats and humans limit the extrapolation of results. To overcome this problem, sensory neurons can be induced from human embryonic fibroblasts, through the transfection with lentiviral vectors of the transcription factor *Brn3a* with either *Ngn1* or *Ngn2* [23]. The pluripotent hematopoietic CD34+ stem cells are also a source for the induction of sensory neurons, which can be available from blood banks or from peripheral blood sampling. The isolated CD34+ stem cells are cultured in the required media and transfected with the lentivirus OCT4 delivery system to produce induced neural progenitor cells (iNPCs). The iNPCs are then cultured in a sensory neuron specification medium, supplemented with brain derived neurotrophic factor, glial derived neurotrophic factor, nerve growth factor, neurotrophin-3 and forskolin, until the desired maturation stage [7]. Likewise, sensory neurons can be differentiated from human induced pluripotent stem cells [6], which has been also utilized as a model for CIPN [24–27].

These "in vitro" models enable the study of the cellular effects of the different cytotoxic drugs and of the effects of potential products directed to protect the neurons of the cytotoxic damage. For this purpose, the cells are cultured with different concentrations of the chemotherapy agent to be studied; after an established incubation period, biochemical and morphological testing can be performed to assess its effects on the concrete functions or structures to which the experiment is directed. These cell cultures enable the study of drugs or natural products with potential properties in reversing the effects of the drugs causing CIPN or with the capability of inducing neuronal regeneration.

6.3 Treatment of CIPN-Related Nerve Damage

At the moment the only clinically available treatments for CIPN are only symptomatic [1], so there is an urgent need for the development of treatments aimed to revert or reduce the neuronal damage. The different cytotoxic drugs causing CIPN affect different cells, organelles, or pathways within the sensory nerve system, resulting in mitochondrial dysfunction, oxidative stress, inflammation, microtubule damage, and alterations in ion channels, along with other effects [10], making the search to uncover CIPN treatments a great challenge. Research can be aimed at a common pathomechanism of damage shared with different drugs or directed to revert the changes induced by a specific drug.

6.3.1 Categorized by Pathomechanism

As chemotherapy targets fast dividing cells and not all chemotherapy agents produce CIPN, there may be additional effects of the cytotoxic drugs on the non-dividing neurons [28]. Most chemotherapy agents do not cross the blood–brain barrier, but they may accumulate in the DRG and nerve terminals, resulting in neuronal body, axonal, or myelin sheath injury [29]. The research toward therapies is aimed at reversing the pathogenic mechanism of the different drugs or in inducing the regeneration of neurons, Schwann cells, or axons.

6.3.1.1 Neurogenesis
The sensory neurons and the supporting glial cells that form the DRG arise from a sub-population of trunk neural crest cell progenitors and the *Notch* signaling pathway is involved in its final differentiation. Some of these cells remain in the undifferentiated stage [30] and express the neural stem cells markers nestin and p75 neurotrophin receptor (p75NTR). The transcription factors involved in its differentiation to neurons or glia could be potential targets in neurogenesis [31]. As seen in the experimental model of peripheral nerve crush injury, the number of DRG neurons increase up to 42%, compared to controls [32]. Alternatively, survival pathways could be activated, as evidenced by the fact that DRG neurons expressing *ptv1* oncogene (plasmacytoma variant translocation 1), a long

non-coding RNA gene, are protected from apoptosis through the activation of the PI3K/AKT pathway [33].

6.3.1.2 Schwann Cell Mechanisms

Schwann cells are essential for the regeneration of peripheral nerves after an injury. In this process Schwann cells halt the production of myelin, digest myelin debris, and facilitate a process of dedifferentiation. These dedifferentiated Schwann cells guide the axon's growth until its completion. After this, the Schwann cells differentiate again and restart the production of myelin [34]. Dynein is a motor protein and regulator of microtubule dynamics, axonal transport, and membrane trafficking. Dynein is essential for the process of Schwann cell dedifferentiation and, consequently, for axon regeneration [35]. Following nerve injury, several pathways are activated in Schwann cells, such as p38, JNK, and ERK, which are involved in the acquisition of the dedifferentiated phenotype of the Schwann cells to start axon recovery [36], resulting in the upregulation of proteins C-Jun and p75NTR, whereas the myelination associated protein EGR2 (early growth response protein 2) becomes downregulated [37]. The involvement of signaling pathways involved in these mechanisms is another focus of research.

6.3.1.3 Axonal Regrowth

The peripheral nervous system, in contrast with the central nervous system, has a capacity to recover after traumatic or toxic injuries. This process involves a series of changes that provides the neuron with the capacity to growth. Axon regeneration is regulated through the activation of several transcription factors, epigenetic changes of chromatin and microRNAs (miRNAs) [38]. Some of the transcribed mRNAs are transported to distal parts of the axon where the translation into proteins occurs, preventing both axon degeneration and neuron apoptosis. One of these retrograde response genes is Bclw (Bcl2l2), which belongs to the Bcl2- family and induces axon survival [39]. Following peripheral nerve injury, the activation of the JNK signaling pathway increases the expression of transcription factors JUN and ATF3, in DRG neurons starting axon regeneration. Other transcription factors induced by peripheral axon injury are members of the SMAD family and STAT3 [38]. Activation of STAT3 happens in DRG neurons after nerve injury by being phosphorylated by cyclin-dependent kinase 5 (Cdk5) [40].

6.3.2   Categorized by Drug

The fact that anticancer chemotherapy targets rapid dividing cells but not all agents produce CIPN supports that different drugs have their own mechanisms of causing neuronal damage [28]. The different gene expression induced by different chemotherapy drugs in normal cells can help in the search for targets in the development of therapies to treat CIPN [41]. As oxaliplatin, paclitaxel, vincristine, cisplatin, and bortezomib are the drugs that commonly cause CIPN in clinical practice, many studies are related to them [42].

6.3.2.1 Oxaliplatin

Animal and "in vitro" studies have shown that the nuclear factor-erythroid-2-related factor 2 (Nrf2) pathway protects from oxaliplatin-induced axonal damage, by stimulating the synthesis of proteins with antioxidant activity. Dimethyl fumarate is a drug used in the treatment of multiple sclerosis that exerts a neuroprotective effect through Nrf2-mediated reduction in oxidative stress. Recent work demonstrated functional and structural improvements with dimethyl fumarate treatment in the rat model of oxaliplatin-induced neuropathy [43]. Another neuroprotective agent, donepezil, an inhibitor of acetylcholinesterase and used for the treatment of Alzheimer's disease, reduced sciatic nerve degeneration and improved mechanical allodynia in rats treated with oxaliplatin, without a reduction in the antitumor efficacy [20]. Oxaliplatin and paclitaxel produce an inflammatory response in DRGs and spinal cord astrocytes with an increased production of inflammatory cytokines (CCL2, CCL3, TNF-α, IL-6, IL1β, and IL-8) and a reduction in the anti-inflammatory cytokines (IL-10 and IL-4). In a rat model of oxaliplatin-induced neuropathy, the selective inhibition of IL-8 receptors improved the results of the behavioral test and reduced the expression of the proteins JAK2 and STAT3, which are associated with oxaliplatin damage [44].

6.3.2.2 Paclitaxel

Oxidative stress produced by the effect of paclitaxel on the mitochondria of DRG neurons and peripheral nerves is one of the pathophysiological mechanisms of CIPN. Melatonin has been shown to be a potent antioxidant that enters the mitochondria. "In vitro" studies showed that melatonin reduces paclitaxel-induced mitochondrial damage. Using the rat model of paclitaxel-induced neuropathy, co-treatment with melatonin improved the results of the behavioral tests and reduced the C-fiber activity-dependent slowing [45]. Paclitaxel-induced apoptosis of DRG neurons is another mechanism involved in CIPN and the tumor suppressor gene *p53* appears to play an essential role in pathways related with DNA-damage and apoptosis. In an "in vitro" study with DRG neurons obtained from neonatal rats treated with paclitaxel and in a mice model of paclitaxel-induced CIPN, duloxetine reduced the expression of p53 and improved thermal and mechanical allodynia. The effect of duloxetine on p53 is through the reduction of oxidative stress [3]. As with oxaliplatin, inflammation in DRGs plays an important role in paclitaxel-induced neuropathy. Pretreatment with an IL-6 neutralizing antibody protects mice from such neuropathy [18].

Membrane drug transporter proteins are also involved in CIPN. These proteins such as ABCB1 and ABCC1 regulate uptake and efflux of drugs and are expressed in the peripheral nervous system [46]. Organic anion-transporting polypeptides (OATPs) are related with the accumulation of paclitaxel in DRG. OATP1B2 knockout mice have a decreased uptake of paclitaxel in DRG. The tyrosine kinase inhibitor nilotinib is a potent inhibitor of OATP1B1 and OATP1B2, protecting mice of paclitaxel induced neuropathy without impairing antitumor activity [47].

6.3.2.3 Vincristine

Axonal degeneration is an active process that is triggered by several transcription factors after a traumatic or toxic lesion. Sterile alpha and TIR motif-containing protein 1 (SARM1) is one of its components. *Sarm1*-knockout mice are protected from vincristine induced neuropathy, when compared with wild-type mice. SARM1 or its down-stream effectors could be potential therapeutic targets for reducing neuropathy [48]. Vincristine also stimulates the immune system, resulting in the consequent release of pro-inflammatory cytokines and neuroinflammation [28]. The anti-diabetes drug metformin reduces the levels of TNF-α, IL-6 and suppress the macrophage activation through the adenosine monophosphate activated protein kinase (AMPK) pathway, preventing mechanical allodynia and numbness in CIPN mice models [29].

6.3.2.4 Cisplatin

Cisplatin targets nuclear and mitochondrial DNA of DRG neurons, causing inter- and intra-strand adducts, inducing DGR-neurons apoptosis and mitochondrial disfunction, with the consequent generation of oxidative stress [49]. Peroxisome proliferator-activated receptor-α (PPAR-α) is a ligand-activated transcription factor of the nuclear hormone receptor superfamily expressed in several cells, including microglia and astroglia. PPAR-α increases mitochondrial and peroxisomal β-oxidation of fatty acids and thus has an important role in oxidation/antioxidant pathway [50]. Stimulation of PPAR-α could increase the levels of endogenous antioxidants reducing the oxidative stress. One stimulator of PPAR-α, undergoing CIPN animal studies, is the endogenous fatty acid, palmitoylethanolamide [49]. "In vitro" studies have shown that cisplatin mediated DRG neurons apoptosis can be prevented with phenoxodiol, an isoflavone analogue, that upregulates the cell-cycle regulator *p21 Waf1/Cip1* stimulating neurite growth [5]. The *sirt2* gene encodes the enzyme NAD-dependent deacetylase sirtuin 2, which results in neurite growth and protects mice from cisplatin-induced neural damage [51].

6.3.2.5 Bortezomib

As described earlier, the drug dimethyl fumarate, used in the treatment of multiple sclerosis, is an antioxidant and neuroprotective agent whose effect is mediated through the upregulation of *Nfr2*. "In vitro" studies using PC12 and rat DRG neurons showed that it reduces the effect of bortezomib, oxaliplatin, and cisplatin on neurite outgrowth, but lacks any protection against apoptosis [52]. Bortezomib alters the energetic metabolism of DRG-neurons, shifting the mitochondrial oxidation to aerobic glycolysis, the so-called Warburg effect. This aerobic glycolysis-phenotype with the consequent overexpression of lactate dehydrogenase A (LDHA) and pyruvate dehydrogenase kinase 1 (PDHK1) contributes to development of CIPN. Studies with a mouse model of bortezomib-induced neuropathy demonstrated that, by inhibition of LDHA and PDHK1 with oxamate and dichloroacetate, respectively, an improvement in the behavioral tests was achieved together with the reversal of the metabolic phenotype [53].

6.4 Treatment of CIPN-Related Pain

There are number of approaches that have been taken to treat CIPN-related pain in animal model systems. Overall, studies suggest that while initial neuropathic pain in CIPN is due to damage to the peripheral sensory nerve fibers, persistent CIPN-related pain is likely due to a combination of peripheral and central pathomechanisms. Supporting this idea is that duloxetine (which appears to act in central nervous system) is the only medication to be shown to be effective in reducing pain from established CIPN in double-blind placebo controlled human clinical trials [54, 55]. Many of the other off-label use of neuropathic pain medications have been tested and shown to provide relief in animal models [56]. The disconnect between successful treatment of CIPN-related pain in animal models versus the failure in human clinical trials is an important point that deserves careful attention.

6.4.1 Categorized by Pathomechanism

The study of pathomechanisms of CIPN-related pain reflects the study of neuro-pathic pain more broadly. As such, many of the pathways discussed below have broad implications for neuropathic pain; however, there are some pathomechanisms that are specific to the CIPN realm, which will be explicitly highlighted. While most of the studies below focused on specific neurotoxic chemotherapy agents, it is unclear how chemotherapy-specific any of the mechanisms below are. For example, a given paper may study a treatment mechanism in cisplatin-induced peripheral neuropathy, but does not explicitly test whether or not the same mechanism is at play in CIPN from other medications. Furthermore the majority of papers either used paclitaxel-, oxaliplatin-, or cisplatin-induced peripheral neuropathy models; bortezomib and vinca alkaloid models are far less represented.

6.4.1.1 Voltage-Gated Ion Channels
Voltage-gated ion channels are a prominent target for CIPN-related pain. Multiple models of CIPN have demonstrated altered voltage-gated ion channel expression that leads to neuronal hyperexcitability and correlates with pain behaviors. Voltage-gated sodium channels have shown increased expression in CIPN [57], especially the Nav1.7-mediated sodium current; blockade of this channel reverses hyperalgesia in a rat model of oxaliplatin-induced peripheral neuropathy [58]. Reduced expression of potassium channels occurs in CIPN models [57, 59, 60], which has been shown to be counteracted by the voltage-gated potassium channel activator retigabine (an FDA-approved epilepsy medication that targets the Kv7 channel) [61]. Voltage-gated T-type calcium channel Cav3.2 expression is increased in paclitaxel-induced peripheral neuropathy models [62]; blockade of this channel or the N-type (Cav2.2) can alleviate CIPN-related pain behaviors [63, 64]. The alpha-2-delta-1 auxiliary subunit for voltage-gated calcium channels, the target of pregabalin and gabapentin, is also upregulated by paclitaxel (PMID

17084535). Finally, the hyperpolarization-activated cyclic nucleotide-gated (HCN) channels have been shown to be upregulated in a rat model of paclitaxel- or oxaliplatin-induced neuropathy [57, 60], and blockade of these channels reduces hyperalgesia and allodynia [60].

6.4.1.2 Neuroinflammation

Neuroinflammation is an often-used, but somewhat nebulous, term that typically refers to the deleterious effects of non-neuronal cells (e.g., immune cells, cytokines, and glial cells) to a neuropathological process (in this case CIPN). Extensive data have established neuroinflammation as playing an important role in CIPN and CIPN-related pain. CIPN is associated with changes in the peripheral immune system, seen as increases in CD4+ and CD8 T-cells [65]. Astrocytosis is seen in the central nervous system with CIPN, which, in part, appears to be mediated by heme oxygenase-1 expression [66], but there are no documented significant changes in microglial activation [65, 67]. Alterations in cytokine levels have been observed in CIPN models, with increased CNS levels of TNF-alpha, IFN-gamma, CCL11, CCL4, CCL3, IL-12p70, and GM-CSF [65]. Blockade of CXCR pathways [68, 69] or MCP-1 [70] can decrease CIPN-related pain behaviors. Increasing evidence also implicates toll-like receptor family activation (a component of the innate immune system) as playing a key role in CIPN-related pain, which can also be beneficially targeted [71–73], noting that data points to sexual dimorphism in this response [71].

6.4.1.3 Sphingosine Metabolism

Sphingosine 1-phosphate is generated via sphingolipid and ceramide metabolism, which can be activated via a number of mechanisms, including bortezomib and paclitaxel. Activation of the sphingosine 1-phosphate receptor in astrocytes has been shown to be important in establishing and maintaining bortezomib and paclitaxel-induced neuropathy in rat models [74, 75]. Importantly, this is an IL-10 dependent mechanism and also exhibits sexual dimorphic response [76]. Accordingly, sphingosine 1-phosphate receptor blockade (via an FDA-approved medication, fingolimod) can both prevent and treat established CIPN in animal models and is being tested in human clinical trials.

6.4.1.4 Endocannabinoids

A number of studies have reported the benefits of cannabinoids for CIPN-related pain syndromes in animal models, which has become more pertinent given the increased legalization of medical and recreational marijuana in many jurisdictions. Endocannabinoids have been implicated in development of CIPN-related pain [77, 78]. Activation of cannabinoid receptors has been shown to reduce CIPN pain behaviors caused by platinates [79–82] and taxanes [80, 83, 84]. The data in these studies is mixed as to whether this effect is mediated primarily by CB1 or CB2 receptors, as well as the relative importance of central versus peripheral cannabinoid receptor activation.

6.4.1.5 Miscellaneous Pathomechanisms

Several pathomechanisms have been explored as a treatment approach for established CIPN, albeit in limited studies. Metalloproteinase 2 and 9 are increased in the DRG of paclitaxel-treated rats, and a study demonstrated reversal of paclitaxel-induced allodynia with intrathecal injection of MMP9 monoclonal antibodies [85]. Histone deacetylase 6 inhibition has been shown to reverse cisplatin-induced allodynia, possibly via improved mitochondrial bioenergetics [86]. The impact of the microbiome has been studied in CIPN. Transferring gut microbiota from a mouse strain that is susceptible to CIPN (C57BL/6) into a resistant strain (129SvEV) can lead to the susceptibility in the 129SvEV strain to paclitaxel-induced neuropathic pain behaviors [87]. It has not been reported whether gut microbiome may be a target for treatment for established CIPN. Finally, an intriguing study demonstrated that voluntary wheel-running decreased paclitaxel-induced allodynia [88].

6.5   Conclusions

There has been considerable laboratory effort made at discovering therapies for established CIPN, and there are a number of promising pathomechanisms that can be further studied in the future. Some of these pathomechanisms are broad and should ameliorate CIPN from varied chemotherapeutic agents, whereas others may be more directed as specific drugs. Finally, it has become clear that in animal models of CIPN there are system level changes due to neurotoxic chemotherapy that may play synergistic or antagonistic roles and will require more sophisticated approaches to elucidate.

References

1. Hershman DL, Lacchetti C, Dworkin RH, Lavoie Smith EM, Bleeker J, Cavaletti G, Chauhan C, Gavin P, Lavino A, Lustberg MB, Paice J, Schneider B, Smith ML, Smith T, Terstriep S, Wagner-Johnston N, Bak K, Loprinzi CL (2014) American Society of Clinical O. prevention and management of chemotherapy-induced peripheral neuropathy in survivors of adult cancers: American Society of Clinical Oncology clinical practice guideline. J Clin Oncol 32(18):1941–1967. Epub 2014/04/16. https://doi.org/10.1200/JCO.2013.54.0914
2. Currie GL, Angel-Scott HN, Colvin L, Cramond F, Hair K, Khandoker L, Liao J, Macleod M, McCann SK, Morland R, Sherratt N, Stewart R, Tanriver-Ayder E, Thomas J, Wang Q, Wodarski R, Xiong R, Rice ASC, Sena ES. Animal models of chemotherapy-induced peripheral neuropathy: A machine-assisted systematic review and meta-analysis. PLoS Biol. 2019;17 (5):e3000243. https://doi.org/10.1371/journal.pbio.3000243. Epub 2019/05/21. PubMed PMID: 31107871; PMCID: PMC6544332
3. Lu Y, Zhang P, Zhang Q, Yang C, Qian Y, Suo J, Tao X, Zhu J (2020) Duloxetine attenuates paclitaxel-induced peripheral nerve injury by inhibiting p53-related pathways. J Pharmacol Exp Ther 373(3):453–462. Epub 2020/04/03. https://doi.org/10.1124/jpet.120.265082
4. Mohiuddin MS, Himeno T, Inoue R, Miura-Yura E, Yamada Y, Nakai-Shimoda H, Asano S, Kato M, Motegi M, Kondo M, Seino Y, Tsunekawa S, Kato Y, Suzuki A, Naruse K, Kato K,

Nakamura J, Kamiya H. Glucagon-like peptide-1 receptor agonist protects dorsal root ganglion neurons against oxidative insult. J Diabetes Res. 2019;2019:9426014. https://doi.org/10.1155/2019/9426014. Epub 2019/03/29. PubMed PMID: 30918901; PMCID: PMC6408997

5. Klein R, Brown D, Turnley AM. Phenoxodiol protects against Cisplatin induced neurite toxicity in a PC-12 cell model. BMC Neurosci. 2007;8:61. https://doi.org/10.1186/1471-2202-8-61. Epub 2007/08/04. PubMed PMID: 17672914; PMCID: PMC1950519

6. Chambers SM, Qi Y, Mica Y, Lee G, Zhang XJ, Niu L, Bilsland J, Cao L, Stevens E, Whiting P, Shi SH, Studer L. Combined small-molecule inhibition accelerates developmental timing and converts human pluripotent stem cells into nociceptors. Nat Biotechnol. 2012;30(7):715–20. https://doi.org/10.1038/nbt.2249. Epub 2012/07/04. PubMed PMID: 22750882; PMCID: 3516136

7. Vojnits K, Mahammad S, Collins TJ, Bhatia M. Chemotherapy-induced neuropathy and drug discovery platform using human sensory neurons converted directly from adult peripheral blood. Stem Cells Transl Med. 2019;8(11):1180–91. https://doi.org/10.1002/sctm.19-0054. Epub 2019/07/28. PubMed PMID: 31347791; PMCID: PMC6811699

8. Hooijmans CR, Draper D, Ergun M, Scheffer GJ. The effect of analgesics on stimulus evoked pain-like behaviour in animal models for chemotherapy induced peripheral neuropathy- a meta-analysis. Sci Rep. 2019;9(1):17549. https://doi.org/10.1038/s41598-019-54152-8. Epub 2019/11/28. PubMed PMID: 31772391; PMCID: PMC6879539

9. Maruta T, Nemoto T, Hidaka K, Koshida T, Shirasaka T, Yanagita T, Takeya R, Tsuneyoshi I. Upregulation of ERK phosphorylation in rat dorsal root ganglion neurons contributes to oxaliplatin-induced chronic neuropathic pain. PLoS One. 2019;14(11):e0225586. https://doi.org/10.1371/journal.pone.0225586. Epub 2019/11/26. PubMed PMID: 31765435; PMCID: PMC6876879

10. Tsubota M, Fukuda R, Hayashi Y, Miyazaki T, Ueda S, Yamashita R, Koike N, Sekiguchi F, Wake H, Wakatsuki S, Ujiie Y, Araki T, Nishibori M, Kawabata A. Role of non-macrophage cell-derived HMGB1 in oxaliplatin- induced peripheral neuropathy and its prevention by the thrombin/thrombomodulin system in rodents: negative impact of anticoagulants. J Neuroinflammation. 2019;16(1):199. https://doi.org/10.1186/s12974-019-1581-6. Epub 2019/11/02. PubMed PMID: 31666085; PMCID: PMC6822350

11. Janes K, Esposito E, Doyle T, Cuzzocrea S, Tosh DK, Jacobson KA, Salvemini D. A3 adenosine receptor agonist prevents the development of paclitaxel-induced neuropathic pain by modulating spinal glial-restricted redox-dependent signaling pathways. Pain. 2014;155 (12):2560–7. https://doi.org/10.1016/j.pain.2014.09.016. Epub 2014/09/23. PubMed PMID: 25242567; PMCID: PMC4529068

12. Slivicki RA, Xu Z, Mali SS, Hohmann AG. Brain permeant and impermeant inhibitors of fatty-acid amide hydrolase suppress the development and maintenance of paclitaxel-induced neuro-pathic pain without producing tolerance or physical dependence in vivo and synergize with paclitaxel to reduce tumor cell line viability in vitro. Pharmacol Res. 2019;142:267–82. https://doi.org/10.1016/j.phrs.2019.02.002. Epub 2019/02/11. PubMed PMID: 30739035; PMCID: PMC6878658

13. Alessandri-Haber N, Dina OA, Joseph EK, Reichling DB, Levine JD. Interaction of transient receptor potential vanilloid 4, integrin, and SRC tyrosine kinase in mechanical hyperalgesia. J Neurosci. 2008;28(5):1046–57. https://doi.org/10.1523/JNEUROSCI.4497-07.2008. Epub 2008/02/01. PubMed PMID: 18234883; PMCID: PMC6671413

14. Lin H, Heo BH, Yoon MH. A new rat model of cisplatin-induced neuropathic pain. Korean J Pain. 2015;28(4):236–43. https://doi.org/10.3344/kjp.2015.28.4.236. Epub 2015/10/27. PubMed PMID: 26495078; PMCID: PMC4610937

15. Zhang M, Du W, Acklin S, Jin S, Xia F. SIRT2 protects peripheral neurons from cisplatin-induced injury by enhancing nucleotide excision repair. J Clin Invest. 2020;130(6):2953–65. https://doi.org/10.1172/JCI123159. Epub 2020/03/07. PubMed PMID: 32134743; PMCID: PMC7260000

16. Duggett NA, Flatters SJL. Characterization of a rat model of bortezomib-induced painful neuropathy. Br J Pharmacol. 2017;174(24):4812–25. https://doi.org/10.1111/bph.14063. Epub 2017/10/04. PubMed PMID: 28972650; PMCID: PMC5727311

17. Boehmerle W, Huehnchen P, Peruzzaro S, Balkaya M, Endres M. Electrophysiological, behavioral and histological characterization of paclitaxel, cisplatin, vincristine and bortezomib-induced neuropathy in C57Bl/6 mice. Sci Rep. 2014;4:6370. https://doi.org/10.1038/srep06370. Epub 2014/09/19. PubMed PMID: 25231679; PMCID: PMC5377307

18. Huehnchen P, Muenzfeld H, Boehmerle W, Endres M. Blockade of IL-6 signaling prevents paclitaxel-induced neuropathy in C57Bl/6 mice. Cell Death Dis. 2020;11(1):45. https://doi.org/10.1038/s41419-020-2239-0. Epub 2020/01/24. PubMed PMID: 31969555; PMCID: PMC6976596

19. Miyano K, Shiraishi S, Minami K, Sudo Y, Suzuki M, Yokoyama T, Terawaki K, Nonaka M, Murata H, Higami Y, Uezono Y. Carboplatin enhances the activity of human transient receptor potential ankyrin 1 through the cyclic AMP-protein kinase A-A-Kinase Anchoring Protein (AKAP) pathways. Int J Mol Sci. 2019;20(13). https://doi.org/10.3390/ijms20133271. Epub 2019/07/07. PubMed PMID: 31277262; PMCID: PMC6651390

20. Kawashiri T, Shimizu S, Shigematsu N, Kobayashi D, Shimazoe T (2019) Donepezil ameliorates oxaliplatin-induced peripheral neuropathy via a neuroprotective effect. J Pharmacol Sci 140(3):291–294. Epub 2019/08/05. https://doi.org/10.1016/j.jphs.2019.05.009

21. Fukuda M, Yamamoto A (2004) Effect of forskolin on synaptotagmin IV protein trafficking in PC12 cells. J Biochem 136(2):245–253. Epub 2004/10/22. https://doi.org/10.1093/jb/mvh116

22. Imai S, Koyanagi M, Azimi Z, Nakazato Y, Matsumoto M, Ogihara T, Yonezawa A, Omura T, Nakagawa S, Wakatsuki S, Araki T, Kaneko S, Nakagawa T, Matsubara K. Taxanes and platinum derivatives impair Schwann cells via distinct mechanisms. Sci Rep. 2017;7(1):5947. https://doi.org/10.1038/s41598-017-05784-1. Epub 2017/07/22. PubMed PMID: 28729624; PMCID: PMC5519765

23. Blanchard JW, Eade KT, Szucs A, Lo Sardo V, Tsunemoto RK, Williams D, Sanna PP, Baldwin KK. Selective conversion of fibroblasts into peripheral sensory neurons. Nat Neurosci. 2015;18(1):25–35. https://doi.org/10.1038/nn.3887. Epub 2014/11/25. PubMed PMID: 25420069; PMCID: PMC4466122

24. Hoelting L, Klima S, Karreman C, Grinberg M, Meisig J, Henry M, Rotshteyn T, Rahnenfuhrer J, Bluthgen N, Sachinidis A, Waldmann T, Leist M. Stem cell-derived immature human dorsal root ganglia neurons to identify peripheral neurotoxicants. Stem Cells Transl Med. 2016;5(4):476–87. https://doi.org/10.5966/sctm.2015-0108. Epub 2016/03/05. PubMed PMID: 26933043; PMCID: PMC4798731

25. Rana P, Luerman G, Hess D, Rubitski E, Adkins K, Somps C (2017) Utilization of iPSC-derived human neurons for high-throughput drug- induced peripheral neuropathy screening. Toxicol In Vitro 45(Pt 1):111–118. Epub 2017/08/28. https://doi.org/10.1016/j.tiv.2017.08.014

26. Wheeler HE, Wing C, Delaney SM, Komatsu M, Dolan ME. Modeling chemotherapeutic neurotoxicity with human induced pluripotent stem cell-derived neuronal cells. PLoS One. 2015;10(2):e0118020. https://doi.org/10.1371/journal.pone.0118020. Epub 2015/02/18. PubMed PMID: 25689802; PMCID: PMC4331516

27. Wing C, Komatsu M, Delaney SM, Krause M, Wheeler HE, Dolan ME. Application of stem cell derived neuronal cells to evaluate neurotoxic chemotherapy. Stem Cell Res. 2017;22:79–88. https://doi.org/10.1016/j.scr.2017.06.006. Epub 2017/06/24. PubMed PMID: 28645005; PMCID: PMC5737666

28. Starobova H, Vetter I. Pathophysiology of chemotherapy-induced peripheral neuropathy. Front Mol Neurosci. 2017;10:174. https://doi.org/10.3389/fnmol.2017.00174. Epub 2017/06/18. PubMed PMID: 28620280; PMCID: PMC5450696

29. Hu LY, Mi WL, Wu GC, Wang YQ, Mao-Ying QL. Prevention and treatment for chemotherapy-induced peripheral neuropathy: therapies based on CIPN mechanisms. Curr Neuropharmacol. 2019;17(2):184–96. https://doi.org/10.2174/

1570159X15666170915143217. Epub 2017/09/20. PubMed PMID: 28925884; PMCID: PMC6343206

30. Wiszniak S, Schwarz Q. Notch signalling defines dorsal root ganglia neuroglial fate choice during early neural crest cell migration. BMC Neurosci. 2019;20(1):21. https://doi.org/10.1186/s12868-019-0501-0. Epub 2019/05/01. PubMed PMID: 31036074; PMCID: PMC6489353

31. Czaja K, Fornaro M, Geuna S. Neurogenesis in the adult peripheral nervous system. Neural Regen Res. 2012;7(14):1047–54. https://doi.org/10.3969/j.issn.1673-5374.2012.14.002. Epub 2012/05/15. PubMed PMID: 25722694; PMCID: PMC4340017

32. Muratori L, Ronchi G, Raimondo S, Geuna S, Giacobini-Robecchi MG, Fornaro M. Generation of new neurons in dorsal root Ganglia in adult rats after peripheral nerve crush injury. Neural Plast. 2015;2015:860546. https://doi.org/10.1155/2015/860546. Epub 2015/02/28. PubMed PMID: 25722894; PMCID: PMC4333329

33. Chen L, Gong HY, Xu L. PVT1 protects diabetic peripheral neuropathy via PI3K/AKT pathway. Eur Rev Med Pharmacol Sci. 2018;22(20):6905–11. https://doi.org/10.26355/eurrev_201810_16160. Epub 2018/11/08

34. Ceci ML, Mardones-Krsulovic C, Sanchez M, Valdivia LE, Allende ML. Axon-Schwann cell interactions during peripheral nerve regeneration in zebrafish larvae. Neural Dev. 2014;9:22. https://doi.org/10.1186/1749-8104-9-22. Epub 2014/10/19. PubMed PMID: 25326036; PMCID: PMC4214607

35. Ducommun Priest M, Navarro MF, Bremer J, Granato M. Dynein promotes sustained axonal growth and Schwann cell remodeling early during peripheral nerve regeneration. PLoS Genet. 2019;15(2):e1007982. https://doi.org/10.1371/journal.pgen.1007982. Epub 2019/02/20. PubMed PMID: 30779743; PMCID: PMC6396928

36. Hung HA, Sun G, Keles S, Svaren J. Dynamic regulation of Schwann cell enhancers after peripheral nerve injury. J Biol Chem. 2015;290(11):6937–50. https://doi.org/10.1074/jbc.M114.622878. Epub 2015/01/24. PubMed PMID: 25614629; PMCID: PMC4358118

37. Wilcox MB, Laranjeira SG, Eriksson TM, Jessen KR, Mirsky R, Quick TJ, Phillips JB. Characterising cellular and molecular features of human peripheral nerve degeneration. Acta Neuropathol Commun. 2020;8(1):51. https://doi.org/10.1186/s40478-020-00921-w. Epub 2020/04/19. PubMed PMID: 32303273; PMCID: PMC7164159

38. Mahar M, Cavalli V. Intrinsic mechanisms of neuronal axon regeneration. Nat Rev Neurosci. 2018;19(6):323–37. https://doi.org/10.1038/s41583-018-0001-8. Epub 2018/04/19. PubMed PMID: 29666508; PMCID: PMC5987780

39. Fukuda Y, Li Y, Segal RA. A mechanistic understanding of axon degeneration in chemotherapy-induced peripheral neuropathy. Front Neurosci. 2017;11:481. https://doi.org/10.3389/fnins.2017.00481. Epub 2017/09/16. PubMed PMID: 28912674; PMCID: PMC5583221

40. Hwang J, Namgung U. Cdk5 phosphorylation of STAT3 in dorsal root ganglion neurons is involved in promoting axonal regeneration after peripheral nerve injury. Int Neurourol J. 2020;24(Suppl 1):S19–27. https://doi.org/10.5213/inj.2040158.080. Epub 2020/06/03. PubMed PMID: 32482054; PMCID: PMC7285696

41. Morales M, Avila J, Gonzalez-Fernandez R, Boronat L, Soriano ML, Martin-Vasallo P. Differential transcriptome profile of peripheral white cells to identify biomarkers involved in oxaliplatin induced neuropathy. J Pers Med. 2014;4(2):282–96. https://doi.org/10.3390/jpm4020282. Epub 2015/01/08. PubMed PMID: 25563226; PMCID: PMC4263976

42. Flatters SJL, Dougherty PM, Colvin LA (2017) Clinical and preclinical perspectives on chemotherapy-induced peripheral neuropathy (CIPN): a narrative review. Br J Anaesth 119 (4):737–749. Epub 2017/11/10. https://doi.org/10.1093/bja/aex229

43. Miyagi A, Kawashiri T, Shimizu S, Shigematsu N, Kobayashi D, Shimazoe T (2019) Dimethyl fumarate attenuates Oxaliplatin-induced peripheral neuropathy without affecting the anti-tumor activity of Oxaliplatin in rodents. Biol Pharm Bull 42(4):638–644. Epub 2019/04/02. https://doi.org/10.1248/bpb.b18-00855

44. Brandolini L, Castelli V, Aramini A, Giorgio C, Bianchini G, Russo R, De Caro C, d'Angelo M, Catanesi M, Benedetti E, Giordano A, Cimini A, Allegretti M. DF2726A, a new IL-8 signalling inhibitor, is able to counteract chemotherapy-induced neuropathic pain. Sci Rep. 2019;9 (1):11729. https://doi.org/10.1038/s41598-019-48231-z. Epub 2019/08/15. PubMed PMID: 31409858; PMCID: PMC6692352

45. Galley HF, McCormick B, Wilson KL, Lowes DA, Colvin L, Torsney C. Melatonin limits paclitaxel-induced mitochondrial dysfunction in vitro and protects against paclitaxel-induced neuropathic pain in the rat. J Pineal Res. 2017;63(4):e12444. https://doi.org/10.1111/jpi.12444. Epub 2017/08/24. PubMed PMID: 28833461; PMCID: PMC5656911

46. Stage TB, Hu S, Sparreboom A, Kroetz DL (2020) Role of drug transporters in chemotherapy-induced peripheral neuropathy. Clin Transl Sci 3:1–8. https://doi.org/10.1111/cts.12915

47. Leblanc AF, Sprowl JA, Alberti P, et al. OATP1B2 deficiency protects against paclitaxel-induced neurotoxicity. J Clin Invest 2018;128(2):816–825. https://doi.org/10.1172/JCI96160

48. Geisler S, Doan RA, Strickland A, Huang X, Milbrandt J, DiAntonio A. Prevention of vincristine-induced peripheral neuropathy by genetic deletion of SARM1 in mice. Brain. 2016;139(Pt 12):3092–108. https://doi.org/10.1093/brain/aww251. Epub 2016/11/01. PubMed PMID: 27797810; PMCID: PMC5840884

49. Lazic A, Popovic J, Paunesku T, Woloschak GE, Stevanovic M. Insights into platinum-induced peripheral neuropathy-current perspective. Neural Regen Res. 2020;15(9):1623–30. https://doi.org/10.4103/1673-5374.276321. Epub 2020/03/27. PubMed PMID: 32209761; PMCID: PMC7437596

50. Tyagi S, Gupta P, Saini AS, Kaushal C, Sharma S. The peroxisome proliferator-activated receptor: A family of nuclear receptors role in various diseases. J Adv Pharm Technol Res. 2011;2(4):236–40. https://doi.org/10.4103/2231-4040.90879. Epub 2012/01/17. PubMed PMID: 22247890; PMCID: PMC3255347

51. Zhao X, Du W, Zhang M, Atiq ZO, Xia F. Sirt2-associated transcriptome modifications in cisplatin-induced neuronal injury. BMC Genomics. 2020;21(1):192. https://doi.org/10.1186/s12864-020-6584-2. Epub 2020/03/04. PubMed PMID: 32122297; PMCID: PMC7053098

52. Kawashiri T, Miyagi A, Shimizu S, Shigematsu N, Kobayashi D, Shimazoe T (2018) Dimethyl fumarate ameliorates chemotherapy agent-induced neurotoxicity in vitro. J Pharmacol Sci 137 (2):202–211. Epub 2018/07/26. https://doi.org/10.1016/j.jphs.2018.06.008

53. Ludman T, Melemedjian OK. Bortezomib-induced aerobic glycolysis contributes to chemotherapy-induced painful peripheral neuropathy. Mol Pain. 2019;15:1744806919837429. https://doi.org/10.1177/1744806919837429. Epub 2019/02/28. PubMed PMID: 30810076; PMCID: PMC6452581

54. Majithia N, Temkin SM, Ruddy KJ, Beutler AS, Hershman DL, Loprinzi CL. National Cancer Institute-supported chemotherapy-induced peripheral neuropathy trials: outcomes and lessons. Support Care Cancer. 2016;24(3):1439–47. https://doi.org/10.1007/s00520-015-3063-4. Epub 2015/12/22. PubMed PMID: 26686859; PMCID: PMC5078987

55. Smith EM, Pang H, Cirrincione C, Fleishman S, Paskett ED, Ahles T, Bressler LR, Fadul CE, Knox C, Le-Lindqwister N, Gilman PB, Shapiro CL, Alliance for clinical trials in O. Effect of duloxetine on pain, function, and quality of life among patients with chemotherapy-induced painful peripheral neuropathy: a randomized clinical trial. Jama. 2013;309(13):1359–67. https://doi.org/10.1001/jama.2013.2813. PubMed PMID: 23549581; PMCID: PMC3912515

56. Xiao W, Naso L, Bennett GJ (2008) Experimental studies of potential analgesics for the treatment of chemotherapy-evoked painful peripheral neuropathies. Pain Med 9(5):505–517. Epub 2008/09/09. https://doi.org/10.1111/j.1526-4637.2007.00301.x

57. Zhang H, Dougherty PM. Enhanced excitability of primary sensory neurons and altered gene expression of neuronal ion channels in dorsal root ganglion in paclitaxel-induced peripheral neuropathy. Anesthesiology. 2014;120(6):1463–75. https://doi.org/10.1097/ALN.0000000000000176. Epub 2014/02/19. PubMed PMID: 24534904; PMCID: PMC4031279

58. Li Y, North RY, Rhines LD, Tatsui CE, Rao G, Edwards DD, Cassidy RM, Harrison DS, Johansson CA, Zhang H, Dougherty PM. DRG voltage- gated sodium channel 1.7 is

upregulated in paclitaxel-induced neuropathy in rats and in humans with neuropathic pain. J Neurosci. 2018;38(5):1124–36. https://doi.org/10.1523/JNEUROSCI.0899-17.2017. Epub 2017/12/20. PubMed PMID: 29255002; PMCID: PMC5792474

59. Thibault K, Calvino B, Dubacq S, Roualle-de-Rouville M, Sordoillet V, Rivals I, Pezet S (2012) Cortical effect of oxaliplatin associated with sustained neuropathic pain: exacerbation of cortical activity and down-regulation of potassium channel expression in somatosensory cortex. Pain 153(8):1636–1647. Epub 2012/06/02. https://doi.org/10.1016/j.pain.2012.04.016

60. Descoeur J, Pereira V, Pizzoccaro A, Francois A, Ling B, Maffre V, Couette B, Busserolles J, Courteix C, Noel J, Lazdunski M, Eschalier A, Authier N, Bourinet E. Oxaliplatin-induced cold hypersensitivity is due to remodelling of ion channel expression in nociceptors. EMBO Mol Med. 2011;3(5):266–78. https://doi.org/10.1002/emmm.201100134. Epub 2011/03/26. PubMed PMID: 21438154; PMCID: PMC3377073

61. Nodera H, Spieker A, Sung M, Rutkove S (2011) Neuroprotective effects of Kv7 channel agonist, retigabine, for cisplatin-induced peripheral neuropathy. Neurosci Lett 505(3):223–227. Epub 2011/09/29. https://doi.org/10.1016/j.neulet.2011.09.013

62. Li Y, Tatsui CE, Rhines LD, North RY, Harrison DS, Cassidy RM, Johansson CA, Kosturakis AK, Edwards DD, Zhang H, Dougherty PM. Dorsal root ganglion neurons become hyperexcitable and increase expression of voltage-gated T-type calcium channels (Cav3.2) in paclitaxel-induced peripheral neuropathy. Pain. 2017;158(3):417–29. https://doi.org/10.1097/j.pain.0000000000000774. Epub 2016/12/03. PubMed PMID: 27902567; PMCID: PMC5303135

63. Cai S, Tuohy P, Ma C, Kitamura N, Gomez K, Zhou Y, Ran D, Bellampalli SS, Yu J, Luo S, Dorame A, Ngan Pham NY, Molnar G, Streicher JM, Patek M, Perez-Miller S, Moutal A, Wang J, Khanna R (2020. Epub 2020/06/17) A modulator of the low-voltage activated T-type calcium channel that reverses HIV glycoprotein 120-, paclitaxel-, and spinal nerve ligation-induced peripheral neuropathies. Pain. https://doi.org/10.1097/j.pain.0000000000001955

64. Bellampalli SS, Ji Y, Moutal A, Cai S, Wijeratne EMK, Gandini MA, Yu J, Chefdeville A, Dorame A, Chew LA, Madura CL, Luo S, Molnar G, Khanna M, Streicher JM, Zamponi GW, Gunatilaka AAL, Khanna R. Betulinic acid, derived from the desert lavender *Hyptis emoryi*, attenuates paclitaxel-, HIV-, and nerve injury-associated peripheral sensory neuropathy via block of N- and T-type calcium channels. Pain. 2019;160(1):117–35. https://doi.org/10.1097/j.pain.0000000000001385. Epub 2018/09/01. PubMed PMID: 30169422; PMCID: PMC6309937

65. Makker PG, Duffy SS, Lees JG, Perera CJ, Tonkin RS, Butovsky O, Park SB, Goldstein D, Moalem-Taylor G. Characterisation of immune and Neuroinflammatory changes associated with chemotherapy-induced peripheral neuropathy. PLoS One. 2017;12(1):e0170814. https://doi.org/10.1371/journal.pone.0170814. Epub 2017/01/27. PubMed PMID: 28125674; PMCID: PMC5268425

66. Shen Y, Zhang ZJ, Zhu MD, Jiang BC, Yang T, Gao YJ (2015) Exogenous induction of HO-1 alleviates vincristine-induced neuropathic pain by reducing spinal glial activation in mice. Neurobiol Dis 79:100–110. Epub 2015/05/10. https://doi.org/10.1016/j.nbd.2015.04.012

67. Robinson CR, Zhang H, Dougherty PM. Astrocytes, but not microglia, are activated in oxaliplatin and bortezomib-induced peripheral neuropathy in the rat. Neuroscience. 2014;274:308–17. https://doi.org/10.1016/j.neuroscience.2014.05.051. Epub 2014/06/07. PubMed PMID: 24905437; PMCID: PMC4099296

68. Zhou L, Hu Y, Li C, Yan Y, Ao L, Yu B, Fang W, Liu J, Li Y (2018) Levo- corydalmine alleviates vincristine-induced neuropathic pain in mice by inhibiting an NF-kappa B-dependent CXCL1/CXCR2 signaling pathway. Neuropharmacology 135:34–47. Epub 2018/03/09. https://doi.org/10.1016/j.neuropharm.2018.03.004

69. Brandolini L, Benedetti E, Ruffini PA, Russo R, Cristiano L, Antonosante A, d'Angelo M, Castelli V, Giordano A, Allegretti M, Cimini A. CXCR1/2 pathways in paclitaxel-induced neuropathic pain. Oncotarget. 2017;8(14):23188–201. https://doi.org/10.18632/oncotarget.15533. Epub 2017/04/21. PubMed PMID: 28423567; PMCID: PMC5410296

70. Zhang H, Li Y, de Carvalho-Barbosa M, Kavelaars A, Heijnen CJ, Albrecht PJ, Dougherty PM. Dorsal root ganglion infiltration by macrophages contributes to paclitaxel chemotherapy-induced peripheral neuropathy. J Pain. 2016;17(7):775–86. https://doi.org/10.1016/j.jpain.2016.02.011. Epub 2016/03/17. PubMed PMID: 26979998; PMCID: PMC4939513

71. Luo X, Huh Y, Bang S, He Q, Zhang L, Matsuda M, Ji RR. Macrophage toll-like receptor 9 contributes to chemotherapy-induced neuropathic pain in male mice. J Neurosci. 2019;39 (35):6848–64. https://doi.org/10.1523/JNEUROSCI.3257-18.2019. Epub 2019/07/05. PubMed PMID: 31270160; PMCID: PMC6733562

72. Li Y, Adamek P, Zhang H, Tatsui CE, Rhines LD, Mrozkova P, Li Q, Kosturakis AK, Cassidy RM, Harrison DS, Cata JP, Sapire K, Zhang H, Kennamer-Chapman RM, Jawad AB, Ghetti A, Yan J, Palecek J, Dougherty PM. The cancer chemotherapeutic paclitaxel increases human and rodent sensory neuron responses to TRPV1 by activation of TLR4. J Neurosci. 2015;35 (39):13487–500. https://doi.org/10.1523/JNEUROSCI.1956-15.2015. Epub 2015/10/02. PubMed PMID: 26424893; PMCID: PMC4588613

73. Li Y, Zhang H, Zhang H, Kosturakis AK, Jawad AB, Dougherty PM. Toll-like receptor 4 signaling contributes to Paclitaxel-induced peripheral neuropathy. J Pain. 2014;15 (7):712–25. https://doi.org/10.1016/j.jpain.2014.04.001. Epub 2014/04/24. PubMed PMID: 24755282; PMCID: PMC4083500

74. Chen Z, Doyle TM, Luongo L, Largent-Milnes TM, Giancotti LA, Kolar G, Squillace S, Boccella S, Walker JK, Pendleton A, Spiegel S, Neumann WL, Vanderah TW, Salvemini D. Sphingosine-1-phosphate receptor 1 activation in astrocytes contributes to neuropathic pain. Proc Natl Acad Sci U S A. 2019;116(21):10557–62. https://doi.org/10.1073/pnas.1820466116. Epub 2019/05/10. PubMed PMID: 31068460; PMCID: PMC6534990

75. Janes K, Little JW, Li C, Bryant L, Chen C, Chen Z, Kamocki K, Doyle T, Snider A, Esposito E, Cuzzocrea S, Bieberich E, Obeid L, Petrache I, Nicol G, Neumann WL, Salvemini D. The development and maintenance of paclitaxel-induced neuropathic pain require activation of the sphingosine 1-phosphate receptor subtype 1. J Biol Chem. 2014;289(30):21082–97. https://doi.org/10.1074/jbc.M114.569574. Epub 2014/05/31. PubMed PMID: 24876379; PMCID: PMC4110312

76. Stockstill K, Wahlman C, Braden K, Chen Z, Yosten GL, Tosh DK, Jacobson KA, Doyle TM, Samson WK, Salvemini D. Sexually dimorphic therapeutic response in bortezomib-induced neuropathic pain reveals altered pain physiology in female rodents. Pain. 2020;161(1):177–84. https://doi.org/10.1097/j.pain.0000000000001697. Epub 2019/09/07. PubMed PMID: 31490328; PMCID: PMC6923586

77. Deng L, Cornett BL, Mackie K, Hohmann AG. CB1 knockout mice unveil sustained CB2-mediated antiallodynic effects of the Mixed CB1/CB2 agonist CP55,940 in a mouse model of paclitaxel-induced neuropathic pain. Mol Pharmacol. 2015;88(1):64–74. https://doi.org/10.1124/mol.115.098483. Epub 2015/04/24. PubMed PMID: 25904556; PMCID: PMC4468646

78. Uhelski ML, Khasabova IA, Simone DA. Inhibition of anandamide hydrolysis attenuates nociceptor sensitization in a murine model of chemotherapy-induced peripheral neuropathy. J Neurophysiol. 2015;113(5):1501–10. https://doi.org/10.1152/jn.00692.2014. Epub 2014/12/17. PubMed PMID: 25505113; PMCID: PMC4346731

79. Mulpuri Y, Marty VN, Munier JJ, Mackie K, Schmidt BL, Seltzman HH, Spigelman I. Synthetic peripherally-restricted cannabinoid suppresses chemotherapy-induced peripheral neuropathy pain symptoms by CB1 receptor activation. Neuropharmacology. 2018;139:85–97. https://doi.org/10.1016/j.neuropharm.2018.07.002. Epub 2018/07/08. PubMed PMID: 29981335; PMCID: PMC6883926

80. King KM, Myers AM, Soroka-Monzo AJ, Tuma RF, Tallarida RJ, Walker EA, Ward SJ. Single and combined effects of Delta(9) – tetrahydrocannabinol and cannabidiol in a mouse model of chemotherapy- induced neuropathic pain. Br J Pharmacol. 2017;174(17):2832–41. https://doi.org/10.1111/bph.13887. Epub 2017/05/27. PubMed PMID: 28548225; PMCID: PMC5554313

81. Vera G, Cabezos PA, Martin MI, Abalo R (2013) Characterization of cannabinoid-induced relief of neuropathic pain in a rat model of cisplatin- induced neuropathy. Pharmacol Biochem Behav 105:205–212. Epub 2013/03/05. https://doi.org/10.1016/j.pbb.2013.02.008

82. Khasabova IA, Khasabov S, Paz J, Harding-Rose C, Simone DA, Seybold VS. Cannabinoid type-1 receptor reduces pain and neurotoxicity produced by chemotherapy. J Neurosci. 2012;32 (20):7091–101. https://doi.org/10.1523/JNEUROSCI.0403-12.2012. Epub 2012/05/18. PubMed PMID: 22593077; PMCID: PMC3366638

83. Segat GC, Manjavachi MN, Matias DO, Passos GF, Freitas CS, Costa R, Calixto JB (2017) Antiallodynic effect of beta-caryophyllene on paclitaxel- induced peripheral neuropathy in mice. Neuropharmacology 125:207–219. Epub 2017/07/22. https://doi.org/10.1016/j. neuropharm.2017.07.015

84. Deng L, Guindon J, Cornett BL, Makriyannis A, Mackie K, Hohmann AG. Chronic cannabinoid receptor 2 activation reverses paclitaxel neuropathy without tolerance or cannabinoid receptor 1-dependent withdrawal. Biol Psychiatry. 2015;77(5):475–87. https://doi.org/10. 1016/j.biopsych.2014.04.009. Epub 2014/05/24. PubMed PMID: 24853387; PMCID: PMC4209205

85. Tonello R, Lee SH, Berta T (2019) Monoclonal antibody targeting the matrix metalloproteinase 9 prevents and reverses paclitaxel-induced peripheral neuropathy in Mice. J Pain 20 (5):515–527. https://doi.org/10.1016/j.jpain.2018.11.003. Epub 2018/11/25. PubMed PMID: 30471427; PMCID: PMC6511475

86. Krukowski K, Ma J, Golonzhka O, Laumet GO, Gutti T, van Duzer JH, Mazitschek R, Jarpe MB, Heijnen CJ, Kavelaars A (2017) HDAC6 inhibition effectively reverses chemotherapy-induced peripheral neuropathy. Pain 158(6):1126–1137. https://doi.org/10.1097/j.pain. 0000000000000893. Epub 2017/03/08. PubMed PMID: 28267067; PMCID: PMC5435512

87. Ramakrishna C, Corleto J, Ruegger PM, Logan GD, Peacock BB, Mendonca S, Yamaki S, Adamson T, Ermel R, McKemy D, Borneman J, Cantin EM (2019) Dominant role of the gut microbiota in chemotherapy induced neuropathic pain. Sci Rep. 9(1):20324. https://doi.org/10. 1038/s41598-019-56832-x. Epub 2020/01/01. PubMed PMID: 31889131; PMCID: PMC6937259

88. Slivicki RA, Mali SS, Hohmann AG (2019) Voluntary exercise reduces both chemotherapy-induced neuropathic nociception and deficits in hippocampal cellular proliferation in a mouse model of paclitaxel-induced peripheral neuropathy. Neurobiol Pain 6:100035. https://doi.org/ 10.1016/j.ynpai.2019.100035. Epub 2019/09/19. PubMed PMID: 31528755; PMCID: PMC6739464

Pharmacological Treatment of Established Chemotherapy-Induced Peripheral Neuropathy

7

Samantha Mayo, Yi Long Toh, Jeong Oh, and Alexandre Chan

Abstract

Pharmacological treatment of chemotherapy-induced peripheral neuropathy (CIPN) is still in its infancy and available options are limited. Both American Society of Clinical Oncology (ASCO) and European Society of Medical Oncology-European Oncology Nursing Society-European Association of Neuro-Oncology (ESMO-EONS-EANO) guidelines recommend the use of duloxetine for treatment of CIPN. The ESMO-EONS-EANO suggest gabapentinoids (pregabalin and gabapentin), tricyclic antidepressants, and opioids may be considered as an option to relieve neuropathic pain where duloxetine cannot be used. The National Comprehensive Cancer Network (NCCN) guidelines do not address CIPN specifically, but consider gabapentinoids (pregabalin and gabapentin) first-line options for cancer-related neuropathic pain. Currently, none of these guidelines recommend the use of any

S. Mayo
Lawrence S. Bloomberg Faculty of Nursing, University of Toronto, Toronto, ON, Canada
e-mail: samantha.mayo@utoronto.ca

Y. L. Toh
Department of Pharmacy, National University of Singapore, Singapore, Singapore
e-mail: tohyilong@u.nus.edu

J. Oh
Clinic for Lasting Effects of Cancer Treatment, Department of General Internal Medicine, MD Anderson Cancer Center, Houston, USA
e-mail: jhoh@mdanderson.org

A. Chan (✉)
Department of Clinical Pharmacy Practice, School of Pharmacy and Pharmaceutical Sciences, University of California, Irvine, Irvine, CA, USA
e-mail: a.chan@uci.edu

© The Author(s), under exclusive license to Springer Nature Switzerland AG 2021
M. Lustberg, C. Loprinzi (eds.), *Diagnosis, Management and Emerging Strategies for Chemotherapy-Induced Neuropathy*,
https://doi.org/10.1007/978-3-030-78663-2_7

155

supplements but they recommend against use of acetyl-l-carnitine due to harm seen in preventions studies. The ESMO guidelines also recommend use of topical menthol but recommend against the use of topical ketamine and amitriptyline. Despite limited options currently available, multiple studies are ongoing and further treatment choices may become available in the future.

Keywords

Chemotherapy-induced peripheral neuropathy · CIPN · Neuropathy · Management · Pharmacological

7.1   Introduction

In other chapters, we have learned that chemotherapy-induced peripheral neuropathy (CIPN) is common and debilitating to many patients and survivors who receive chemotherapy. Over the years, a number of agents have been tried for treatment of symptoms associated with CIPN. Although evidence-based guidelines are available, many of these agents are used in clinical practice primarily due to its extrapolated data from pharmacologic studies of more common nonchemotherapy-induced neuropathic pain syndromes, such as diabetic neuropathy. In this chapter, we are providing an update on the literature related to the various categories of CIPN treatment, with a specific focus on studies that were published over the past five years. The categories of agents that will be covered in this chapter include the serotonin–norepinephrine reuptake inhibitors, anticonvulsants, opioids, tricyclic antidepressants supplements, and topical agents (Tables 7.1 and 7.2). Promising therapies that are currently under trials will also be covered.

7.2   Serotonin–Norepinephrine Reuptake Inhibitors

7.2.1   Duloxetine

Evidence Duloxetine belongs to the drug class of serotonin–norepinephrine reuptake inhibitor (SNRI) with known efficacy in treating neuropathic pain such as diabetic neuropathy and chemotherapy-induced peripheral neuropathy (CIPN). An SNRI helps to decrease pain transmission by inhibiting the reuptake of neurotransmitters and increasing their synaptic concentrations. The main evidence cited for efficacy of duloxetine in treatment of CIPN stems from a randomized, placebo-controlled, crossover trial conducted by Smith et al. [1], which had reported a moderately large effect size of 0.51. Patients were treated with duloxetine via a regimen consisting of 30 mg daily for the first week and 60 mg daily for 4 additional weeks. Patients receiving duloxetine showed significantly greater decrease in pain score compared to those who had received placebo, with a mean change of -1.06 vs -0.34 on the Brief Pain Inventory-Short Form, $p = 0.003$. They had also reported greater decrease of pain interference with daily function and an improvement in pain-related quality of life. In a secondary analysis by Smith et al [2], patients with higher emotional functioning were found to more likely respond to duloxetine treatment (OR = 4.04, 95% CI = 0.99–16.31). An improvement in emotional functioning with the use of duloxetine was also reported in another study involving breast cancer patients [3]. This underlines that specific sub-groups of patients may stand to benefit more from the use of duloxetine and how managing distress may help to optimize management for painful neuropathy. In a pilot study conducted by Hirayama et al., duloxetine administration was found to decrease mean visual analog scale scores for both numbness and pain, when compared to Vitamin B12. It may be of further interest to investigate if the extent of duloxetine's effectiveness in treating CIPN is dependent on the type of chemotherapeutic drug as this study focused on Japanese patients treated with paclitaxel, oxaliplatin, bortezomib or vincristine [4]. In another comparative trial against venlafaxine and placebo, duloxetine was reported to demonstrate effectiveness in treating established CIPN. In the duloxetine group, cranial neuropathy grade in patients decreased significantly throughout the study period and beneficial effects were observed on motor, sensory and neuropathic pain grade as well, with lower frequency of patients reporting higher pain grade in the aforementioned aspects [5].

Conversely, patients receiving duloxetine reported more adverse side effects experienced such as fatigue, insomnia, and nausea, resulting in an 11% dropout rate compared to 1% in placebo group in trial by Smith et al. The use of duloxetine also warrants consideration of drug-interaction-risks, with it being a moderate CYP2D6 inhibitor. Keeping in mind the adverse effect profile of duloxetine, future trials may be needed to compare different dosing and duration of duloxetine in order to optimize its effective dose in treating CIPN. Despite duloxetine being recommended for treatment of CIPN, the incidence of duloxetine dispensing shows an underutilization in the USA. According to a retrospective claims study conducted by Gewandter et al, the most commonly dispensed drug after initiating neurotoxic chemotherapy was found to be gabapentin, in 7.1% of patients compared to 0.78% for the dispensing of duloxetine in patients undergoing neurotoxic chemotherapy [6]. Other factors such as patient-related factors and cost could also decide on the choice of treatment for CIPN.

Guidelines Recommendation To date, duloxetine remains one of the few pharmacological options recommended by guidelines for treatment of CIPN. Duloxetine has been approved to treat diabetic neuropathy and other neuropathic pain by the FDA. According to the American Society of Clinical Oncology (ASCO) guidelines published in 2014, duloxetine is given a moderate recommendation for use as a pharmacological treatment option for CIPN [7]. In the 2020 ASCO guideline update, data from 3 additional trials were considered and duloxetine remains the only agent recommended for use to treat patients with established painful CIPN. However, the limited amount of benefit from its use is also noted [8]. Similarly, under the European Society for Medical Oncology (ESMO) 2018 guideline for management of cancer pain in adult patients, duloxetine is also given a strong recommendation as single agent for neuropathic pain first-line treatment, along with gabapentin, pregabalin, and tricyclic antidepressants ($\leq$75 mg/day) [9]. In the updated ESMO 2020 guideline for systemic anticancer therapy-induced peripheral and central neurotoxicity, duloxetine is given grade B recommendation with level I evidence for treatment of neuropathic pain [10]. The National Comprehensive Cancer Network (NCCN) 2020 guideline for adult cancer pain included a supporting statement for use of duloxetine with a starting dose of 20–30 mg daily, increase to 60 mg daily as tolerated, as an adjuvant analgesic for neuropathic pain [11].

Future Directions Although duloxetine had demonstrated efficacy in treating CIPN based on pain score measures, its use had been limited to cases of paclitaxel and oxaliplatin-induced peripheral neuropathy so far. The magnitude of benefit may be considered modest in a clinical setting. Future studies may be required to see if its effect extends to neuropathy induced by other chemotherapeutic agents and to optimize its effective dosage.

7.2.2   Venlafaxine

Evidence Venlafaxine belongs to the same drug class of SNRI as duloxetine. Its efficacy in prevention of CIPN was supported in a randomized, placebo-controlled trial conducted by Durand et al. [12]. Patients were randomized to receive either venlafaxine 50 mg 1 h prior to oxaliplatin infusion and venlafaxine extended release 37.5 mg twice daily from day 2 to 11 or placebo. In the venlafaxine arm, the proportion of patients who experienced complete relief of acute neurotoxicity as measured on the Neuropathic Pain Symptom Inventory was significantly higher at 31.3% vs 5.3% in patients who received placebo ($p = 0.03$). In terms of adverse effect profile, a higher frequency of emesis was observed with the use of venlafaxine. In another trial, comparing venlafaxine extended release at dose of 37.5 mg twice daily against placebo, no difference in the motor and autonomic subscales measured on the European Organization for Research and Treatment of Cancer Quality of Life Questionnaire-Chemotherapy-induced   peripheral   neuropathy   (EORTC QLQ-CIPN20) was shown between both arms while sensory subscale data favored the placebo arm [13]. Depending on the instrument used and timing of oxaliplatin administration, results may differ, and future studies would be required before drawing an inference on venlafaxine's efficacy in treating CIPN.

In a double-blinded clinical trial conducted by Farschian et al. which provided direct comparison between the effects of duloxetine and venlafaxine on CIPN over 4 weeks, findings espoused the use of duloxetine over the latter [5]. It was reported while decreased neuropathy was observed in both groups, duloxetine had a more pronounced effect on reducing the grade of motor neuropathy and neuropathic pain severity than venlafaxine. It would be useful to have these findings validated in independent patient cohorts with larger sizes.

Guidelines Recommendation Despite venlafaxine's ability to potentially act as an agent for the treatment of established CIPN, the ASCO guideline does not recommend the routine use of venlafaxine in clinical practice [7, 8]. The ESMO 2018 guideline does not mention the use of venlafaxine for treating cancer-related neuropathic pain [9]. However, under the ESMO 2020 guideline, venlafaxine is considered as a pharmacological intervention for treatment of neuropathic pain with a grade C of recommendation and level II evidence [10]. In the National Comprehensive Cancer Network (NCCN) 2020 guideline for adult cancer pain, the use of venlafaxine (with a starting dose of 37.5 mg daily, titrated up to 75-225 mg daily) is supported as an adjuvant analgesic for neuropathic pain [11].

Future Directions The mechanism of action may be imperative in understanding whether venlafaxine may be effective in treating CIPN as analgesic effectiveness is not found to be dependent on its anti-depressant activity. Venlafaxine's efficacy against oxaliplatin-induced neurotoxicity is hypothesized to be due to its ability to modulate oxidative stress in the nervous system. Although venlafaxine belongs to the same drug class of SNRI as duloxetine, it was not found to be as effective. Compared to other SNRIs, venlafaxine has a higher affinity for 5-HT transporter but lower affinity for norepinephrine transporter. A recent meta-analysis suggests SNRI as a promising treatment option, with improvement in CIPN shown (standardized mean difference = 2.20, 95% CI = 0.90–3.49) [14]. Future treatments for established CIPN may require a more targeted approach, using drugs tailored to the nature of the CIPN induced.

7.3 Anticonvulsants

Evidence The use of anticonvulsant agents for the treatment of CIPN is an area of interest given their effectiveness in the treatment of neuropathic pain in other non-cancer contexts, such as diabetic neuropathy and post-herpetic neuralgia. The analgesic effects of gabapentinoids, such as gabapentin and pregabalin, are attributed to their binding to the alpha-2-delta subunit of presynaptic calcium channels, which reduces the release of excitatory neurotransmitters [15]. Despite a similar mechanism of action, structural differences between the compounds account for pregabalin exhibiting more rapid absorption and bioavailability at lower doses than gabapentin [16]. In the context of neuropathic pain, the maximum dosage of gabapentin is 3600 mg/day, divided into three doses, whereas for pregabalin it is 600 mg/day [17].

In the context of CIPN, small single-arm studies have reported improvements in CIPN symptoms at dosages of gabapentin at a maximum 900 mg/day divided into three doses [18] and pregabalin at a target dose of 450 mg/day, divided into three doses [19]. However, in these studies, a large proportion of patients did not stay on drug, with 7/20 (35%) [18] and 8/23 (35%) [19] patients, respectively, stopping the drug due to no benefit or adverse effects.

Moreover, randomized studies of gabapentinoids for the treatment of established CIPN have been unable to provide evidence of effectiveness. A double-blind randomized cross-over trial of gabapentin (at a maximum dose of 2700 mg/day) was unable to demonstrate benefit on CIPN-related pain (measured by the numeric rating scale and ECOG neuropathy scale) when compared to placebo, in a sample of 115 mixed cancer patients treated with a variety of neurotoxic chemotherapies [20]. A small double-blind randomized cross-over trial of 26 mixed cancer patients with CIPN after treatment with oxaliplatin, docetaxel, or paclitaxel chemotherapy, tested a 4-week treatment with pregabalin (600 mg/day) and found no significant difference between pregabalin and placebo in reducing average daily pain from baseline (average daily pain: 22.5% vs 10.7%, $P = 0.23$, or worst pain: 29.2% vs 16.0%, $p = 0.13$) [21]. Compared to the duloxetine, pregabalin was more effective at improving pain and insomnia domains of QOL as measured by the EORTC QLQ-C30, though global QOL improved for both groups [3]. Finally, an early RCT was unable to demonstrate any benefit of lamotrigine on CIPN pain (target dose of 300 mg daily) as compared to placebo in a sample of 131 mixed cancer patients [22].

Guideline Recommendations In the 2020 practice guidelines from ESMO-EONS-EANO, anticonvulsants are recognized as having potential for symptom control in CIPN, despite limited evidence to support efficacy, in cases of duloxetine failure or presence of contraindications [10]. These guidelines provide the following suggested doses, as tolerated: gabapentin at a target dose of 2700 mg daily, pregabalin at a target dose of 300 mg daily, and lamotrigine at a starting dose of 25 mg/day up to a target dose of 300 mg/day [10].

In the latest update of the ASCO CIPN guideline, no recommendation was made regarding the use of gabapentin or pregabalin, as a consequence of low levels of evidence to support its benefit [23]. NCCN recognizes anticonvulsants as an option for first-line adjuvant analgesics for cancer-related neuropathic pain, though not specific to CIPN [24]. The NCCN guidelines also highlight that titration rate and/or maximum dose may require adjustment for patients who are elderly, medically frail, or have renal insufficiency and note that pregabalin is more efficiently absorbed through the GI tract than gabapentin [24].

Future Directions Other anticonvulsants have been investigated for their effectiveness in managing CIPN. Lacosamide is an anti-epileptic drug with additional anticonvulsant effects through its inhibition of neuronal voltage-gated sodium channel activation. One case report describes a 52-year old male patient with metastatic, high grade urothelial carcinoma that experienced painful peripheral neuropathy after MVAC chemotherapy (methotrexate, vincristine, doxorubicin, and cisplatin) that was uncontrolled with combination gabapentin, morphine, and oxycodone-acetaminophen. Treatment with lacosamide at a dose of 100 mg twice daily was accompanied with immediate pain improvement and management of symptoms over the subsequent chemotherapy cycle appeared to coincide with lacosamide administration [25]. Recent pre-clinical data suggest that lacosamide may have comparable effects on paclitaxel-induced peripheral neuropathy, but with less motor adverse effects related to motor functioning. Lacosamide was not effective in fibromyalgia and chronic neuropathic pain in a 2012 Cochrane review [26], but a recent RCT reported benefit in reducing pain compared to placebo in non-cancer small fiber peripheral neuropathy [27].

7.4   Opioids

Evidence   The evidence related to the use of opioids specific to the treatment of CIPN is limited to few single-arm studies. In an open-label, single-arm study, 46 hematological cancer patients with uncontrolled bortezomib-induced peripheral neuropathy pain were treated with controlled-released oxycodone (mean daily dose of 24.28 mg). After two weeks, there was a significant reduction in pain intensity on the NRS, from 7.6 at baseline to 1.3 on day 14. For most participants (38/46), the CR oxycodone was added to their previously established anticonvulsant treatment, but response did not differ from those not receiving a concurrent anticonvulsant drug. Side effects were reported in 26/46 patients, with the most common being grade 1–2 constipation in 12 patients (26%) [28]. To reduce the risk of opioid-induced constipation, another single-arm study tested an oxycodone/naloxone combination in 72 Korean patients of mixed cancer diagnoses with uncontrolled CIPN. Oxycodone/naloxone, starting at 20/10 mg/day and titrated up to 80/40 mg/day, was added to the existing treatment with gabapentin or pregabalin and a 21.4% reduction in NRS score was observed after 4 weeks (23.3 vs. 1.29, $p<0.0001$) [29]. The combination of tramadol/acetaminophen, administered as one tablet every 6 h, significantly reduced VAS scores (3.1 vs. 2.1, $p<0.001$) after 24 h in a sample of 96 patients with colorectal and gastric carcinomas with mild to moderate oxaliplatin-induced chemotherapy [30]. However, findings from this single-arm study also highlight potential for variability in analgesic response; the benefit of tramadol/acetaminophen varied based on mu-opioid receptor gene (OPRM1) A118G polymorphism, with reduced response in participants with G allele variants [30].

There is a lack of randomized controlled trials testing the effect of opioids specific to the treatment of CIPN. One open-label randomized controlled trial compared pregabalin to transdermal fentanyl for neuropathic pain due to cancer or its treatment and found that fentanyl alone did not have a benefit over pregabalin. In the fentanyl group, 36.7% of participants achieved at least a 30% reduction in VAS compared to the 73.3% in the pregabalin group ($p<0.0001$) [31]. However, the proportion of the sample with CIPN in this trial was not reported, making the applicability of these findings to CIPN unclear.

Guidelines Recommendation   In the 2020 practice guidelines from ESMO-EONS-EANO, opioids are referred to as a salvage option for CIPN given the evidence of efficacy in the treatment of neuropathic pain related to other causes, but recognizing the lack of evidence in the treatment of CIPN in particular [10]. The 2020 update of the ASCO guideline on prevention and management of CIPN in survivors of adult cancers does not address the use of opioids, as only randomized trials were eligible for inclusion in the evidence review [23].

7.5   Tricyclic Antidepressants

Evidence Nortriptyline and amitriptyline, tricyclic antidepressants (TCA) that inhibit the reuptake of the biogenic amines—mostly norepinephrine and serotonin are effective in the treatment of neuropathic pain. Evidence is well established in the non-CIPN neuropathic pain setting, with a Cochrane review reported that TCAs were effective for the achievement of at least moderate pain relief [32].

In the setting of CIPN, there is minimal and mixed evidence. One RCT involving 51 patients with cisplatin-induced peripheral neuropathy investigated nortriptyline at 25 mg daily with increasing doses at weekly intervals of 25 mg (maximum target dose, 100 mg daily). Patients received either nortriptyline or placebo in two 4-week phases, separated by a 1-week washout period. There were no significant differences in paresthesias between groups in the first treatment period. Although nortriptyline appeared to have a modest benefit in the second treatment period, overall, there was no significant difference between groups [33]. In another RCT, amitriptyline ($n = 44$) at dosages of 10–50 mg daily for the treatment of CIPN from a variety of chemotherapeutic agents failed to improve sensory neuropathic symptoms [34].

Guidelines Recommendation In the ASCO 2014 and 2020 guidelines, recommendation for the use of TCA for treatment of CIPN is inconclusive. The two trials that informed the recommendation possessed limited statistical power which limited the generalizability of the data. However, it is also acknowledged that the potential of harms and benefits for TCAs is generally low, suggesting that they could be viable options that may be offered for patients despite not yet having been proven to be helpful for CIPN.

7.6   Supplements

7.6.1   Acetyl-L-Carnitine

Evidence Acetyl-L-carnitine (ALC) has shown a neuroprotective effect in diabetic-related neuropathy possibly thru neuroprotective effect mediated by neuronal nerve growth factor, regulation of acetyl-CoA, and acetylation of tubulin. Initially, two small treatment studies of CIPN from paclitaxel and cisplatin have suggested potential benefit from ALC [17]. However, a large double-blind randomized prevention trial showed harm from ALC supplementation [35]. Follow-up long term prevention study showed further persistence of worse CIPN up to 2 years of discontinuation [36]. In view of the harm shown in these prevention studies, ALC cannot be recommended for treatment of CIPN.

7.6.2 Glutamate/Glutamine

Evidence Glutamate was thought to induce local nerve growth factor release and aid in the assembly microtubule [37]. Initial studies suggested that glutamate ameliorate modestly both human [38] and experimental neuropathy induced by vincristine [39], paclitaxel and cisplatin [39]. More recently, in a randomized control trial of 49 patients between 4 and 19 years old who developed vincristine-induced neuropathy, glutamine group had lower neuropathy scores (National Cancer Institute Common Terminology Criteria for Adverse Events v.3) than placebo group on day 21. Neuropathy scores were not statistically different on day 42 after 21 days washout period [40]. However, glutamate supplementation has failed to prevent peripheral neurotoxicity of paclitaxel and current models for neuropathic pain from oxaliplatin seem to be associated with excessive activation of glutamate receptors in the spinal cord with increased amount of synaptically released glutamate [41]. Due to these findings and concerns, we do not have enough evidence to recommend glutamine for the treatment of CIPN.

7.6.3 Kampo

Evidence The Kampo (a traditional Japanese herbal medicine) formulas are made of different herbal components to treat peripheral neuropathies. Early studies have suggested potential benefit of Kampo formula containing Goshajinkigan in CIPN prevention studies [42]. A retrospective review of a database of 24 ambulatory patients with cancer in Japan who had developed neuropathy after chemotherapy and treated with diverse Kampo formulas (mostly containing commonly Goshajinkigan, hachimijiogan, and keishibukuryogan) showed beneficial outcomes in 80.0%. A reduction of $\geq 50\%$ in numbness and pain was observed in 37.8% [43]. However, as further prevention studies failed to show benefit, and the only study with beneficial results used diverse formulas, further studies are needed before we can recommend Kampo formulas for treatment of CIPN [44].

7.6.4 Guidelines Recommendation

The current ASCO Guideline Update does not provide any official recommendations for any supplements, and recommendations were made against ALC [23]. Similarly, in the Society for Integrative Oncology guidelines Clinical Recommendation for treatment of Neuropathy, ALC is not recommended to treat neuropathy because of harm [45]. Furthermore, the guidelines stated that there is currently insufficient evidence to form a clinical recommendation for omega 3 fatty acids and vitamin E. ESMO guidelines do not recommend any supplements for treatment of CIPN in adults but state glutamine has modest evidence for efficacy in children [10].

7.6.5   Future Directions

Based on current evidence, no supplements can be recommended as a treatment option in usual clinical practice setting. However, several supplements have been shown to be potentially beneficial in CIPN in recent studies and should be monitored for results of further clinical research. In a study for acupuncture, methylcobalamin only control group also showed significant improvements in pain scores for CIPN but needs further studies in placebo-controlled setting [46]. Retrospective studies in CIPN suggested improvement of numbness and pain with Goshajinkigan formula but further prospective studies are needed to evaluate its effectiveness [43].

7.7   Topical Treatment

7.7.1   Baclofen, Amitriptyline, and Ketamine Topical Gel

Evidence   Topical amitriptyline and ketamine were shown to decrease neuropathic pain in patients diabetic neuralgia [47]. A trial of a pluronic lecithin organogel containing baclofen 10 mg, amitriptyline HCL 40 mg, and ketamine 20 mg twice daily for 4 weeks was studied in 150 patients with CIPN using EORTC Quality of Life Questionnaire–Chemotherapy-Induced Neuropathy 20. After 4 weeks there was a nonsignificant trend toward benefit in sensory subscale scores as well as significant improvement in motor subscale scores [48].

7.7.2   Topical Amitriptyline and Ketamine

Evidence   A trial for patients with CIPN a combination of 4% amitriptyline/2% ketamine preparation was studied in a double-blind randomized placebo-controlled trial involving 462 patients. However, topical amitriptyline/ketamine showed no effect on 6-week CIPN scores for pain, numbness, and tingling in an intention to treat analysis [49].

7.7.3   High Concentration Topical Amitriptyline

Evidence   Case reports of high concentration topical amitriptyline (5–10%) suggested benefit in neuropathies of diverse etiologies. A recent pilot study of 44 patients with CIPN of hands and feet evaluated the use of 10% amitriptyline cream twice daily. VAS pain score decreased at least 3 points in all patients after 1 week of treatment. After 4 weeks of topical amitriptyline, mean VAS pain score decreased from 7 to 2. Twenty percent of the patients stopped treatment after 1 month with no worsening of symptoms after initial relief of pain. However, further studies are needed in placebo-controlled environment [50].

7.7.4 Capsaicin Patch

Evidence *Capsaicin* 8% patch has been used in the past for the treatment of other neuropathic pains. A study of 16 patients with CIPN from taxanes, protease inhibitors, and platinum compounds treated with one single 30 minutes application of capsaicin 8% patch. Patients had significant reduction numerical pain rating scale for spontaneous, touch-evoked, and cold-evoked pain 3 months after initial application [51]. Further placebo-controlled studies are needed to ensure the improvements were not spontaneous over time.

7.7.5 Topical Citrullus colocynthis (Bitter Apple)

Evidence Citrullus colocynthis extract has been shown to decrease pain in diabetic neuropathy [52]. However, a placebo-controlled RCT of C. colocynthis twice daily for 2 months in breast cancer patients with CIPN failed to demonstrate any improvement in FACT/GOG-Ntx scores in sensory or functional domains [53].

7.7.6 Topical Menthol

Evidence Menthol is a topical activator of transient receptor potential melastatin 8 (TRPM8), a cation channel present on sensory neurons and has a potential to produce analgesia in CIPN. Early case reports have suggested benefit in bortezomib- and carboplatin-induced neuropathy [54, 55]. A more recent open-label proof of concept study evaluated the use of topical menthol in 1% aqueous cream twice daily for 4–6 weeks in patients with chronic neuropathic pain. Of the 51 participants only 35 (69%) had CIPN. Thirty-one of thirty-eight patients (82%) who completed 4–6 week assessment had statistically significant improvement of their pain scores using Brief Pain Inventory as well as in Quantitative Sensory Testing, although it was not statistically significant in all items. However, as there were no separate analysis for patients with CIPN, further studies are needed in subjects with chemotherapy-induced neuropathy [56].

7.7.7 Guidelines Recommendation

In the latest ASCO guideline update, there is no official recommendations for any topical treatment outside of clinical trials. There are also no recommendations made for topical gel treatment containing baclofen, amitriptyline HCL, plus/minus ketamine [8]. ESMO guidelines rated grade of recommendation B (generally recommended) to topical menthol based on prospective cohort studies; however, this was a mixed cohort with only 69% having CIPN. ESMO also rated grade C (optional) for topical baclofen, amitriptyline, and ketamine gel and capsaicin patches based on level II and III evidence, respectively. However, ESMO rated grade D (generally not recommended) to topical ketamine and amitriptyline based on level I evidence [10].

7.7.8 Future Directions

Based on current evidence, no topical formulation can be recommended as a treatment option in usual clinical practice setting. Topical gabapentin gel was studied in rats with cisplatin-induced neuropathy in a prevention study and showed benefit in the alleviation of neuropathic hypoalgesia [57] but treatment studies in human are needed to explore its viability. Topical menthol may be beneficial but research is needed in CIPN specific target population [56]. High concentration topical amitriptyline [50] and capsaicin 8% patch needs further studies in placebo-controlled setting and topical baclofen and amitriptyline, and ketamine needs further studies to confirm beneficial results observed in initial studies. However, topical menthol, high dose amitriptyline, combination baclofen/amitriptyline/ketamine, and capsaicin patch could be an option in patients who have significant symptoms and are refractory to other recommended treatment options.

7.8 Other Agents under Investigation

Currently, a number of agents are undergoing clinical trials for evaluating their efficacy in treatment of CIPN. We have extracted information from ClinicalTrial.gov on all the registered trials that are currently recruiting or about to recruit, and these agents include TRK-750, dextromethorphan, nicotinamide riboside, calmangafodipir, hemp-based cannabidiol, minocycline, and intravenous lidocaine [58]. Results of these studies will be able to inform the future directions on the management of CIPN.

7.9 Summary

To date, the options of effective pharmacological treatment for established CIPN have remained scarce. The adverse impact on patient's quality of life presents an unmet clinical need to be bridged. Recommendations by international guidelines or evidence generated from robust clinical trials for treatment of CIPN are limited, other than for the use of duloxetine. Given that the observed effect for drugs which commonly work against neuropathic pain might not necessarily work in CIPN, there is a need for enhanced understanding and appreciation of the patient-related factors and biological mechanisms underlying CIPN. The manifestation of CIPN symptoms is likely due to a combination of multiple factors and in this regard, a combination with non-pharmacological interventions may also be helpful to mitigate the adverse side effects. The future carries hope for the realization of a potential treatment as ongoing research shows promise in clarifying novel agents and/or tailored interventions specific to the mechanisms of action (Tables 7.1 and 7.2).

Table 7.1 Relevant past literature and the most updated literature in the past 6 years

Agent	Authors	Study Design	Primary Endpoint	Measure of Endpoint	Participants	Interventions	Results
Serotonin–norepinephrine reuptake inhibitors							
Duloxetine	Yang 2011 [59]	Open-label pilot study	Efficacy of in treatment of CIPN	Visual analog scale (VAS) and CT-CAE	$N = 39$	Starting dose of 30 mg of duloxetine daily and increased to 60 mg daily for 12 weeks	63.3% of patients who completed treatment had a VAS score improvement
	Smith 2013 [1]	Double-blind, placebo-controlled, crossover trial	Effect of duloxetine in reducing average pain score	Brief pain inventory-short form	$N = 231$	30 mg of duloxetine daily for 1 week and then 60 mg for 4 additional weeks	Mean difference in average pain score of 0.73 (95% CI 0.26–1.20) for Duloxetine group compared to placebo
	Hirayama 2015 [4]	Open-label, cross-over trial	Effectiveness of duloxetine in decreasing severity of CIPN	Visual analog scale	$N = 34$	20 mg of duloxetine daily for 1 week and 40 mg for 3 additional weeks	Significant differences between duloxetine group with respect to numbness and pain at 4 weeks after, compared to vitamin B12 group
	Otake 2015 [60]	Retrospective study	Efficacy of duloxetine for CIPN	NCI-CTCAE	$N = 25$	20–40 mg of duloxetine daily as maintenance dose	Duloxetine has beneficial effects on symptoms of CIPN in women with gynecologic malignancies (56%)

(continued)

Table 7.1 (continued)

Agent	Authors	Study Design	Primary Endpoint	Measure of Endpoint	Participants	Interventions	Results
	Kanbayashi 2017 [61]	Retrospective	Effect of duloxetine 2 weeks after administration	Subjective improvement based on patient's complaints	$N = 74$	2 weeks of duloxetine administration (dose not specified)	Predictors for effect of duloxetine include body height, history of docetaxel use, and site of symptom
	Avan 2018 [3]	Double-blind trial	Mean global health status after 6 week treatment period	EORTC QLQ-C30 for QOL	$N = 108$	30 mg of duloxetine daily for 1 week, followed by 30 mg bd until 6 weeks	Global health status did not differ between duloxetine and pregabalin but emotional functioning improved more favorably with duloxetine ($p < 0.001$)
	Farshchian 2018 [5]	Double-blind trial	Pain reduction of medication compared to placebo	Standard grading by radiation therapy oncology group (RTOG)	$N = 156$	30 mg of duloxetine daily for 4 weeks compared to 37.5 mg of venlafaxine daily and placebo	Both duloxetine and venlafaxine showed beneficial effects on cranial, motor, sensory neuropathy and neuropathic pain
Venlafaxine	Kus 2016 [62]	Retrospective case-control study	Rate of more than 75% symptomatic relief of acute neuropathy	NCI-CTCAE Grading scale, numeric rating scale and NPSI v 4.03	$N = 206$	75 mg of venlafaxine (extended release) daily	Venlafaxine has clinical activity against taxane and oxaliplatin-induced acute

						neurosensory. Achieved more than 75% symptomatic relief in first, second, and third visits in 52.5%, 58.3%, and 45.2% of patients in venlafaxine arm
Zimmerman 2016 [13]	Double-blind, placebo-controlled pilot study	Neurotoxicity based on sensory subscale of EORTC QLQ-CIPN20	EORTC QLQ-CIPN20	$N = 50$	37.5 mg of venlafaxine extended release twice daily	Not conclusive to support the use of venlafaxine for preventing oxaliplatin-induced neuropathy
Anticonvulsants						
Gabapentin	Rao 2007 [20]	Double-blind, placebo-controlled, cross-over trial	Patient-reported "average" pain over a particular day	NRS and ENS	$N = 155$, any cancer type, variety of chemotherapies	Gabapentin increased to target dose of 2700 mg daily, for 6 weeks	Negative
	Magnowska 2019 [18]	Single-arm	Neuropathy symptoms, neuropathic pain, neurological deficits, quality of life	Neuropathy symptoms scale, McGill's neuropathic pain scale, neurologic deficit scale, visual analogue scale	$N = 20$, ovarian cancer, paclitaxel and carboplatin chemotherapy	Gabapentin up to 300 mg × 3 times daily (900 mg daily), for one month	Positive

(continued)

Table 7.1 (continued)

Agent	Authors	Study Design	Primary Endpoint	Measure of Endpoint	Participants	Interventions	Results
Pregabalin	Salahifar 2020 [63], Avan 2018 [3]	Double-blind, RCT (vs. duloxetine)	Average pain score, sensory neuropathy grade after 6-week treatment period, quality of life	VAS, NCI–CTCAE v.4.03 and patient neurotoxicity questionnaire, EORTC QLQ-C30	N = 82, breast cancer, paclitaxel or docetaxel chemotherapy	*Pregabalin:* 75 mg/day for 1 week, followed by 75 mg BID for 5 weeks. *Compared to Duloxetine:* 30 mg/day for 1 week, followed by 30 mg BID for 5 weeks.	Suggestion that pregabalin was better than duloxetine. Same trial published twice, in different ways.
	Hinkner 2019 [21]	Double-blind, randomize d, placebo-controlled, cross-over trial	Change in pain intensity from baseline to the average of daily pain on days 24 to 28 (i.e., last 5 days) of each treatment phase	NRS	N = 26, oxaliplatin, docetaxel, or paclitaxel chemotherapy	Pregabalin increased up to 600 mg/day for 4 weeks, followed by placebo, and vice versa	Negative
	Saif 2010 [19]	Single-arm	Sensory neuropathy	NCI-CTCAE	N = 23, oxaliplatin-induced neuropathy	Pregabalin increased to target dose of 150 mg three times daily	Positive, chronic neuropathy

Lamotrigine	Rao 2008 [22]	RCT	Pain intensity	NRS	N = 131, variety of chemotherapies	Lamotrigine 25 mg × 2 weeks, 25 mg twice daily × weeks, 50 mg twice daily × weeks, 100 mg twice daily × 2 weeks, 150 mg twice daily ×2 weeks	Negative
Opioids							
Opioids	Kim 2018 [29]	Single-arm	Change in pain intensity score over a 4-week period	NRS	N = 72, any cancer type, variety of chemotherapies, on concomitant gabapentin or pregabalin	Starting dose of oxycodone/naloxone 20/10 mg/day in addition to gabapentin (≥900 mg/day) or pregabalin (≥300 mg/day). Dose titration up to 80/40 mg/day, as per investigator discretion	Suggestion of benefit
	Raptis 2014 [31]	Open-label RCT	30% decrease in VAS score compared to baseline	VAS	N = 120, mixed diagnoses, with either CIPN or non-CIPN neuropathic cancer pain	Increasing doses of either oral pregabalin (starting at 75 mg/day up to 600 mg/day, as appropriate) or transdermal	Negative

(continued)

Table 7.1 (continued)

Agent	Authors	Study Design	Primary Endpoint	Measure of Endpoint	Participants	Interventions	Results
						fentanyl (starting at 25mcg/hr. every 72 h up to 150 mcg/h., as appropriate) for 28 days	
	Liu 2012 [30]	Single-arm	Pain intensity	VAS	N = 96, oxaliplatin neuropathy	Ultracet (each table contains tramadol 37.5 mg and acetaminophen 325 mg). One tablet administered every 6 h	Suggestion of benefit
	Cartoni 2012 [28]	Open-label single-arm	Pain intensity	NRS	N = 46, bortezomib neuropathy, with or without concomitant anticonvulsant use	Controlled-released oxycodone, starting at 10 mg q12hr and titrated to optimal dose, up to maximum daily dose of 80 mg, over 14 days	Suggestion of benefit
Tricyclic antidepressants							
Nortriptyline	Hammack 2002 [33]	Double-blind placebo-controlled trial	Change in the ordinal pain/paresthesia severity score from baseline	VAS scores and verbal descriptor scale	N = 51, cisplatin-induced neuropathy	Nortriptyline 25 mg daily, Increasing weekly to max target dose 100 mg daily	Negative

	Study	Type	Measure	Assessment	N	Intervention	Result
	Kautio 2008 [34]	Double-blind placebo-controlled trial	Relief of neuropathic symptoms with amitriptyline compared with placebo and the change in the effect of pain on daily activities score	NCI-CTCAE, patient-reported global improvement of neuropathic symptoms using a numeric scale	N = 44, variety of chemotherapies	Amitriptyline 10–50 mg daily	Negative
Supplements							
Glutamine	Sands 2017 [40]	Double-blind, randomized placebo-controlled study	Sensory and motor neuropathy scores after 6 weeks	NCI-CTCAE v3	N = 49	Glutamine 0.6 g/m² twice daily	Glutamate group had statistically significant lower sensory scores (but not motor) in week 3 but not on week 6 on patient with vincristine-induced peripheral neuropathy
Kampo formulas	Kimata 2016 [43]	Retrospective case series	Pain, numbness, and pins-and-needles	11-point numerical rating scale for pain intensity	N = 24	Variable formulas (most commonly Goshajinkigan, hachimijiogan, and keishibukuryogan)	Reduction of symptoms was observed in 80.0% of patients (>50% reduction in 37.8%)

(continued)

Table 7.1 (continued)

Agent	Authors	Study Design	Primary Endpoint	Measure of Endpoint	Participants	Interventions	Results
Topical agents							
Topical baclofen, amitriptyline, ketamine	Barton 2011 [48]	Randomize d placebo-controlled study	Sensory neuropathy after 4 weeks	Sensory subscale of EORTC QLQ-CIPN20	$N = 150$	Pluronic lecithin organogel with baclofen 10 mg, amitriptyline HCL 40 mg, and ketamine 20 mg	Treatment group showed a trend towards improvement in sensory subscale and significant improvement on motor subscale
Topical amitriptyline and ketamine	Gewandter 2014 [49]	Double-blind, randomize d placebo-controlled study	Pain, numbness, or tingling after 6 weeks	11-point numerical rating scale for pain, numbness, or tingling	$N = 458$	2% ketamine plus 4% amitriptyline cream (KA)	There were no significant differences in 6-week pain, numbness, or tingling between KA and placebo group
Topical amitriptyline (high concentration)	Rossignol 2019 [50]	Open-label, single-arm, prospective clinical trial	Neuropathic pain after 4 weeks	Visual analogue scale (VAS) for pain	$N = 44$	10% amitriptyline cream	VAS pain score went down from 7 (4–9) to 2 (0–4) after 4 weeks
Topical capsaicin	Anand 2019 [51]	Open-label, single-arm, prospective clinical trial	Neuropathic pain after 12 weeks	11-point numerical rating scale for pain, numbness, or tingling	$N = 16$	Capsaicin 8% patch	Substantial reduction in scores for spontaneous, touch, and cold evoked pain was observed

Citrullus colocynthis (bitter apple)	Rostami 2019 [53]	Double-blind, randomize d placebo-controlled study	Neuropathy scores on FACT/GOG-Ntx	Functional assessment of Cancer therapy/ gynecologic oncology group (FACT/ GOG)-neurotoxicity (Ntx) score	N = 34	Topical C. colocynthis oil twice daily on plantar and dorsal surface of hands and feet	No significant improvement was observed on FACT/GOG-Ntx total score or sensory and motor domains
Menthol	Fallon 2015 [9]	Open-label, single-arm, prospective clinical trial	Change in neuropathic pain after 4–6 weeks	Brief pain inventory	N = 38	1% menthol aqueous cream	82% had substantial improvement in pain scores (BPI Score 47 to 34)

Table 7.2 Future direction table (Case reports/series or Abstracts)

Agent	Authors	Study design	Primary endpoint	Measure of endpoint	Participants	Interventions	Results
Lacosamide	Ibrahim 2015 [25]	Case study	Pain	Not reported	52-year old male patient with metastatic, high grade urothelial carcinoma and Uncontrolled severe CIPN	Lacosamide 100 mg twice a day	Immediate reduction in pain *in a single patient*
Duloxetine	Takenaka 2012 [64]	Case report	Effect of addition of duloxetine to gabapentin	Visual analog scale (VAS) for pain	N = 1	Addition of 20 mg of duloxetine to 50 mg of pregabalin daily. Dose of duloxetine up-titrated to 30 mg daily	VAS improve from 45 mm to 20 mm to 5 mm (after 5 months)
Duloxetine	Battaglini 2018 [65]	Double-blind, placebo-controlled, cross-over trial	Effect of duloxetine on patient-reported outcome	FACT/GOG-Ntx	N = 48	30 mg of duloxetine daily for 1 week followed by 60 mg daily for 6 weeks. During eighth week, dose will be down-titrated to 30 mg	Trial protocol

References

1. Smith EM, Pang H, Cirrincione C, Fleishman S, Paskett ED, Ahles T et al (2013) Effect of duloxetine on pain, function, and quality of life among patients with chemotherapy-induced painful peripheral neuropathy: a randomized clinical trial. JAMA 309(13):1359–1367. https://doi.org/10.1001/jama.2013.2813. PubMed PMID: 23549581; PubMed Central PMCID: PMCPMC3912515
2. Smith EM, Pang H, Ye C, Cirrincione C, Fleishman S, Paskett ED et al (2017) Predictors of duloxetine response in patients with oxaliplatin-induced painful chemotherapy-induced peripheral neuropathy (CIPN): a secondary analysis of randomised controlled trial - CALGB/alliance 170601. Eur J Cancer Care (Engl) 26(2):e12421. https://doi.org/10.1111/ecc.12421. PubMed PMID: 26603828; PubMed Central PMCID: PMCPMC4879099
3. Avan R, Janbabaei G, Hendouei N, Alipour A, Borhani S, Tabrizi N et al (2018) The effect of pregabalin and duloxetine treatment on quality of life of breast cancer patients with taxane-induced sensory neuropathy: A randomized clinical trial. J Res Med Sci 23:52. https://doi.org/10.4103/jrms.JRMS_1068_17. PubMed PMID: 30057636; PubMed Central PMCID: PMCPMC6040148
4. Hirayama Y, Ishitani K, Sato Y, Iyama S, Takada K, Murase K et al (2015) Effect of duloxetine in Japanese patients with chemotherapy-induced peripheral neuropathy: a pilot randomized trial. Int J Clin Oncol 20(5):866–871. Epub 2015/03/13. PubMed PMID: 25762165. https://doi.org/10.1007/s10147-015-0810-y
5. Farshchian N, Alavi A, Heydarheydari S, Moradian N (2018) Comparative study of the effects of venlafaxine and duloxetine on chemotherapy-induced peripheral neuropathy. Cancer Chemother Pharmacol 82(5):787–793. https://doi.org/10.1007/s00280-018-3664-y. PubMed PMID: 30105459
6. Gewandter JS, Kleckner AS, Marshall JH, Brown JS, Curtis LH, Bautista J et al (2020) Chemotherapy- induced peripheral neuropathy (CIPN) and its treatment: an NIH Collaboratory study of claims data. Support Care Cancer 28(6):2553–2562. https://doi.org/10.1007/s00520-019-05063-x. PubMed PMID: 31494735; PubMed Central PMCID: PMCPMC7060096
7. Hershman DL, Lacchetti C, Dworkin RH, Lavoie Smith EM, Bleeker J, Cavaletti G et al (2014) Prevention and management of chemotherapy-induced peripheral neuropathy in survivors of adult cancers: American society of clinical oncology clinical practice guideline. J Clin Oncol 32 (18):1941–1967. https://doi.org/10.1200/JCO.2013.54.0914. PubMed PMID: 24733808
8. Loprinzi CL, Lacchetti C, Bleeker J, Cavaletti G, Chauhan C, Hertz DL et al (2020) Prevention and management of chemotherapy-induced peripheral neuropathy in survivors of adult cancers: ASCO guideline update. J Clin Oncol. https://doi.org/10.1200/jco.20.01399. PubMed PMID: 32663120
9. Fallon M, Giusti R, Aielli F, Hoskin P, Rolke R, Sharma M et al (2018) Management of cancer pain in adult patients: ESMO clinical practice guidelines. Annals Oncol 29:iv166–iv191. https://doi.org/10.1093/annonc/mdy152
10. Jordan B, Margulies A, Cardoso F, Cavaletti G, Haugnes HS, Jahn P et al (2020. Epub 2020/08/03. PubMed PMID: 32739407) Systemic anticancer therapy-induced peripheral and central neurotoxicity: ESMO-EONS-EANO clinical practice guidelines for diagnosis, prevention, treatment and follow-up. Ann Oncol. https://doi.org/10.1016/j.annonc.2020.07.003
11. NCCN Clinical practice guidelines in oncology-adult cancer pain (Version 1). 2020. Epub April 8
12. Durand JP, Deplanque G, Montheil V, Gornet JM, Scotte F, Mir O et al (2012) Efficacy of venlafaxine for the prevention and relief of oxaliplatin-induced acute neurotoxicity: results of EFFOX, a randomized, double-blind, placebo-controlled phase III trial. Ann Oncol 23 (1):200–205. https://doi.org/10.1093/annonc/mdr045. PubMed PMID: 21427067
13. Zimmerman C, Atherton PJ, Pachman D, Seisler D, Wagner-Johnston N, Dakhil S et al (2016) MC11C4: a pilot randomized, placebo-controlled, double-blind study of venlafaxine to prevent oxaliplatin-induced neuropathy. Support Care Cancer 24(3):1071–1078. https://doi.org/10.

1007/s00520-015-2876-5. PubMed PMID: 26248652; PubMed Central PMCID: PMCPMC4939800

14. Song SY, Ko YB, Kim H, Lee GW, Yang JB, Chang HK et al (2020) Effect of serotonin-norepinephrine reuptake inhibitors for patients with chemotherapy-induced painful peripheral neuropathy: A meta-analysi. Medicine (Baltimore) 99(1):e18653. https://doi.org/10.1097/MD.0000000000018653. PubMed PMID: 31895829; PubMed Central PMCID: PMCPMC6946453

15. Calandre EP, Rico-Villademoros F, Slim M (2016) Alpha2delta ligands, gabapentin, pregabalin and mirogabalin: a review of their clinical pharmacology and therapeutic use. Expert Rev Neurother 16(11):1263–1277. https://doi.org/10.1080/14737175.2016.1202764

16. Bockbrader HN, Wesche D, Miller R, Chapel S, Janiczek N, Burger P (2010) A comparison of the pharmacokinetics and pharmacodynamics of pregabalin and gabapentin. Clin Pharmacokinet 49(10):661–669. Epub 2010/09/08. PubMed PMID: 20818832. https://doi.org/10.2165/11536200-000000000-00000

17. Pachman DR, Watson JC, Lustberg MB, Wagner-Johnston ND, Chan A, Broadfield L et al (2014) Management options for established chemotherapy-induced peripheral neuropathy. Support Care Cancer 22(8):2281–2295. Epub 2014/06/01. PubMed PMID: 24879391. https://doi.org/10.1007/s00520-014-2289-x

18. Magnowska M, Iżycka N, Kapoła-Czyż J, Romała A, Lorek J, Spaczyński M et al (2018) Effectiveness of gabapentin pharmacotherapy in chemotherapy-induced peripheral neuropathy. Ginekol Pol 89(4):200–204. Epub 2018/05/22. PubMed PMID: 29781075. https://doi.org/10.5603/GP.a2018.0034

19. Saif MW, Syrigos K, Kaley K, Isufi I (2010) Role of pregabalin in treatment of oxaliplatin-induced sensory neuropathy. Anticancer Res 30(7):2927–2933. Epub 2010/08/05. PubMed PMID: 20683034

20. Rao RD, Michalak JC, Sloan JA, Loprinzi CL, Soori GS, Nikcevich DA et al (2007) Efficacy of gabapentin in the management of chemotherapy-induced peripheral neuropathy: a phase 3 randomized, double-blind, placebo-controlled, crossover trial (N00C3). Cancer 110(9):2110–2118. Epub 2007/09/14. PubMed PMID: 17853395. https://doi.org/10.1002/cncr.23008

21. Hincker A, Frey K, Rao L, Wagner-Johnston N, Ben Abdallah A, Tan B et al (2019) Somatosensory predictors of response to pregabalin in painful chemotherapy-induced peripheral neuropathy: a randomized, placebo-controlled, crossover study. Pain 160(8):1835–1846. https://doi.org/10.1097/j.pain.0000000000001577. Epub 2019/07/25. PubMed PMID: 31335651; PubMed Central PMCID: PMCPMC6687437 the end of this article

22. Rao RD, Flynn PJ, Sloan JA, Wong GY, Novotny P, Johnson DB et al (2008) Efficacy of lamotrigine in the management of chemotherapy-induced peripheral neuropathy: a phase 3 randomized, double- blind, placebo-controlled trial, N01C3. Cancer 112(12):2802–2808. Epub 2008/04/23. PubMed PMID: 18428211. https://doi.org/10.1002/cncr.23482

23. Loprinzi CL, Lacchetti C, Bleeker J, Cavaletti G, Chauhan C, Hertz DL et al (2020. Epub 2020/07/15. PubMed PMID: 32663120) Prevention and management of chemotherapy-induced peripheral neuropathy in survivors of adult cancers: ASCO guideline update. J Clin Oncol. https://doi.org/10.1200/jco.20.01399

24. Swarm RA, Paice JA, Anghelescu DL, Are M, Bruce JY, Buga S et al (2019) Adult Cancer pain, version 3.2019, NCCN clinical practice guidelines in oncology. J Natl Compr Cancer Netw 17(8):977–1007. Epub 2019/08/08. PubMed PMID: 31390582. https://doi.org/10.6004/jnccn.2019.0038

25. Ibrahim SA, Albany Z, Albany C (2015) Significant response to lacosamide in a patient with severe chemotherapy-induced peripheral neuropathy. J Community Support Oncol 13(5):202–204. Epub 2015/06/02. PubMed PMID: 26029937. https://doi.org/10.12788/jcso.0136

26. Hearn L, Derry S, Moore RA (2012) Lacosamide for neuropathic pain and fibromyalgia in adults. Cochrane Database Syst Rev 2:Cd009318. Epub 2012/02/18. PubMed PMID: 22336864. https://doi.org/10.1002/14651858.CD009318.pub2

27. de Greef BTA, Hoeijmakers JGJ, Geerts M, Oakes M, Church TJE, Waxman SG et al (2019) Lacosamide in patients with Nav1.7 mutations-related small fibre neuropathy: a randomized controlled trial. Brain 142(2):263–275. https://doi.org/10.1093/brain/awy329
28. Cartoni C, Brunetti GA, Federico V, Efficace F, Grammatico S, Tendas A et al (2012) Controlled- release oxycodone for the treatment of bortezomib-induced neuropathic pain in patients with multiple myeloma. Support Care Cancer 20(10):2621–2626. Epub 2012/06/16. PubMed PMID: 22699304. https://doi.org/10.1007/s00520-012-1511-y
29. Kim BS, Jin JY, Kwon JH, Woo IS, Ko YH, Park SY et al (2018) Efficacy and safety of oxycodone/naloxone as add-on therapy to gabapentin or pregabalin for the management of chemotherapy-induced peripheral neuropathy in Korea. Asia Pac J Clin Oncol 14(5):e448–ee54. Epub 2017/12/28. PubMed PMID: 29280313. https://doi.org/10.1111/ajco.12822
30. Liu YC, Wang WS (2012) Human mu-opioid receptor gene A118G polymorphism predicts the efficacy of tramadol/acetaminophen combination tablets (ultracet) in oxaliplatin-induced painful neuropathy. Cancer 118(6):1718–1725. Epub 2011/08/13. PubMed PMID: 21837673. https://doi.org/10.1002/cncr.26430
31. Raptis E, Vadalouca A, Stavropoulou E, Argyra E, Melemeni A, Siafaka I (2014) Pregabalin vs. opioids for the treatment of neuropathic cancer pain: a prospective, head-to-head, randomized, open-label study. Pain Pract 14(1):32–42. Epub 2013/03/08. PubMed PMID: 23464813. https://doi.org/10.1111/papr.12045
32. Saarto T, Wiffen PJ (2007) Antidepressants for neuropathic pain. Cochrane Database Syst Rev 4:CD005454. Epub 2007/10/19. PubMed PMID: 17943857. https://doi.org/10.1002/14651858.CD005454.pub2
33. Hammack JE, Michalak JC, Loprinzi CL, Sloan JA, Novotny PJ, Soori GS et al (2002) Phase III evaluation of nortriptyline for alleviation of symptoms of cis-platinum-induced peripheral neuropathy. Pain 98(1–2):195–203. Epub 2002/07/06. PubMed PMID: 12098632. https://doi.org/10.1016/s0304-3959(02)00047-7
34. Kautio AL, Haanpaa M, Saarto T, Kalso E (2008) Amitriptyline in the treatment of chemotherapy- induced neuropathic symptoms. J Pain Symptom Manag 35(1):31–39. Epub 2007/11/06. PubMed PMID: 17980550. https://doi.org/10.1016/j.jpainsymman.2007.02.043
35. Hershman DL, Unger JM, Crew KD, Minasian LM, Awad D, Moinpour CM et al (2013) Randomized double-blind placebo-controlled trial of acetyl-L-carnitine for the prevention of Taxane-induced neuropathy in women undergoing adjuvant breast Cancer therapy. J Clin Oncol 31(20):2627–2633. https://doi.org/10.1200/jco.2012.44.8738
36. Hershman DL, Unger JM, Crew KD, Till C, Greenlee H, Minasian LM et al (2018) Two-year trends of Taxane-induced neuropathy in women enrolled in a randomized trial of acetyl-L-carnitine (SWOG S0715). JNCI: Journal of the National Cancer Institute 110(6):669–676. https://doi.org/10.1093/jnci/djx259
37. Stubblefield MD, Vahdat LT, Balmaceda CM, Troxel AB, Hesdorffer CS, Gooch CL (2005) Glutamine as a neuroprotective agent in high-dose paclitaxel-induced peripheral neuropathy: a clinical and Electrophysiologic study. Clin Oncol 17(4):271–276. https://doi.org/10.1016/j.clon.2004.11.014
38. Jackson DV, Wells HB, Atkins JN, Zekan PJ, White DR, Richards F et al (1988) Amelioration of vincristine neurotoxicity by glutamic acid. Am J Med 84(6):1016–1022. https://doi.org/10.1016/0002-9343(88)90306-3
39. Boyle FM, Wheeler HR, Shenfield GM (1999) J Neuro-Oncol 41(2):107–116. https://doi.org/10.1023/a:1006124917643
40. Sands S, Ladas EJ, Kelly KM, Weiner M, Lin M, Ndao DH et al (2016) Glutamine for the treatment of vincristine-induced neuropathy in children and adolescents with cancer. Support Care Cancer 25(3):701–708. https://doi.org/10.1007/s00520-016-3441-6
41. Loven D, Levavi H, Sabach G, Zart R, Andras M, Fishman A et al (2009) Long-term glutamate supplementation failed to protect against peripheral neurotoxicity of paclitaxel. Eur J Cancer Care 18(1):78–83. https://doi.org/10.1111/j.1365-2354.2008.00996.x

42. Nishioka M, Shimada M, Kurita N, Iwata T, Morimoto S, Yoshikawa K et al (2011) The Kampo medicine, Goshajinkigan, prevents neuropathy in patients treated by FOLFOX regimen. Int J Clin Oncol 16(4):322–327. https://doi.org/10.1007/s10147-010-0183-1

43. Kimata Y, Ogawa K, Okamoto H, Chino A, Namiki T (2016) Efficacy of Japanese traditional (Kampo) medicine for treating chemotherapy-induced peripheral neuropathy: A retrospective case series study. World J Clin Cases 4(10):310. https://doi.org/10.12998/wjcc.v4.i10.310

44. Kuriyama A, Endo K (2017) Goshajinkigan for prevention of chemotherapy-induced peripheral neuropathy: a systematic review and meta-analysis. Support Care Cancer 26(4):1051–1059. https://doi.org/10.1007/s00520-017-4028-6

45. Greenlee H, DuPont-Reyes MJ, Balneaves LG, Carlson LE, Cohen MR, Deng G et al (2017) Clinical practice guidelines on the evidence-based use of integrative therapies during and after breast cancer treatment. CA Cancer J Clin 67(3):194–232. https://doi.org/10.3322/caac.21397.. PubMed PMID: 28436999

46. Han X, Wang L, Shi H, Zheng G, He J, Wu W et al (2017) Acupuncture combined with methylcobalamin for the treatment of chemotherapy-induced peripheral neuropathy in patients with multiple myeloma. BMC Cancer 17(1):40. https://doi.org/10.1186/s12885-016-3037-z

47. Lynch ME, Clark AJ, Sawynok J (2003) A pilot study examining topical amitriptyline, ketamine, and a combination of both in the treatment of neuropathic pain. Clin J Pain 19 (5):323–328. https://doi.org/10.1097/00002508-200309000-00007

48. Barton DL, Wos EJ, Qin R, Mattar BI, Green NB, Lanier KS et al (2010) A double-blind, placebo- controlled trial of a topical treatment for chemotherapy-induced peripheral neuropathy: NCCTG trial N06CA. Support Care Cancer 19(6):833–841. https://doi.org/10.1007/s00520-010-0911-0

49. Gewandter JS, Mohile SG, Heckler CE, Ryan JL, Kirshner JJ, Flynn PJ et al (2014) A phase III randomized, placebo-controlled study of topical amitriptyline and ketamine for chemotherapy-induced peripheral neuropathy (CIPN): a University of Rochester CCOP study of 462 cancer survivors. Support Care Cancer 22(7):1807–1814. https://doi.org/10.1007/s00520-014-2158-7

50. Rossignol J, Cozzi B, Liebaert F, Hatton S, Viallard M-L, Hermine O et al (2019) High concentration of topical amitriptyline for treating chemotherapy-induced neuropathies. Support Care Cancer 27(8):3053–3059. https://doi.org/10.1007/s00520-018-4618-y

51. Anand P, Elsafa E, Privitera R, Naidoo K, Yiangou Y, Donatien P et al (2019) Rational treatment of chemotherapy-induced peripheral neuropathy with capsaicin 8% patch: from pain relief towards disease modification. J Pain Res 12:2039–2052. https://doi.org/10.2147/jpr.s213912

52. Heydari M, Homayouni K, Hashempur MH, Shams M (2016) Topical Citrullus colocynthis (bitter apple) extract oil in painful diabetic neuropathy: a double-blind randomized placebo-controlled clinical trial. J Diabetes 8(2):246–252. Epub 2015/03/25. PubMed PMID: 25800045. https://doi.org/10.1111/1753-0407.12287

53. Rostami N, Mosavat SH, Heydarirad G, Arbab Tafti R, Heydari M (2019) Efficacy of topical Citrullus colocynthis (bitter apple) extract oil in chemotherapy-induced peripheral neuropathy: a pilot double-blind randomized placebo-controlled clinical trial. Phytother Res 33 (10):2685–2691. Epub 2019/08/03. PubMed PMID: 31373112. https://doi.org/10.1002/ptr.6442

54. Colvin LA, Johnson PR, Mitchell R, Fleetwood-Walker SM, Fallon M (2008) From bench to bedside: a case of rapid reversal of bortezomib-induced neuropathic pain by the TRPM8 activator, menthol. J Clin Oncol 26(27):4519–4520. Epub 2008/09/20. PubMed PMID: 18802169. https://doi.org/10.1200/JCO.2008.18.5017

55. Storey DJ, Colvin LA, Mackean MJ, Mitchell R, Fleetwood-Walker SM, Fallon MT (2010) Reversal of dose-limiting carboplatin-induced peripheral neuropathy with TRPM8 activator, menthol, enables further effective chemotherapy delivery. J Pain Symptom Manag 39(6):e2–e4. https://doi.org/10.1016/j.jpainsymman.2010.02.004

56. Fallon MT, Storey DJ, Krishan A, Weir CJ, Mitchell R, Fleetwood-Walker SM et al (2015) Cancer treatment-related neuropathic pain: proof of concept study with menthol—a TRPM8 agonist. Support Care Cancer 23(9):2769–2777. https://doi.org/10.1007/s00520-015-2642-8

57. Shahid M, Subhan F, Ahmad N, Sewell RDE (2019) Efficacy of a topical gabapentin gel in a cisplatin paradigm of chemotherapy-induced peripheral neuropathy. BMC Pharmacol Toxicol 20(1):51. https://doi.org/10.1186/s40360-019-0329-3

58. U.S. National Library of Medicine (2020) ClinicalTrials.gov (August 25, 2020). Available from: https://clinicaltrials.gov/.

59. Yang YH, Lin JK, Chen WS, Lin TC, Yang SH, Jiang JK et al (2012) Duloxetine improves oxaliplatin- induced neuropathy in patients with colorectal cancer: an open-label pilot study. Support Care Cancer 20(7):1491–1497. https://doi.org/10.1007/s00520-011-1237-2. PubMed PMID: 21814779

60. Otake A, Yoshino K, Ueda Y, Sawada K, Mabuchi S, Kimura T et al (2015) Usefulness of duloxetine for paclitaxel-induced peripheral neuropathy treatment in gynecological cancer patients. Anticancer Res 35(1):359–363

61. Kanbayashi Y, Inagaki M, Ueno H, Hosokawa T (2017) Predictors of the usefulness of duloxetine for chemotherapy-induced peripheral neuropathy. Med Oncol 34(8):137. https://doi.org/10.1007/s12032-017-0995-1. PubMed PMID: 28687964

62. Kus T, Aktas G, Alpak G, Kalender ME, Sevinc A, Kul S et al (2016) Efficacy of venlafaxine for the relief of taxane and oxaliplatin-induced acute neurotoxicity: a single-center retrospective case- control study. Support Care Cancer 24(5):2085–2091. https://doi.org/10.1007/s00520-015-3009-x. PubMed PMID: 26546457

63. Salehifar E, Janbabaei G, Hendouei N, Alipour A, Tabrizi N, Avan R (2020) Comparison of the efficacy and safety of Pregabalin and duloxetine in Taxane-induced sensory neuropathy: a randomized controlled trial. Clin Drug Investig 40(3):249–257. Epub 2020/01/12. PubMed PMID: 31925721. https://doi.org/10.1007/s40261-019-00882-6

64. Takenaka M, Iida H, Matsumoto S, Yamaguchi S, Yoshimura N, Miyamoto M (2013) Successful treatment by adding duloxetine to pregabalin for peripheral neuropathy induced by paclitaxel. Am J Hosp Palliat Care 30(7):734–736. https://doi.org/10.1177/1049909112463416. PubMed PMID: 23064035

65. Battaglini E, Park SB, Barnes EH, Goldstein D (2018) A double blind, placebo controlled, phase II randomised cross-over trial investigating the use of duloxetine for the treatment of chemotherapy- induced peripheral neuropathy. Contemp Clin Trials 70:135–138. https://doi.org/10.1016/j.cct.2018.04.011. PubMed PMID: 29680317

Systematic Review of Exercise for Prevention and Management of Chemotherapy-Induced Peripheral Neuropathy

8

Ian R. Kleckner, Susanna B. Park, Fiona Streckmann, Joachim Wiskemann, Sara Hardy, and Nimish Mohile

Abstract

Chemotherapy-induced peripheral neuropathy (CIPN) is a highly prevalent and dose-limiting toxicity of many widely used chemotherapy regimens for the treatment of common cancers including lung, breast, prostate, gastrointestinal, and blood cancers. Symptoms include numbness, tingling, pain, and cramping in the hands and feet, as well as impaired balance and gait that collectively increase the risk of falls and compromise activities of daily living. Among the extremely limited treatment options for CIPN, exercise has emerged as a promising intervention based on a growing body of studies. Here, we review preclinical and clinical evidence on the use of exercise and related modalities for the prevention, treatment, and management of CIPN. We identified 2 studies in rodents plus 23 studies in humans, including 15 randomized studies (10 comparing

I. R. Kleckner (✉) · S. Hardy · N. Mohile
University of Rochester Medical Center, Rochester, NY, USA
e-mail: Ian_Kleckner@URMC.Rochester.edu; Sara_Hardy@URMC.Rochester.edu;
Nimish_Mohile@URMC.Rochester.edu

S. B. Park
Brain and Mind Centre, Faculty of Medicine and Health, University of Sydney, Sydney, Australia
e-mail: Susanna.Park@Sydney.edu.au

F. Streckmann
Department of Sport, Exercise and Health, University of Basel, Basel, Switzerland

Department of Oncology, University Hospital Basel, Basel, Switzerland

German Sport University Cologne, Cologne, Germany
e-mail: fiona.streckmann@unibas.ch

J. Wiskemann
Heidelberg University Clinic and National Center for Tumor Diseases, Heidelberg, Germany
e-mail: Joachim.Wiskemann@nct-heidelberg.de

exercise vs. non-exercise control), plus 19 pre-registered studies. The 10 randomized studies collectively suggest that exercise is beneficial for the treatment and prevention of CIPN with little to no side effects. However, these studies tend to be either rigorous yet small or large yet simple and exploratory, with no Phase III randomized studies published or pre-registered. Next, we discuss biological and psychosocial mechanisms by which exercise might exert its effects. We are optimistic for the trajectory of this work including seeking definitive answers to *whether* exercise is beneficial, *what dose* of exercise is needed, *how* it exerts its effects mechanistically, and how to best disseminate exercise to patients in the real world.

| Keywords |

Exercise · Chemotherapy · Neuropathy · CIPN · Review · Mechanism

8.1   Introduction

Chemotherapy-induced peripheral neuropathy (CIPN) is a highly prevalent and severe toxicity of many widely used chemotherapy drugs including platinum-based agents (oxaliplatin, cisplatin, carboplatin), taxanes (paclitaxel, docetaxel), vinca alkaloids, bortezomib, and thalidomide analogs [1, 2], as well as certain immunotherapies [3]. These drugs are used to treat many cancers including lung, breast, prostate, gastrointestinal, cervical, ovarian, testicular, blood, and bone marrow cancers. CIPN can involve acute symptoms that present in the hours and days after an infusion [4–6] plus ongoing symptoms that affect on average 58–78% of patients one month after completion of chemotherapy [7]. The prevalence of CIPN is approximately 6–54% six months post-chemotherapy [7], and many patients develop a chronic CIPN [8, 9]. CIPN can become so severe that it gives oncologists cause to reduce chemotherapy dose or terminate neurotoxic chemotherapy altogether, and it reduces adherence to at-home chemotherapy [10], which may compromise anti-cancer treatment [11]. CIPN is also stressful on the healthcare system—medical claims data suggest that CIPN is under-diagnosed [12] and that patients with CIPN typically require 12 more outpatient visits, 3 more hospital days, and $17,000 USD more in medical expenses than matched patients without CIPN [13].

CIPN includes patient-reported symptoms, clinical signs, and mechanistic features resulting from damage, dysfunction, and death of peripheral neurons and downstream sequelae. The symptoms of CIPN are primarily felt in the hands and feet with some combination of numbness, tingling, shooting or stabbing pains, burning pain, cramping, and hypersensitivity to cold (e.g., cold weather, touching something cold, or pain in the throat from drinking a cold beverage) [2, 14]. The consequences of CIPN include loss of tactile or vibration sensitivity, walking gait and balance problems (i.e., postural instability; especially with eyes closed) [15, 16], increased risk of falls [17, 18], compromised participation in activities of daily living, occasional changes in peripheral sensory nerve conduction (e.g., reduced sensory nerve action potential amplitudes) and, in rare cases, damage to the autonomic nervous system leading to impaired organ function (e.g., constipation, orthostatic hypotension, sexual dysfunction) [2]. The pathophysiological mechanisms underlying the development of CIPN are varied and agent-dependent but include neuronal loss in the dorsal root ganglion, axonal degeneration, oxidative stress and mitochondrial dysfunction, interruption of axonal transport, neuroinflammation, and excitability alterations [1, 19, 20]. There is no gold-standard assessment for CIPN [21, 22], but it is recommended to include both patient-reported outcome measures (e.g., CIPN-20; [23]) and clinical grading scales [21, 22], e.g., the Total Neuropathy Score (TNS) [24]. Diagnosis depends on patient history, type and dose of chemotherapy, and symptoms [16, 25].

There are only minimally effective treatments for CIPN, despite over 20 years of research and over 48 RCTs testing drugs to treat or prevent CIPN [25–27]. The 2020 American Society for Clinical Oncology (ASCO) Guidelines for CIPN concluded no recommended methods to prevent CIPN, and only one established method to treat CIPN: a moderate recommendation of the drug duloxetine to treat CIPN-related pain [25]. However, in the most definitive RCT of duloxetine ($N = 231$), CIPN pain was only mildly improved with this drug [28]. Duloxetine also has poor adherence of 30–38% [29], perhaps due to its side effects such as constipation and dizziness [30]. As of yet there are no recommended supplements, integrative therapies [31], devices, or behavioral interventions for CIPN [25] due to lack of multiple definitive Phase III randomized controlled trials (RCTs). Therefore, research on promising treatments for CIPN is a high-priority area of inquiry to ultimately identify and optimize additional treatments for CIPN [32]. One of the most promising treatments for CIPN is exercise, as shown by a growing body of studies [33, 34].

Here we review preclinical and clinical evidence on the use of exercise, physical therapy, and occupational therapy for the prevention, treatment, and management of CIPN. There have been two excellent and recent systematic reviews of studies investigating exercise for CIPN [33, 34], and so we only briefly review these existing published studies and then extend beyond these reviews in a few unique ways. First, we will review preclinical studies of exercise for CIPN. Then we present pre-registered studies of exercise for CIPN to get a sense of the future literature. Next, we discuss biological and psychosocial mechanisms by which exercise might exert its effects on CIPN. We conclude with implications for future research on the use of exercise for preventing and/or treating CIPN.

8.2 What Is Exercise, Physical Therapy, and Occupational Therapy?

Definitions *Physical activity* is any bodily movement produced by skeletal muscles that results in energy expenditure. Physical activity can be categorized into occupational, sports, conditioning, household, or other activities [35]. *Exercise* is a subset of physical activity that is planned, structured, and repetitive and has as a final or an intermediate objective to improve or maintain physical fitness [35]. *Physical therapy* in our chapter here is defined as a branch of passive rehabilitative measures (e.g., manual therapy, massage, traction, ultrasound, electrical stimulation) to help patients regain or improve their physical abilities. *Occupational therapy* helps people with injuries do what they want and need to do via therapeutic use of daily activities, thus enabling patients to live life to its fullest, including activities of daily living in the occupational, recreational, and household setting (American Occupational Therapy Association).

In studies or prescriptions of exercise, it is important to consider the dose, which comprises the frequency (how often an exercise session is performed; e.g., 3 sessions per week), intensity (based on percent maximum heart rate, percent maximum force production, or perceived exertion), type (aerobic, resistance, mixed), and duration (minutes per session). It is also important to consider principles of exercise training including (1) specificity, (2) progression, (3) overload, (4) initial values, (5) diminishing returns, and (6) reversibility [36]. Typically, dose is progressively increased over several weeks and may be periodized into cycles of higher doses (a larger stimulus to ultimately drive physiological adaptation) alternated with lower doses to overcome training adaptation barriers (i.e., the principle of diminishing returns).

Exercise is effective for treating a variety of clinical problems [37] including cardiovascular disease [38], depression [39], diabetic neuropathy [40, 41], and neuropathic pain [42]. There is also a strong body of evidence suggesting that exercise treats cancer- and cancer treatment-related side effects such as fatigue, cardiovascular toxicity, quality of life, physical function, and others [43–48].

Although the literature has suggested that exercise and physical therapy can help patients with CIPN since the mid-2000s [49–51], this is a relatively new area of research, when compared to exercise for cancer-related fatigue or cancer-related cardiovascular disease, which both have been tested via multiple Phase III RCTs [47, 48]. The body of research on exercise and CIPN includes several correlational studies in humans suggesting that CIPN is associated with lower levels of physical activity (e.g., [52–56]), and that exercise adherence is associated with better psychological outcomes especially in patients with worse CIPN [57]. However, these correlations do not reveal whether CIPN reduces physical activity and/or if a reduction in physical activity worsens CIPN. In the following sections, we review the two preclinical studies of exercise for CIPN, followed by 23 clinical studies of exercise, physical therapy, and occupational therapy for the treatment or prevention of CIPN, followed by 19 pre-registered studies that are planned or in progress, suggesting where this body of research is moving in the coming years. In a separate chapter, we prepared a set of suggestions on how to use exercise for the prevention or treatment of CIPN in a clinical setting [58].

8.3 Preclinical Studies on Exercise for CIPN (Table 8.1)

Our literature review identified two preclinical studies of CIPN and exercise (Table 8.1). In the first study, Park et al. randomized 32 mice to exercise or control with infusions of paclitaxel or vehicle (control) [59]. The exercise consisted of treadmill running starting 1 week before paclitaxel (3 injections across 5 days) for 60 min/session each day for 4 weeks. They found that daily treadmill exercise partially prevented paclitaxel-induced thermal hypoalgesia, reductions in nerve fiber density, and detyrosinated tubulin in peripheral nerves, which appears to be

Table 8.1 Preclinical studies testing exercise for the treatment or prevention of CIPN

Citation	Sample and study design	Chemotherapy regimen	Exercise protocol	Effects of exercise on CIPN
Park et al 2015 [59]	32 AJ mice age 6 weeks *Randomization:* • Control − paclitaxel. • Control + paclitaxel. • Exercise − paclitaxel. • Exercise + paclitaxel. *Assessments* • Pre-intervention. • Post-intervention (4 weeks).	• Paclitaxel 25 mg/kg every other day for 3 injections into the tail vein	• Treadmill exercise starting 1 week before paclitaxel for 60 min/session, 7 sessions/week for 4 weeks • 5-min warm up at 6 m/min 50 min running at 10 m/min 5 min cool down at 6 m/min	• Partially reduced axonal degeneration (nerve fiber. density), thermal hypoalgesia • Prevented detyrosinated tubulin in nerves as seen. in paclitaxel treated mice
Slivicki et al 2019 [60]	Mice C57BL/6 J age 12–14 weeks *Randomization* • Free access to running. Wheel • No access. *Experiments* • During onset of CIPN. • Prior to paclitaxel vs. vehicle • After establishment of. CIPN *Assessments* • 6 times over a 3-week period.	• Paclitaxel 4 mg/kg every other day for 4 injections intraperitoneally	• Access to running wheels that measured number of revolutions • Paclitaxel did not affect amount of voluntary running	• Voluntary running delayed and partially prevented CIPN • Voluntary running reduced established CIPN (mechanical and cold allodynia) • Voluntary running did not alter mechanical or cold responsivity in non-paclitaxel-treated mice • Voluntary running reduced paclitaxel-induced reductions in cell hippocampal proliferation (Ki67) and also increased cellular survival (BrdU)

related to neuronal dysfunction in CIPN [59]. In the second study, Slivicki et al. randomized mice to a free access running wheel or no running wheel in three different experiments: before paclitaxel, upon the onset of CIPN signs due to paclitaxel, and after establishment of CIPN due to paclitaxel [60]. In all cases, paclitaxel was delivered every other day for a total of 4 injections. They found that voluntary running was beneficial under all conditions, both in delaying and partially preventing CIPN signs, and in reducing established CIPN signs as measured by mechanical and cold allodynia tests. The wheel running did not alter mechanical or cold allodynia in non-paclitaxel treated mice. The voluntary running also had beneficial effects on the brain, in terms of mitigating paclitaxel-induced reductions in hippocampal neural proliferation and cell survival.

Taken together, these studies suggest that aerobic exercise (either mandatory or voluntary) is helpful for preventing or reducing paclitaxel-induced CIPN signs and related biomarkers. Given the number of human studies of exercise for CIPN, and the large number of rodent studies of CIPN [20], it is surprising that there are only two studies of exercise for CIPN in rodents.

Clearly, these two studies pave the way for future work in non-human animals to evaluate different chemotherapy agents (e.g., oxaliplatin, bortezomib), exercise doses, and mechanistic measures that are difficult to assess in humans, in order to gain more insight into how exercise affects CIPN.

8.4   Human Studies on Exercise for CIPN (Table 8.2)

Our literature search identified 23 interventional studies of exercise that included a measure of CIPN;[1] details of these studies are provided in Table 8.2. First, we give a broad overview of these studies, then delve into details of their methods and key findings, with a focus on the randomized studies because their results provide the strongest tests for the potential benefits or harms of exercise; finally we discuss a few noteworthy studies in detail. The interventions studied included various modalities, such as aerobic, resistance, balance, stretching, vibration therapy, yoga, and dance. Fifteen of these studies were randomized and 8 were non-randomized. The control conditions were typically usual care (10 studies) and in other cases were a different exercise condition or a physical therapy condition that lacked the experimental

---

[1]We searched found studies in two ways: (1) PubMed search for (exercise[Title/Abstract] OR exercises [Title/Abstract] OR yoga[title/abstract] OR "physical therapy" [Title/Abstract] OR "occupational therapy" [Title/Abstract] OR "training"[Title/Abstract])AND (chemotherapy[Title/Abstract] OR oxaliplatin [Title/Abstract] OR carboplatin[Title/Abstract] OR cisplatin[Title/Abstract] OR paclitaxel[Title/Abstract] OR docetaxel[Title/Abstract] OR vincristine[Title/Abstract] OR vinblastine[Title/Abstract] OR thalidomide[Title/Abstract] OR bortezomib [Title/Abstract]) AND (neuropathy[Title/Abstract] OR allodynia[Title/Abstract] OR hyperalgesia[Title/Abstract]), and (2) references to other studies within the published papers (which only revealed one more study). We excluded studies that used passive devices for therapy (e.g., electrical nerve stimulation, heat therapy, cryotherapy).

Table 8.2 Studies testing exercise and related interventions for treatment or prevention of CIPN

Citation	Sample and study design	Prevention vs. treatment and type of chemotherapy	Exercise protocol and adherence	Effects of exercise on CIPN
Andersen Hammond et al. 2020 [63]	*48 patients analyzed* • Stage I–III breast cancer. • Winnipeg, Manitoba, Canada *2-arm randomization* • Physical therapy home program • Usual care. *Assessed 5 times* • Typically, 2 weeks.before chemotherapy • Mid-taxane chemotherapy (7–18 weeks depending on chemo regimen). • Post-chemotherapy. • 3 months post-chemotherapy. • 6 months post-chemotherapy.	*CIPN prevention* *Standard adjuvant taxane chemotherapy* • 4 infusions of TC q3w • 3 infusions of AC q3w, then 3 infusions of docetaxel q3w.	*Home exercise and education program at the start of chemotherapy throughout and after chemotherapy until CIPN symptoms subsided* • Four visits to the PT to develop.home program • Nerve gliding exercises (5–10 min/session, 3 sessions/day). • Education on management of CIPN (compression gloves, heated mittens, resting wrist splints, desensitization, stereognosis exercises). • Stretching for neck and upper.limb • Follow-up phone call from PT at 6 weeks. • No other CIPN interventions. were offered *Adherence* • Not assessed.	*Primary outcome (three patient-reported outcomes)* • Numerical pain rating scale (specific to CIPN). • Disability of arm, shoulder, and hand (DASH). • Leeds assessment for neuropathic symptoms and signs (S-LANSS). → Exercise helped with all 3 outcomes (trend or significant p-values) with biggest benefit post-chemo and 3 months post-chemo *Other outcomes* • Vibration testing on index finger. • Pressure algometry on quadriceps. • Hand grip strength. • Temperature detection thresholds. • Patient-reported exercise level. → No benefit of exercise for other outcomes → Non-randomized subgroup analysis: participants reporting general exercise preserved vibration and heat pain thresholds compared to more sedentary participants (significant p-values)

(continued)

Table 8.2 (continued)

Citation	Sample and study design	Prevention vs. treatment and type of chemotherapy	Exercise protocol and adherence	Effects of exercise on CIPN
Bland et al. 2019 [61]	*27 patients analyzed* • Women with stage I–III breast cancer. • Vancouver, British, Columbia, Canada *2-arm randomization* • Immediate exercise during taxane chemo, followed by no exercise post-chemo. • Usual care during chemo, followed by exercise post-chemo. *Assessed 4 times* • Before taxane chemotherapy • Pre-cycle 4 (before the final taxane cycle). • End of chemotherapy. • 10–15 weeks post-chemotherapy.	*CIPN prevention* *Taxane chemotherapy* • Typically, 4 infusions of docetaxel or paclitaxel, q2w or q3w.	*Supervised aerobic, resistance, and balance training 3x/week for the duration of taxane chemo plus 2–3 weeks (8–12 weeks total)* • 3 sessions/week supervised • 2 sessions/week at home • Aerobic: walking, treadmill, cycling, elliptical • Resistance: 5 exercises with machines, free weights, or bands, abdominal exercises. • Balance: single leg standing. • Dose periodized with lower intensity in the days post-infusion • Aerobic and resistance exercise progressed in volume and intensity. *Adherence* • Duration 10 weeks long. • > 75% for attendance to supervised sessions • > 78% for home-based sessions • No adverse events.	*Primary outcome* • Patient-reported CIPN severity (CIPN-20). → No significant differences in CIPN-20 between groups at any time point *Other outcomes* • Vibration sensation. • Temporal summation of pain (pinprick) on lower limb. • Chemotherapy completion rate. → Pre-cycle 4: exercise reduced patient-reported moderate to severe numbness in the toes or feet (exercise: $n = 1$, 9%, control: $n = 7$, 50%, $P = 0.04$) and impaired vibration sense in the feet (exercise: $n = 2$, 18%, control: $n = 10$, 83%, $P < 0.01$) → End of chemotherapy: no differences between groups for moderate to severe numbness in the toes or feet ($P = 1.0$) or impaired vibration sense in the feet ($P = 0.71$). → More exercise participants received ≥85% relative dose intensity (exercise: $n = 12$, 100%, control: $n = 10$, 67%, $P < 0.05$)

Cammisuli et al. 2016 [128]	*7 patients analyzed* • Diagnosis of CIPN and at least one fall referred to rehabilitation after completion of chemotherapy. • Genoa, Italy. *Non-randomized intervention* • Balance training. *Assessed 2 times* • Pre-intervention. • Post-intervention (4 weeks).	*CIPN treatment* *Various chemotherapies* • Paclitaxel, cisplatin, oxaliplatin • Sample was 9 months–8 years post-chemo.	*Supervised rehabilitation* • 60 min/session, 3 sessions/week, 4 weeks duration, 3 exercises (4 min each): Rhythmic weight shifting. • Tandem walk. • Step up/over. *Adherence* • 100% • No adverse events.	*Primary outcome* • Force plate to assess postural sway velocity with firm or foam. surface, and with eyes open or eyes closed → No difference from pre- to post-intervention ($p > 0.457$) *Other outcomes* • Limits of stability test, moving center of gravity to various targets to assess balance and control. • Tandem walk. • Step up/over. • Berg Balance Scale (14 tasks testing balance). → Pre- vs. post-treatment improvement in balance as measured by static-dynamic posturography ($p = 0.004$) and Berge balance scale ($p < 0.002$).
Clark et al. 2012 [64]	*26 patients analyzed* • Fairfax, Virginia, USA. *4-arm randomization* • Yoga. • Reiki. • Meditation. • Control (education). *Assessed 2 times* • Pre-intervention. • Post-intervention (6 weeks).	*CIPN treatment* *Various chemotherapies* • At least 3 months.duration CIPN • Sample was 3 months–11 years post-chemo (mean 2.8 years).	*Supervised yoga for 6 weekly sessions* • Breathing and body scanning, breathing and postures, gentle stretching, relaxation. *Adherence* • 6 of the 7 participants attended at least 4 of the 6 sessions	*Primary outcome* • Psychological distress (Brief Symptom Inventory). *Other outcomes* • Patient-reported CIPN severity (FACT-Ntx). • Mindfulness. → FACT-NTX improved in yoga but reduced in control (effect size Cohen's d = 0.62)

(continued)

Table 8.2 (continued)

Citation	Sample and study design	Prevention vs. treatment and type of chemotherapy	Exercise protocol and adherence	Effects of exercise on CIPN
Courneya et al. 2014 [74]	*301 patients analyzed* • Women with stage I–IIIc breast cancer. • Starting adjuvant chemo. • 3 sites: Edmonton, Ottawa, and Vancouver, Canada. *3-arm randomization* • Standard dose of aerobic. exercise • High dose of aerobic. exercise • Combined aerobic and. resistance exercise *Assessed 4 times* • Baseline. • During chemo. • During chemo. • 3–4 weeks post-chemo.	*CIPN prevention* *Various chemotherapies* • Some included taxane and some did not. • Mostly FEC-D (docetaxel; 33%), TC (docetaxel; 23%), and TAC (docetaxel; 10%), and AC-T (AC followed by paclitaxel; 7%).	*Supervised exercise 3 sessions/ week for the duration of their chemotherapy (12–26 weeks)* *Standard dose aerobic* • 25–30 min/session • Cycling, treadmill, elliptical, rower, or combination • Progressed from 60–75% VO_{2peak} over 6 weeks. *High dose aerobic* • 50–60 min/session • Otherwise same as standard. *Combined aerobic and resistance* • 50–60 min/session • Same as standard plus 2 sets of 10–12 reps of 9 strengths exercises at 60–75% 1RM. • Exercises spanned the entire. body *Adherence* • 88% of sessions attended.	*Primary outcome* • Patient-reported depression (CES-D). *Other outcomes* • Patient-reported taxane/ neuropathy symptoms (FACT-taxane). → Did not report main effects of the 3 interventions on CIPN → Body mass index moderated the effects of the exercise interventions on bodily pain ($p = 0.038$), taxane/ neuropathy symptoms ($p = 0.013$), aerobic fitness ($p = 0.041$), muscular strength ($p = 0.007$), and fat mass ($p = 0.005$). → For CIPN, healthy weight patients responded better to the higher-dose exercise interventions than overweight/obese patients. → For CIPN, premenopausal, younger, and fitter patients achieved greater benefits from the higher-dose exercise interventions compared to the lower-dose exercise intervention.

Dhawan et al. 2020 [65]	45 patients analyzed • Receiving paclitaxel or carboplatin • Had CIPN. • Delhi, India. 2-arm randomization • Exercise. • Usual care. Assessed 2 times • Pre-intervention. • Post-intervention (10 weeks).	CIPN treatment Currently receiving paclitaxel and carboplatin • On average 10. months duration	Home-based training for 30 min/session, 7 sessions/week, 10 weeks duration • Face-to-face instruction. Strength and balance exercises • Ankle, hip, and leg raise, finger and wrist movements, knee extension and flexion, etc. • One legged stand, toe stand, tandem forward walking Adherence phone calls every 15 days and during medical visits Adherence • 68% of patients adhered for 10 weeks • No exercise-related adverse events.	Primary outcome • CIPN Assessment Tool. Other outcomes • Nerve conduction velocity (NCV) test. • Leeds Assessment of Neuropathic Symptoms and Signs pain scale. → Exercise reduced neuropathic pain scores ($p < 0.0001$) improved Functional QOL ($p = 0.0002$), Symptom QOL ($P = 0.0003$), Global Health Status QOL ($P = 0.004$) scores. → Effects of exercise on NCV not reported
Fernandes & Kumar 2016 [129]	25 patients analyzed • With CIPN. • Belagavi, India (assumed, not reported) Non-randomized intervention • Lower-body exercise. Assessed 2 times • Pre-intervention. • Post-intervention (3 weeks).	CIPN treatment Various chemotherapies • Agents and cancer types not reported	Closed kinematic chain exercises were administered for a total of 15 sessions over 3 weeks. • Warm up. • Toe and heel raises, shifting weight between feet while standing, unipedal toe and heel raises, unipedal balance, etc. • Total number of sets increased over the weeks of the intervention.	Primary outcome • Modified Total Neuropathy Score (mTNS), assessed in a lab/clinic. → mTNS score decreased (improved) from 13.88 at baseline to 6.5 at the end of intervention ($p < 0.01$) Other outcomes • Berg Balance Scale (BBS). → BBS increased (improved) from 26 at baseline to 42 after intervention ($p < 0.01$).

(continued)

Table 8.2 (continued)

Citation	Sample and study design	Prevention vs. treatment and type of chemotherapy	Exercise protocol and adherence	Effects of exercise on CIPN
Galantino et al. 2019 [130] and Galantino et al. 2019 [131]	*8–10 patients analyzed* • With CIPN. • Stockton, New Jersey, USA *Non-randomized intervention* • Somatic yoga and.meditation *Assessed 2 times* • Pre-intervention. • Post-intervention (8 weeks).	*CIPN treatment* *Various chemotherapies* • Not reported but from breast, colon, ovarian, pancreatic cancers. • Time since last chemo ranged <1 year ago to over 10 years ago.	*Supervised somatic yoga and meditation twice a week for 8 weeks for 1.5 h, with home program and journaling.* *Adherence* • Attended 61–81% of sessions. • No adverse events.	*Primary outcomes* • Sit and Reach (SR), Functional Reach (FR), and Timed Up and Go (TUG). → Significant improvements in flexibility (SR; $P = 0.006$); balance (FR; $P = 0.001$) and fall risk (TUG; $P = 0.004$). *Secondary outcomes* • Patient-reported Patient Neurotoxicity Questionnaire (PNQ). • Functional Assessment of Cancer Therapy-Neurotoxicity (FACT-GOG-NTX). • Brief Pain Inventory. • Fear of falling. → PNQ improved significantly ($P = 0.003$) with other measures improving non-significantly. → FACT-Ntx improved significantly (from 88.88 to 106.88, standard deviation = 20.03; $p = 0.039$) → Fear of falling improved (from 39.26 to 34.38, standard deviation = 6.081; $p = 0.058$) → Pain worsened, Brief Pain Inventory pain severity (from 3.50 to 3.75, $p = 0.041$)

			Biomarkers/mechanisms
Henke et al. 2014 [75]	29 patients analyzed • Lung cancer. • No requirement of CIPN. • Berlin, Germany. *2-arm randomization* • Special physiotherapeutic training (with resistance training). • Conventional physiotherapy (breathing techniques and manual therapy). *Assessed 2 times* • Pre-intervention. • Post-intervention (after 3 cycles of chemotherapy).	*Patients were not selected based on CIPN.* During platinum-based chemotherapy	*Supervised endurance training 8 min/session, 5 sessions/week* • Walking (6 min) and stair walking (2 min). • 55–75% of HR reserve Supervised breathing techniques 5 sessions/week • Active cycle of breathing to reduce airway obstruction Supervised strength training every other day • For training trunk stability, legs, arms, and abdominals • Resistance bands. • 50% maximum load Adherence • Patients who attended <75% of sessions were not analyzed further	*Biomarkers/mechanisms* • Salivary cortisol. • Bioesthesiometer (vibration sensation). → The salivary cortisol decreased non-significantly (0.131 to 0.118, $P = 0.12$) → Bioesthesiometer sensation showed trend improvement in feet. *Primary outcome* • Barthel Index (independence in activities of daily living). → Barthel index improved after exercise vs. control *Other outcomes* • Patient-reported quality of life (EORTC QLQ C30/LC-13), includes neuropathy severity question. • Physical function assessments: 6-Min Walk Test (6MWT), stair walking, the Modified Borg Scale (BBS), and muscle strength. → Exercise improved neuropathy severity (single question on QLQ) → Exercise improved 6MWT, stair walking, strength capacity, and in the patient's dyspnea perception during submaximal walking activities

(continued)

Table 8.2 (continued)

Citation	Sample and study design	Prevention vs. treatment and type of chemotherapy	Exercise protocol and adherence	Effects of exercise on CIPN
Kleckner et al. 2018 [66]	*355 patients analyzed* • Any cancer type. • No requirement of CIPN. • 20 sites across the USA. *2-arm randomization* • Walking and resistance exercise. • Usual care. *2 assessments* • Pre-intervention/chemo. • Post-intervention (6 weeks).	*CIPN prevention* During first 6 weeks of taxane, platinum, or vinca alkaloid chemotherapy	*Home-based walking and resistance training for 6 weeks* • Face-to-face instruction followed by at-home unsupervised exercise. • Daily walking goal using pedometer, progressed 5–20%/week from baseline. • Daily resistance band training using 16 upper- and lower-body exercises progressing in sets, reps, and resistance level. *Adherence* • Daily steps increased only in exercisers. • 77% of exercisers completed some resistance training, and 50% of resistance training sessions were completed	*Primary outcome* • Patient-reported fatigue severity (Brief Fatigue Inventory). *Other outcomes* • Patient-reported neuropathy severity from single-item ratings of numbness/tingling (0–10) and hot/coldness in hands/feet (0–10). → Exercise reduced CIPN symptoms of hot/coldness in hands/feet (-0.46 units, $p = 0.045$) and numbness and tingling (-0.42 units, $p = 0.061$) compared to the control. → Exercise reduced CIPN symptoms more for patients who were older ($p = 0.086$), male ($p = 0.028$), or had breast cancer ($p = 0.076$).
Kneis et al. 2019 [76]	*41 patients analyzed* • With CIPN. • After completion of anti-cancer therapy (any cancer). • Baden-Württemberg, Germany. *2-arm randomization* • Endurance and balance training	*CIPN treatment Various types of chemotherapy* • For treatment of breast, gastrointestinal, gynecological, lung, lymphoma, and multiple myeloma. • Range 1–167 weeks since completing treatment.	*Supervised training 30–60 min/session, 2 sessions/week for 12 weeks* *Both groups* • 30 min moderate intensity aerobic training *Only intervention group* • Extra 30 min of balance training	*Primary outcome* • Force plate measures: center of force sway path. → No effect of balance training on any outcomes (sway path in semi-tandem stance) *Other outcomes* • Vibration sense on knuckle and patella using tuning fork.

	• Control: endurance training *Assessed 2 times* • Pre-intervention. • Post-intervention (12 weeks).	• 3–8 exercises for 20–30 sec each progressively increasing difficulty by reducing support surface, visual input, etc. *Adherence* • 90% of patients attended ≥70% of sessions	• Patient-reported CIPN severity (CIPN-20). • Maximum standing vertical jump height and power (from force plate) • VO2max using cycle ergometer. → Analysis of 37/41 patients with training compliance ≥70% revealed: balance training reduced sway path during semi-tandem stance, improved duration standing on one leg on instable surface (11 s vs 0 s) and reported decreased motor symptoms (−8 vs. −2 points). → Both groups reported reduced total CIPN-20 (IG: −1, CG: −6) and sensory CIPN-20 (IG: −7, CG: −7) → Only the CG exhibited better vibration sense → Maximum power output during cardiopulmonary exercise test increased in both groups → Only the CG improved their jump height (2 cm vs. 1 cm).
Kneis et al. 2020 [132]	*8 patients with CIPN and 15 healthy subjects* • Severe CIPN symptoms. • Controls matched to participant's age, weight, and height.	*CIPN treatment various chemotherapies* • Bortezomib, carboplatin, cisplatin, paclitaxel, docetaxel, and vincristine.	*Supervised training 30–60 min/session, 2 sessions/week for 12 weeks* *Exercise balance training* • 30 min moderate intensityaerobic training • 30 min of balance training	*Primary outcome* • Postural sway: spontaneous sway and perturbed stance on motion platform with eyes open and eyes closed. → No pre-post difference in center of pressure deviation or CF

(continued)

Table 8.2 (continued)

Citation	Sample and study design	Prevention vs. treatment and type of chemotherapy	Exercise protocol and adherence	Effects of exercise on CIPN
	Non-randomized intervention • Exercise balance training *Assessed 2 times* • Pre-intervention. • Post-intervention (12 weeks). • Controls only assessed once		• 3–8 exercises for 20–30 sec each progressively increasing difficulty by reducing support surface, visual input, etc. *Adherence* • 70% of sessions attended • No adverse events.	→ Post-intervention, sway increased (bad), and PHASE decreased (good)
McCrary et al. 2019 [79]	*29 patients analyzed* • With CIPN. • >3 months post-treatment • Sydney, Australia. *Non-randomized intervention* • Exercise. *Assessed 3 times* • Baseline. • Pre-intervention (8 weeks). • Post-intervention (16 weeks).	*CIPN treatment various chemotherapies* • Bortezomib, cisplatin, paclitaxel, oxaliplatin, vincristine.	*Supervised resistance training, balance training, and cardiovascular exercises 1 h/ session, 3 sessions/week, 8 weeks duration* • Half the sessions were supervised and half were at home • Weeks 1–4: 2 sessions/week face-to-face (1 at home), and weeks 5–8: 3 sessions/week face-to-face (2 at home). *Resistance training* • Chest press, row, leg press, deadlift, shoulder press, pull down, squat, one-leg deadlift. *Balance training* • Tandem standing, tandem walk, single leg standing, etc.	*Primary outcome* • Clinical assessment of CIPN severity using the Total Neuropathy Score–clinical (TNSc; muscle weakness, pinprick sensitivity, vibration sensitivity, tendon reflex). → CIPN severity reduced post-exercise compared to pre-exercise ($p = 0.001$) *Other outcomes* • Patient-reported CIPN severity (CIPN-20). • 6 min walk test • Standing balance (postural sway) without a force plate with eyes open and closed on stable and unstable surfaces • Five-times-sit-to-stand test. • Overall disability via the CIPN Rasch Built Overall Disability.

		Cardiovascular training • Waling, incline walking, or cycling at RPE 13–15 *Adherence* • 83%, with 98% of supervised sessions attended and 67% home exercise completed	Score (CIPN-R-ODS) • Motor and Sensory nerve conduction (tibial, sural) and nerve excitability studies (median nerve). → Objective and patient-reported CIPN, dynamic balance, standing balance in eyes open conditions, mobility and quality of life were improved from pre- to post-exercise ($p < 0.05$), with no changes over the control period ($p > 0.21$). → No changes were observed in sensory or motor neurophysiologic parameters (i.e., nerve conduction; $p > 0.23$).
Moonsammy et al. 2013 [133]	*19 patients* • Ovarian cancer. • No requirement for CIPN • During or after treatment. • Toronto, Ontario, Canada *Non-randomized intervention* • Exercise plus cognitive behavioral therapy *Assessed 3 times* • Baseline. • 3 months • 6 months.	*Patients were not selected based on CIPN* Platinum-based chemotherapy	*Home-based unsupervised aerobic and resistance exercise, 30–60 min/session, 3–5 sessions/week for 6 months* • Equipment: stability ball, yoga mat, resistance bands (3 tension levels). • Individualized program from an exercise physiologist • Aerobic: brisk walking (60–70% HR max, or 4–7/10 RPE). • Resistance: 10 upper- and lower-body exercises. • Progressive increase in sets and. reps and RPE over the 6 months	*Primary outcome* • Unclear. *Other outcomes* • FACT-GOG-Ntx. → No effects reported on CIPN symptoms (FACT-GOG-Ntx) • Increase in peak VO2 from baseline to 6 months ($p = 0.015$).

(continued)

Table 8.2 (continued)

Citation	Sample and study design	Prevention vs. treatment and type of chemotherapy	Exercise protocol and adherence	Effects of exercise on CIPN
			• Aerobic and resistance training on alternate days Cognitive behavioral therapy (CBT) was also used 1 h/session, 1 session every 2 weeks for 6 months	
Schönsteiner et al. 2017 [77]	*131 patients* • With CIPN (CTC grade II–III) • Any cancer (solid and blood) • Ulm, Germany. *2-arm randomization* • Whole-body vibration (WBV) plus integrated program (massage, mobilization, physical exercises). • Integrated program. *Assessed 4 times* • Pre-intervention. • Mid-intervention (after 8 sessions). • Post-intervention (after 15 sessions). • Follow-up (1 month post-intervention).	*CIPN treatment various chemotherapy regimens* • Platinum, taxane, vinca alkaloids, bortezomib.	*Supervised therapy 15 sessions on a biweekly basis* Integrated program (all participants) • Massage and passive mobilization 30 min/side • At-home exercises for posture and transport movements. • Patients encouraged to walk as frequently and as long as possible Whole-body vibration (only experimental group) • 18 min including warm up, different positions, and cool down *Adherence* • 67% of patients in experimental arm completed it • 77% of patients in the control arm completed it.	*Primary outcome* • Chair-rising test. → Experimental arm (WBV) improved chair-rising test score at follow-up *Other outcomes* • NCI common terminology criteria (CTC) scale. • FACT-GOG-Ntx. • Physical examination (tendon reflex, tuning fork sensitivity, thermal discrimination, light vs. pinprick touch discrimination) • Quantitative sensory testing on the right foot: cold detection threshold, warm detection threshold, thermal sensory limen, cold pain threshold, heat pain threshold, pressure pain threshold, mechanical pain sensitivity, wind-up ratio, mechanical detection threshold.

			→ No significant effect of WBV on FACT-GOG-Ntx → WBV improved results of quantitative sensory testing compared to control ($p = 0.03$)
Schwenk et al. 2016 [67]	**CIPN treatment** Various chemotherapies used to treat lung, multiple myeloma, breast, colorectal, melanoma, bladder, prostrate, pancreas, ovarian, and chronic lymphoid leukemia.	*Supervised training 45 min/session, 2 sessions/week for 4 weeks* • Used a special computer setup with wearable sensors to receive computer game input based on participant movement • Ankle reaching tasks, virtual obstacle tasks, balance exercises *Adherence* • Unclear.	*Primary outcome* • Unclear. *Other outcomes* • CIPN severity (VPT score). • Neuropathy-related pain. • Fear of falling (Falls-Efficacy-Scale International). • BMI. • Balance measures using 3 wearable sensors (feet close together eyes open and closed, semi-tandem eyes open) to assess center of mass via sway in various directions. → Exercise improved nearly all balance/sway measures ($p < 0.035$, except ankle sway) → No significant effects of exercise on other balance/sway measures with eyes closed nor on gait speed (direction suggests exercise helped but effects were too small to be significant) → No significant effects of exercise on fear of falling
	22 patients • With CIPN. • Any cancer. • Tucson, Arizona, USA. *2-arm randomization* • Intervention: interactive game-based balance training including repetitive weight shifting and virtual obstacle crossing tasks. • Control: usual care. *Assessed 2 times* • Pre-intervention. • Post-intervention (4 weeks).		

(continued)

Table 8.2 (continued)

Citation	Sample and study design	Prevention vs. treatment and type of chemotherapy	Exercise protocol and adherence	Effects of exercise on CIPN
				Predictors of training response → Patients with worse baseline balance or worse baseline CIPN • (numbness in feet, pain) or worse fear of falling showed greater improvements in balance
Streckmann et al. 2014 [68]	*61 patients* • Lymphoma. • No requirement for CIPN • Freiburg, Germany. *2-arm randomization* • Sensorimotor + endurance + strength training. • Usual care. *Assessed 4 times* • Pre-chemotherapy. • 12 weeks • 24 weeks • 36 weeks.	*CIPN prevention* Chemotherapy not specified (but used to treat lymphoma)	*Supervised, 1 h/session, 2 sessions/week for 36 weeks* • Aerobic exercise: 10–30 min of cycling or walking (60–80% HR max). • Sensorimotor training: 4 posture stabilization tasks that progressed in difficulty. • Strength training: 4 resistance band exercises *Adherence* • 65% sessions attended overall • 99% of sensorimotor sessions attended	*Primary outcome* • Quality of life (EORTC QLQ-C30). *Other outcomes* • Peripheral deep sensitivity using tuning fork (toe and ankle). • Activity level via patient-reported logbook. • Balance control on static surface (force plate) eyes closed bipedal and monopedal stance for 20 sec intervals • Balance control on dynamic surface (foam pad on top of force. • Balance control following mechanical perturbation (oscillating platform) • Incremental step test for lactic acid threshold. → Exercise improved peripheral deep sensitivity

				→ Exercise increased physical activity outside the intervention → Exercise improved balance on stable surface via monopedal sway but not via other measures → Exercise improved balance on dynamic surface via monopedal sway → Exercise improved balance from mechanical perturbation via sway on bipedal and monopedal tasks → Exercise reduced maximum lactate and improved performance on incremental step test
Streckmann et al. 2019 [78]	*30 patients (+10 healthy controls)* • With CIPN for 1–5 years. • Cologne, Germany. *3-arm randomization* • Sensorimotor training. (SMT) • Whole-body vibration. (WBV) training • Oncological control group. Study also recruited 10 age- and gender-matched healthy controls *Assessed 2 times* • Pre-intervention. • Post-intervention (6 weeks).	*CIPN treatment* Various neurotoxic chemotherapies • Taxane, platinums, and vinca alkaloids	*Supervised training 2 sessions/ week for 6 weeks* Sensorimotor training • Progressively more difficult balance exercises on unstable surfaces. Whole-body vibration training • 4 progressing sets of vibration exercises 30–60 sec Adherence • 97.5% • No adverse events.	*Primary outcome* • Clinical test battery for CIPN: (1) achilles and patella tendon reflexes (hammer reflex), (2) peripheral deep sensitivity (128 Hz tuning fork), (3) light-touch perception in legs and feet, (4) sense of position change in first toe with eyes closed, and (5) lower leg strength (knee and ankle dorsiflexion). *Other outcomes* • Patient-reported CIPN severity (FACT-GOG-Ntx). • Nerve conduction velocity and amplitude (compound muscle

(continued)

Table 8.2 (continued)

Citation	Sample and study design	Prevention vs. treatment and type of chemotherapy	Exercise protocol and adherence	Effects of exercise on CIPN
				action potentials, distal motor latency, conduction velocity from tibial nerve). • Balance control on force plate (sway via monopedal and bipedal stance) • Patient-reported neuropathic pain via Pain-DETECT. → Both exercise modalities (SMT and WBV) improved sensory and associated motor symptoms vs. control. Significant differences were found for the tendon reflexes, peripheral deep sensitivity, and pain, with trend difference for subjective improvement of symptoms. → SMT group yielded superior results for tendon reflexes and a tendency regarding the subjective report of symptoms. → WBV was superior regarding pain.
Stuecher et al. 2019 [69]	*44 patients randomized* • No requirement for CIPN • Stage III–IV gastrointestinal cancer.	*CIPN prevention* Platinum-based chemotherapy	*Home-based walking intervention for 12 weeks* • Exercise counseling to develop program.	*Primary outcome* • Short physical performance battery (SBBP; balance, gait speed, chair stand)

Study	Sample/design	CIPN type	Intervention	Outcomes
	• Frankfurt, Germany. *2-arm randomization* • Home-based exercise. • Usual care (wait-list control). *3 assessments* • Pre-chemotherapy. • After two chemotherapy cycles. • After 12 weeks.		• Initially, moderate intensity walking for 20 min/session, 3 sessions/week for 2–3 weeks. • Then increase to 150 min/week at a moderate intensity. • Pedometer to measure daily steps. • Study coordinator called weekly to enhance adherence	→ Exercise attenuated worsening in SBBP scores at 12 weeks compared to baseline *Other outcomes* • Walking speed. • Postural control. • Strength of knee extensors. • Peripheral deep sensitivity (i.e., vibration sensitivity). • Body composition. → Exercise improved postural sway (T0 to T1; T0 toT2) and lean body mass (T1 to T2; T0 to T2) compared to the control group ($p < 0.05$). → No significant effects of exercise on gait speed, peripheral neuropathy, and strength ($p > 0.05$).
Vollmers et al. 2018 [70]	*36 patients* • Female breast cancer. • Receiving paclitaxel for 12 weeks. • No requirement for CIPN. • Germany. *2-arm randomization* • Exercise intervention. • Usual care (written instructions for exercise) *3 assessments* • Pre-intervention /	*CIPN prevention* *Paclitaxel*	2 sessions/week for 18 weeks (12 weeks paclitaxel +6 more after) • Physical training and sensorimotor exercises • Moderate to difficult effort. (RPE 13–15/20). • Not clear if it is supervised or home-based *Adherence* • Not reported.	*Primary outcome* • Sway areas (on force plate). • Fullerton Advanced Balance Scale score. → Exercise improved balance per sway area and Fullerton scale over multiple time points *Other outcomes* • Upper- and lower-body strength (handgrip, chair-rising test). • Patient-reported CIPN severity (CIPN-20).

(continued)

Table 8.2 (continued)

Citation	Sample and study design	Prevention vs. treatment and type of chemotherapy	Exercise protocol and adherence	Effects of exercise on CIPN
	pre-paclitaxel. • Post-paclitaxel (12 weeks). • Post-intervention (18 weeks).			→ Exercise improved handgrip strength, which decreased in controls → No significant effect of exercise on chair-rising test → No significantly effect of exercise on CIPN-20
Wonders et al. 2013 [134]	*6 patients* • Female breast cancer survivors. • With CIPN or currently receiving paclitaxel, docetaxel, or vinorelbine. • Cincinnati, Ohio, USA. *Non-randomized intervention* • Home-based exercise. *2 assessments* • Pre-intervention. • Post-intervention (10 weeks).	*No requirement of CIPN* *Paclitaxel, docetaxel, vinorelbine*	*Home-based exercise program for 10 weeks* • Pedometer to measure daily. steps • Three resistance bands. • Moderate intensity (55–65% HR max). • Progression: week 1 brisk walk 10 min/session, 2 sessions/week. By week 10, 30 min/session, 5 sessions/ week (i.e., 150 min/week). • Resistance training at least 3 sessions/week *Adherence* • 8 of 14 patients dropped out (42% remained) • Mean steps increased from 4441 ± 322 to 14,673 ± 4356 steps/day.	*Primary outcome* • Unclear. *Other outcomes* • Leeds Assessment of Neuropathic Symptoms and Signs. → After the exercise, patients reported reduced prevalence of unpleasant skin sensations (pre to post change: $n = 6$ to 3), abnormally sensitive to touch ($n = 6$ to 1), and coming on suddenly in bursts for no apparent reason ($n = 5$ to 4) → It was also determined that troublesome symptoms were significantly reduced after 10-weeks of home-based exercise ($p = 0.05$).

Study	Population	Design/Assessments	Intervention	Outcomes
Worthen-Chaudhari et al. 2019 [135]	*No requirement of CIPN Various chemotherapies* • Mostly for breast cancer (12/22)	*22 patients* • Any cancer. • No requirement for CIPN • Columbus, Ohio, USA. *Non-randomized intervention* • Partnered Tango. *3 Assessments* • Pre-intervention. • Mid-intervention (4 weeks). • Post-intervention (8 weeks).	*Supervised 1 h/session, 2 sessions/week for 8 weeks* Partnered Tango • Warm up. • Walking forward, backward, and side-to-side • More steps and embellishments involving horizontal rotation *Adherence* • Over half of participants attended over half the classes • Those who enrolled with a companion ($n = 9$) attended more sessions than those who did not ($n = 13$)	*Primary outcome* • Center of pressure from force plate. → Participants with demonstrated deficits ($n = 9$) improved in 3 center of pressure measures at midpoint (i.e., medial-lateral sway, ellipse area, medial-lateral velocity), retaining improvement in 2 CoP measures at endpoint (i.e., medial-lateral sway, ellipse area). *Other outcomes* • Unclear/none.
Zimmer et al. 2018 [71]	*No requirement of CIPN About half of patients were receiving oxaliplatin chemotherapy*	*30 patients* • Stage IV colorectal cancer receiving outpatient palliative treatment. • No requirement for CIPN • Essen/Cologne, Germany. *2-arm randomization* • Exercise intervention. • Usual care (written instructions for exercise) *3 Assessments* • Pre-intervention. • Post-intervention (8 weeks). • Follow-up (12 weeks).	*Supervised exercise training 60 min/session, 2 sessions/week for 8 weeks* • Balance training, coordination.training • Endurance training (e.g., cycling) at 60–70% HR max • Resistance training on machines for upper- and lower-body at 60–80% 1RM. • Cool down. *Adherence* • 80% completed the study • 88% sessions attended.	*Primary outcome* • Patient-reported well-being FACT-GOG-Ntx (sum of physical well-being plus functional well-being plus neurotoxicity subscales). → Exercise attenuated CIPN severity (severity was stable in exercisers, and worsened in controls) *Other outcomes* • Balance using the GG-Reha (28 tasks in 3 blocks) for static balance, dynamic balance, unstable static balance.

(continued)

Table 8.2 (continued)

Citation	Sample and study design	Prevention vs. treatment and type of chemotherapy	Exercise protocol and adherence	Effects of exercise on CIPN
				• Strength estimates of 1-rep maximum for bench press, leg press, and lat pulldown. • Endurance capacity (6-min walk test). → Exercise improved balance (GGT-Reha static balance) and strength → No significant effects of exercise on other measures of balance from GGT-Reha

component of interest (e.g., endurance training with vs. without whole-body vibration therapy). Eleven studies were designed for the treatment of existing CIPN, 7 studies for prevention of future CIPN (e.g., exercise starting before or with neurotoxic chemotherapy), and 5 studies were mixed. Sample sizes ranged from 7 to 355 (mean ± SD = 69 ± 93 patients). Nearly half of the publications (11 of 23) were published in 2019 or 2020, illustrating that this is a very rapidly growing area of research.

In terms of chemotherapy regimens, most (10) studies recruited patients receiving various neurotoxic chemotherapy regimens, 5 studies focused on platinum-based chemotherapy, 4 studies focused on taxane-based chemotherapy, 1 study focused on combined taxane/platinum, and 1 study focused on vinca alkaloids. In terms of cancer types, the majority of studies allowed any cancer type (13 studies), whereas 5 studies focused on breast cancer, 2 studies focused on gastrointestinal cancers, and single studies focused on lung cancer, ovarian cancer, and lymphoma.

Details of Exercise Regimens There were several different exercise modalities tested, including aerobic (11 studies), resistance (11 studies), balance/sensorimotor (11 studies, e.g., tandem walk, standing on one foot, standing on unstable surfaces), physical therapy (3 studies, e.g., stretching, nerve gliding, symptom management), whole-body vibration (2 studies), yoga (2 studies), and dance (1 study). Many studies used some combination of aerobic, resistance, and balance/sensorimotor exercises. Most of the interventions were supervised (15 studies), and many were home-based (7 studies) usually with an initial face-to-face instructional session, and others were combined (1 study) or not specified (1 study). The length of the interventions ranged from 4–36 weeks (mean ± SD = 11 ± 8 weeks). The length of the intervention was typically fixed but, in some studies, it matched the chemotherapy treatment and therefore could differ by patient. In terms of exercise principles [36], nearly all studies followed baseline testing of abilities, some studies used periodization (e.g., some type of progression over the weeks of the intervention), and one study purposefully reduced the exercise dose following a chemotherapy infusion (Bland et al., 2019 [61]).

CIPN Outcome Assessments The 23 studies utilized a wide range of outcomes, which we grouped into five categories: patient-reported CIPN, clinical assessments of CIPN, balance measures, physical function assessments, and chemotherapy completion. First, 17 studies assessed patient-reported CIPN severity, e.g., using the CIPN-20, FACT-GOG-Ntx, single-symptom numerical rating scales (NRS), Brief Pain Inventory, or Fear of Falling neuropathy instruments. Second, 12 studies used clinical assessments of CIPN signs such as vibration testing with a tuning fork (also called peripheral deep sensitivity), quantitative sensory testing (e.g., temperature discrimination), nerve conduction, and composite measures such as the Total Neuropathy Score (TNS; © Johns Hopkins University) modified or clinical version. Third, 10 studies included balance measures such as using a force plate (center of pressure, sway), timed standing on one leg, or the Berg Balance Scale. Fourth, 12 studies used physical functional assessments such as handgrip strength, standing vertical jump, maximum strength (1-rep max test for leg press or chest press), chair-rising test, VO_{2max}, or 6-min walk. Finally, 1 study assessed chemotherapy completion.

Studies typically included multiple measures, with the average $\pm$ SD number of measures being 4.7 $\pm$ 3.5 (range 1–15), and the number of categories of measures (of the 5 indicated above) being 2.3 $\pm$ 1.0 (range 1–4). On average, studies included 1.5 $\pm$ 0.9 patient-reported measures of CIPN (range 1–4), 3.3 $\pm$ 3.6 clinical assessments of CIPN (range 1–13), 1.8 $\pm$ 1.1 measures of balance (range 1–4), and 2.1 $\pm$ 1.4 measures of physical function (range 1–5). Because CIPN is not a simple unitary phenomenon, it is important to include multiple measures of its different signs and symptoms [21, 22], as nearly all these 23 exercise studies have done. However, very few studies appeared to account for the chemotherapy dose received, which is important to assess because differences in dose reductions across intervention vs. control groups would likely yield differences in CIPN severity that might mask intervention effects [62].

Results of ten randomized studies of exercise vs. non-exercise control (Andersen Hammond et al. 2020 [63]; Bland et al. 2019 [61]; Clark et al. 2012 [64]; Dhawan et al. 2020 [65]; Kleckner et al. 2018 [66]; Schwenk et al. 2016 [67]; Streckmann et al. 2014 [68]; Stuecher et al. 2019 [69]; Vollmers et al. 2018 [70], and Zimmer et al. 2018 [71]). These ten studies have the potential to suggest whether it is better to exercise or not to treat or prevent CIPN, with the caveat that usual care control groups do not account for non-specific intervention effects including patient expectancy of benefit and behavioral artifacts (e.g., the Hawthorne effect) [72]. In the following text, we place more emphasis on results from a study's primary outcome because those results are less likely to be biased if the primary outcome is selected before data collection [73]. In contrast, results found with non-primary outcomes have a greater risk for bias (e.g., false positive) because they might only be published because they show a benefit of exercise on CIPN. Results from non-primary outcomes are not necessarily incorrect but should be considered as more hypothesis-generating results that can inform the design of more definitive future studies [73].

We identified several randomized studies suggesting beneficial effects of exercise vs. non-exercise control on CIPN symptom severity or functional balance measure: 6 of those studies found benefits of exercise on the study's primary outcome (Andersen Hammond et al. 2020; Dhawan et al. 2020; Stuecher et al. 2019; Vollmers et al. 2018 and Zimmer et al. 2018) and 8 of those studies found benefits of exercise on a non-primary outcome (Bland et al. 2019; Clark et al. 2012; Kleckner et al. 2018; Schwenk et al. 2016; Streckmann et al. 2014 and Zimmer et al. 2018). Five studies found undetectable or no effects of exercise vs. non-exercise control: 1 study on the primary outcome (Bland et al. 2019) and 4 studies on other outcomes (Andersen Hammond et al. 2020; Schwenk et al. 2016; Stuecher et al. 2019 and Vollmers et al. 2018).

No randomized studies suggested that exercise was worse than non-exercise control. Taken together, most studies found exercise to be beneficial compared to no exercise, some studies found exercise to not be beneficial, with no studies finding exercise to be harmful. However, there is significant heterogeneity in exercise dose, CIPN outcome measures, and populations studied, and the studies were not definitive in nature (not large Phase III RCTs designed to assess CIPN).

Results of five randomized studies comparing different doses or modalities of exercise (Courneya et al. 2014 [74]; Henke et al. 2014 [75]; Kneis et al. 2019 [76]; Schönsteiner et al. 2017 [77]; Streckmann et al. 2019 [78]). These five studies have the potential to suggest which dose of exercise (frequency, intensity, type, duration) is most beneficial for CIPN. First, in a study of 301 women with breast cancer, Courneya et al. 2014, found that premenopausal, younger, and fitter patients achieved benefit on CIPN from the higher-dose aerobic exercise interventions compared to the lower-dose aerobic exercise intervention (60 min/session vs. 30 min/session, both for 3 sessions/week for at least 12 weeks during chemotherapy). These results suggest that exercise dose should be tailored to each patient's individual abilities. Second, Henke et al. (2014) found that adding resistance training to standard physiotherapy (endurance training, breathing exercises, and manual therapy) improved patient-reported neuropathy severity and physical function (6-min walk test, strength, etc.) in 29 patients with lung cancer during platinum-based chemotherapy. These results suggest that resistance training can be additionally beneficial on top of existing endurance training. Third, Kneis et al. (2019) found that adding balance training to an endurance training program did not affect measures of balance and eliminated the beneficial effects on a sign of CIPN (vibration sensation) and physical function (jump height) in 41 patients with CIPN. These results suggest that the balance training was not rigorous enough or that the endurance training already elicited beneficial effects on balance, and perhaps that the dual-modality exercise intervention is asking too much of participants. The latter suggestion is consistent with a recent meta-analysis of exercise for cancer-related fatigue suggesting that exercise alone and psychological interventions alone are each more effective than the combination of exercise plus psychological interventions [47]. Fourth, Schönsteiner et al. (2017) reported that adding whole-body vibration therapy to an integrated program (massage, mobilization, physical exercises) improved physical function (chair-rising test) and CIPN signs from quantitative sensory testing, with no significant effects on patient-reported CIPN severity in 131 patients with CIPN. Finally, Streckmann et al. (2019) compared sensorimotor (balance) training to whole-body vibration training to an oncological control group as well as to healthy age- and gender-matched controls, for reference values in a total of 40 individuals. They found that both exercise conditions improved CIPN but that sensorimotor training was better for improving tendon reflexes, peripheral deep sensitivity (i.e., vibration sensitivity), and patient-reported CIPN severity, whereas whole-body vibration training was better for improving pain.

Taken together, these results suggest that some exercise modalities are better than others for treating certain symptoms of CIPN (e.g., pain, vibration sensitivity), sometimes (but not always) adding more exercise modalities can be additionally beneficial, and that the dose of exercise that best treats CIPN may depend on the individual's baseline fitness level or other factors.

Results of eight non-randomized studies (Cammisuli et al. 2016; Fernandes et al. 2016; Galantino et al. 2019; Kneis et al. 2020; McCrary et al. 2019; Moonsammy et al. 2013; Wonders et al. 2013; Worthen-Chaudhari et al. 2019). These studies have the potential to suggest feasibility and provide qualitative feedback on novel exercise interventions or populations. Results from these types of studies are critical to help optimize interventions and provide pilot data to obtain future funding for subsequent randomized trials. However, because these studies lack randomization it is not possible to attribute any changes in CIPN to the exercise interventions themselves. These studies suggest the feasibility of seldom-used physical activity modalities for CIPN such as combined yoga/meditation, partnered Tango, and combined aerobic exercise, resistance exercise, and cognitive behavioral therapy, as well as various combinations of balance, aerobic, and strength training. The partnered Tango intervention by Worthen-Chaudhari et al. (2019) is particularly interesting because dance is a form of physical activity or exercise that strongly leverages psychosocial mechanisms—namely, dance can be incredibly fun, socially oriented, and culturally relevant, thereby increasing adherence. Indeed, Worthen-Chaudhari et al. found greater adherence by patients who attended with a companion. These types of findings are important to help broaden our understanding and optimization of the use of exercise.

Results on Predictors of the Effects of Exercise on CIPN Three studies included data suggesting factors that moderate the effects of exercise on CIPN. First, Courneya et al. (2014) found that healthy weight patients had greater reductions in CIPN from the higher-dose exercise interventions than overweight/obese patients in a study of 301 women with breast cancer during chemotherapy. Second, Kleckner et al. (2018) found that exercise reduced CIPN symptoms more for patients who were older or had breast cancer (compared to other cancer types, primarily colorectal) in 355 patients receiving chemotherapy (mostly breast cancer patients). Third, Schwenk et al. (2016) reported that patients with worse baseline balance, fear of falling, or CIPN (numbness in feet, pain) showed greater improvements in balance in 22 patients with CIPN (mixed cancer types). Typically, these types of moderating analyses require larger sample sizes and are simply exploratory analyses that are hypothesis-generating and require tests for replication in future studies. However, the Courneya and Kleckner studies both invite the same hypothesis that lower doses of exercise are effective for patients who are older, whereas younger, fitter patients require or can tolerate a higher dose of exercise to better reduce CIPN.

Highlighting Key Studies in Detail Next, we focus in on three separate studies to see results, strengths, and limitations considering published recommendations for the design of CIPN clinical trials [62] and principles of exercise interventions [36]. We hope this provides the reader with an idea of how to interpret the primary literature of exercise for CIPN with three examples: (1) a smaller non-randomized study, (2) a larger randomized study that is exploratory, and (3) a smaller randomized study comparing multiple exercise interventions.

Highlighted Study 1. McCrary et al. (2019) [79] Design. This is a non-randomized study conducted in Sydney, Australia, using supervised and home-based exercise in 29 patients with CIPN at least 3 months post-treatment with a mixed patient population including multiple neurotoxic chemotherapy types. The exercise included resistance training (8 upper- and lower-body exercises), balance (tandem walk, single leg standing, etc.), and cardiovascular training (walking or cycling) at a moderate intensity (rating of perceived exertion [RPE] 13–15 out of 20) for a total of 1 h/session, 3 sessions/week for 8 weeks (half of the sessions were supervised, half were home-based). Although this study is not randomized, it used an 8-week control period before the intervention. Patients were assessed 3 times: at baseline (0 weeks), pre-intervention (8 weeks), and post-intervention (16 weeks) using a wide array of outcomes including clinical assessments of CIPN (primary outcome: TNS-clinical version TNSc), neurophysiological measures (nerve conduction and excitability studies), patient-reported CIPN (CIPN-20, CIPN R-ODS, and SF36 QoL), functional tests (6-min walk, five times sit-to-stand), balance tests (postural sway). **Results.** Adherence was good, at 83% (98% for supervised, 67% for home-based). The exercise appeared to improve CIPN severity because it decreased from pre- to post- intervention (TNSc $p = 0.001$) with no significant change in the control period. Many of the other outcomes were improved as well from pre- to post-intervention (with no significant changes in the control period) including patient-reported CIPN severity and balance, but there were no observed changes in neurophysiological outcomes. **Strengths.** The major strengths of this study include the use of a wide range of CIPN outcome measures, the combined supervised plus home-based exercise program, and the heterogeneous sample. These are very useful design features for a smaller Phase I study to investigate which outcomes show the greatest sensitivity to change, the number and type of outcome measures patients are willing to complete, how patients adhere to and enjoy the intervention and, by recruiting a diverse sample, how to improve the next study to fit the needs of a diverse group of patients. **Limitations.** The major limitations of this study are its non-randomized nature and small sample size, but these limitations are appropriate for a study of this type (i.e., a pilot or Phase I study). Indeed, larger sample sizes or randomization might be considered an inappropriate use of resources at this phase because it limits the number of patients who receive the experimental intervention, thus limiting the possibility for patient feedback on that intervention. The use of a non-exercise control period before the intervention is a good way to allow all patients to receive the intervention while obtaining an estimate of the effects of exercise vs. no exercise. **Overall impression.** This study suggests that 8 weeks of combined supervised plus at-home resistance, balance, and cardiovascular exercise is beneficial for the treatment of CIPN assessed in multiple ways and sets the stage for a larger follow-up Phase II randomized study, which is currently ongoing (Goldstein & Park, ACTRN12618001422213; Table 8.3).

Highlighted Study 2. Kleckner et al. (2018) [66] Design. This is a 2-arm randomized study conducted across 20 sites in the United States using home-based exercise compared to usual care in 355 patients starting neurotoxic chemotherapy

Table 8.3 Unpublished studies ongoing or proposed to test the effects of exercise or related interventions for treatment or prevention of CIPN

Citation	Status	Sample and study design	Prevention vs. treatment and type of chemotherapy	Exercise protocol	Primary outcome
Galantino NCT03786055	Completed	18 patients *Non-randomized* • Somatic yoga and meditation Pomona, New Jersey, USA	Treatment of CIPN Various neurotoxic chemotherapy	Participants will engage in 16 sessions of somatic yoga and meditation with appropriate props as needed over 8 weeks and continue with a home practice. Application of somatic yoga and meditation throughout activities of daily living is reinforced. All sessions are facilitated by trained yoga therapists.	Physical function (sit and reach, forward reach, timed up and go) Secondary includes patient-reported pain, CIPN, etc.
Goldstein & Park ACTRN12618001422213	Recruiting	96 patients *Randomized* • Clinic exercise. program • Usual care then home exercise program Sydney, Australia	Treatment of CIPN Various chemotherapy regimens	Group 1: 8-week clinic-based exercise program with 3 sessions/week individualized cardiovascular, resistance, balance, and stretching exercises. Group 2: 8-weeks of usual care then an at-home exercise intervention completed 3 sessions/week for 8 weeks	Primary outcome: Total Neuropathy Score-clinical version Secondary outcomes: 6-min walk test, timed up and go test, balance and gait analyses, dynamometer, grooved pegboard, CIPN-20, CIPN-R-ODS, SF-36, IPAQ, Falls questionnaire, corneal confocal microscopy

Kanzawa-Lee & Smith NCT03515356	Completed	54 patients • Gastrointestinal cancer. *Randomized* • Exercise. • Education control Ann Arbor, Michigan, USA	Prevention of CIPN Oxaliplatin chemotherapy	Physical activity education pamphlet and 8-weeks of motivational enhancement therapy- and a home-based aerobic walking intervention. Motivational interviewing will be delivered with concurrent feedback and motivational techniques in 30–45-min sessions at intervention orientation (T1), 2 weeks (T3), and 4 weeks (T4).	Patient-reported CIPN severity (CIPN-20) at 8 weeks
Kleckner NCT03021174	Completed	19 patients *Randomized* • Exercise. • Nutrition education control Rochester, New York, USA	Prevention of CIPN During neurotoxic chemotherapy (various agents)	Face-to-face instruction and a prescription for an at-home progressive walking and resistance exercise program for 12 weeks using 16 upper- and lower-body exercises with resistance bands and using a wrist-worn activity tracker	Feasibility (percent patients eligible, consented, completed intervention and assessments) Secondary outcomes include CIPN-20, finger and toe tactile sensitivity, blood measures of inflammation, and brain MRI scans
Kleckner NCT03858153	Recruiting	80 patients • Women with breast cancer *Randomized* • Exercise.	Prevention of CIPN During taxane chemotherapy	Face-to-face instruction and a prescription for an at-home progressive walking and resistance exercise program for	Patient-reported CIPN severity (CIPN-20) at 6 weeks Secondary outcomes include finger and toe

(continued)

Table 8.3 (continued)

Citation	Status	Sample and study design	Prevention vs. treatment and type of chemotherapy	Exercise protocol	Primary outcome
		• Nutrition education control Rochester, New York, USA		12 weeks using 16 upper- and lower-body exercises with resistance bands and using a wrist-worn activity tracker	tactile sensitivity, blood measures of inflammation, and brain MRI scans
Kleckner NCORP study; NCT04888988	Recruiting	120 patients with CIPN within 9 months of chemotherapy *Randomized* • Exercise. • Usual care. Multiple community oncology clinics across the United States in the NCI NCORP Network	Treatment of CIPN During or after neurotoxic chemotherapy	Face-to-face instruction and a prescription for an at-home progressive walking and resistance exercise program for 6 weeks using 16 upper- and lower-body exercises with resistance bands and using a wrist-worn activity tracker	Patient-reported CIPN severity (CIPN-20) at 6 weeks Secondary outcomes include finger and toe tactile sensitivity, blood measures of inflammation, and perhaps brain MRI scans
Knoerl NCT03824860	Recruiting	50 patients *Randomized* • Yoga. • Usual care Boston, Massachusetts, USA	Treatment of CIPN Various neurotoxic chemotherapy Regimens • At least 3 months after completion.	Participants will attend at least one group class per week (in-person or Zoom) and practice one self-guided yoga video class at home on their own per week, over eight weeksParticipants may choose to attend "Flow Yoga" and/or "Chair Flow Yoga" classes.	Feasibility (accrual, adherence, outcome measures)

Koltyn NCT04075097	Completed	6 patients • Breast cancer. *Randomized, crossover study* • Walking. • Iyengar yoga Madison, Wisconsin, USA	Treatment of CIPN using acute exercise or yoga	These classes will be videotaped and made available to participants electronically. Classes consist of guided breathing exercises, upper and lower extremity stretching structured postures and movements to improve balance and strength	Change in Neuropathic Pain Sensations as determined by visual analog scale. Time Frame: This will be measured 2 times during each experimental session-once before the assigned exercise task (i.e., baseline) and once upon task completion (i.e., approx. 50 mins later), approximately week 1 and week 2 on study. Other outcomes include change in cold pain threshold. Time Frame: This will be measured 2 times during each experimental session—

Outcomes are measured before and after the participants complete a single session of either 44 min of moderate aerobic exercise (i.e., walking on a treadmill) or 44 min of yoga

(continued)

Table 8.3 (continued)

Citation	Status	Sample and study design	Prevention vs. treatment and type of chemotherapy	Exercise protocol	Primary outcome
					once before the assigned exercise task (i.e., baseline) and once upon task completion (i.e., approx. 50 mins later), approximately week 1 and week 2 on study.
Nagaraj & D'Errico NCT03272919	Recruiting	40 patients • Breast or gynecologic cancer *Randomized* • Intraneural facilitation (INF) • Exercise Loma Linda, California, USA	Prevention of CIPN Taxane or platinum chemotherapy	Intraneural facilitation utilizes three holds or positions, which widens tiny openings in arteries surrounding nerves and improves blood flow to targeted nerves. The improved blood flow in these nerves stimulates healing and reduces or even stops nerve pain. Standardized program of muscle stretching and strengthening exercise under the supervision of treating physical therapist.	Severity of patient-reported CIPN via the PQAS Other outcomes include distress, MSNI physical screening of CIPN, and CTCAE graded CIPN

Najafi NCT02773329	Completed	35 patients *Randomized* • Intervention. with game-based exercise • Intervention without game-based exercise Houston, Texas, USA	Treatment of CIPN During neurotoxic chemotherapy (various agents)	Game-based exercise: This intervention includes interactive game-based balance training including repetitive weight shifting and virtual obstacle crossing tasks. Wearable sensors will provide real-time visual/auditory feedback from foot and ankle position and allowed perception of motor-errors during each motor-action. Two sessions/week for 6 weeks. Intervention without game-based exercise Subjects are asked to perform non-technology-based foot and ankle exercises, which include body weight shifting and obstacle crossing tasks. Two sessions/week for 6 weeks.	Gait speed and balance (body sway) change at 6 weeks
Najafi NCT02165163	Completed	20 patients • Any cancer. *Randomized* • Exercise.	Treatment of CIPN Various chemotherapy regimens	Balance training will be conducted individually two times per week for 4 weeks. Each training session will include	Postural Balance during quiet standing for 30 seconds will be assessed using validated wearable sensor

(continued)

Table 8.3 (continued)

Citation	Status	Sample and study design	Prevention vs. treatment and type of chemotherapy	Exercise protocol	Primary outcome
		• Usual care Tucson, Arizona, USA		virtual reality tasks such as "ankle reaching" and "obstacle crossing" using a virtual obstacle shown on a computer screen. Each session will last 30–45 min.	technologies (BalanSens™)
Ozdemir NCT03798379	Completed	60 participants Non-randomized Hitit University, Turkey	Prevention of CIPN	Strengthening, balance, and aerobic exercises were explained to the patients examined before chemotherapy sessions and demonstrated by applying on the patient.	Balance at 3 months (center of gravity, postural control along with static and dynamic stability)
Ryan NCT02991677	Recruiting	60 patients • Any cancer. *Randomized* • Aerobic exercise. • Resistance. Exercise • Usual care Baltimore, Maryland	Treatment of CIPN Prior oxaliplatin, docetaxel, or paclitaxel 6–12 months prior	Aerobic exercise intervention: exercise physiologist supervised walking or running on the treadmill 3 times weekly for 12 weeks Resistance training: exercise physiologist supervised upper and lower extremity resistive training 3 times weekly for 12 weeks	Thermal, mechanical, and vibration sensation by quantitative sensory testing

Streckmann et al. 2021	Planned	128 patients • Children and adolescents (6–18 years) Randomized • Sensorimotor training • Usual care. *Multicenter trial* Basel, Switzerland; Bern, Switzerland; Aarau, Switzerland; St. Gallen, Switzerland; Heidelberg, Germany; Freiburg, Germany	Prevention of CIPN (sensory and motor dysfunctions) scheduled to receive chemotherapy containing either a platinum derivate or vinca alkaloid	Patients in the intervention group will perform a standardized, age-adjusted, playful sensorimotor training (SMT) program twice a week for the duration of their medical therapy, in addition to usual care, while the control group receives treatment as usual.	Ped-mTNS score (assesses the change in the severity of CIPN symptoms). Data will be assessed at 3–4 time points: Prior to chemotherapy (baseline T0), after 12 weeks (T1), after completion of therapy for children that are treated >3 months (Tp), and after 12 months follow-up (T3). Additionally, status of CIPN reported symptoms will be monitored every 6 weeks.
Streckmann et al. 2019 [136] NCT03032718	Completed	44 patients *Randomized* • Whole-body vibration therapy. • Usual care Basel, Switzerland and Cologne, Germany	Treatment of CIPN Various neurotoxic chemotherapy	Patients in the intervention group will receive a defined exercise program twice a week for twelve weeks. The vibration exercises will take place on a side-alternating vibration platform (GalileoTM, Pforzheim, Germany ®) according to the previously determined optimal (highest neuromuscular response)	Patient-reported CIPN (FACT-Ntx)

(continued)

Table 8.3 (continued)

Citation	Status	Sample and study design	Prevention vs. treatment and type of chemotherapy	Exercise protocol	Primary outcome
				setting for each individual. Each session will last 15–30 min, leaving sufficient time for regeneration. Training will consist of four vibration exercises, chosen from a standardized pool of exercises with increasing difficulty to allow for individual, optimal progression.	
Streckmann et al. 2018 [137] DRKS00006088	Completed	158 patients *Randomized* • Sensorimotor training • Whole-body vibration training. • Usual care Cologne, Germany and Eschweiler, Germany	Prevention of CIPN Oxaliplatin or vinca-alkaloid chemotherapy	Sensorimotor training consists of progressively more difficult balance exercises on progressively instable surfaces. Each patient performs 4 exercises per session following a standardized protocol. Each exercise is performed three times for 20 s, allowing a 40 s rest between each set and a 3-min rest between each exercise, to avoid	Primary endpoint is the time to incidence of neurologically confirmed CIPN. Secondary endpoints are pain, maintenance of the functionality of sensory as well as motor nerve fibers as well as the level of physical activity

(continued)

neuronal fatigue. Patients are asked to stand barefoot or in socks, their foot in a previously acquired "short-foot-position," knees slightly flexed (30°), and to maintain balance. Vibration training takes place on a vibration platform (Galileo™, Pforzheim, Germany)®. Each training session consists of four sets of 30 s to 1 min vibration. The frequency of the vibrating platform ranges between 18 and 35 Hz with a 2 mm amplitude. Between sets, the patients rest for at least 1 min to avoid fatigue. Patients are asked to stand on the platform barefoot and on their forefeet or if they are too instable, an 80/20% distribution of weight on the forefeet rather than the heels.

Table 8.3 (continued)

Citation	Status	Sample and study design	Prevention vs. treatment and type of chemotherapy	Exercise protocol	Primary outcome
Visovsky NCT00869804	Terminated due to low accrual	19 patients *Randomized* • Exercise. • Education control Omaha, Nebraska, USA	Prevention of CIPN Taxane chemotherapy	The intervention will consist of a tailored home-based program of both aerobic (walking, using pedometer) and strength training exercises for upper and lower extremities using resistance power bands.	Effect size for reduction of neuropathic symptoms
Winters-Stone NCT04170075	Recruiting	30 patients *Randomized* • Whole-body vibration (WBV). therapy • Usual care Portland, Oregon, USA	Treatment of CIPN Various chemotherapy regimens	Twice daily 10-min WBV training sessions, 7 days a week. Each WBV session will consist of a series of timed stands on the vibration platform (Marodyne LiV). During a timed stand, participants will perform slow controlled weight shifting exercises and gentle squats. The vibration frequency will be set at 30 Hz and the amplitude set at 50–200 microns, for a total body acceleration of 0.4 g ± 20%.	Accrual, adherence, compliance, adverse events

| Wiskemann NCT02871284 | Terminated due to low accrual | 243 (termination at 159 patients) *Randomized* • Sensorimotor training. • Resistance training. • Usual care Heidelberg, Germany | Prevention of CIPN All neurotoxic treatments eligible | Sensorimotor training: 3 sessions/week for 35 min/session. During an introductory one-to-one training, the patients received a catalog of exercises, including 45 illustrated exercise cards, and necessary training materials (e.g., Airex balance pad). The patients exercise either at home or in an open supervised training session. Each exercise carried out 3 × 30 sec with at least 30 sec pause between sets. Resistance training: machine-based exercises 2 sessions/week for 45 min/session, and a 15 min home-based training once a week. After two familiarization sessions, a one-repetition-maximum strength test (1RM) was conducted at each resistance machine. Its results were used to define initial training weights based on current guidelines (70–80% 1RM). | Total Neuropathy Score (TNS) reduced |

(mostly taxane chemotherapy for breast cancer). This study is an exploratory secondary analysis in that the parent trial was designed to study fatigue, not CIPN. The exercise intervention involved a face-to-face meeting to train the participant in how to conduct the exercise intervention, including assessment of baseline physical activity (daily steps), a plan to complete low-moderate intensity walking each day, and a plan to increase daily steps by 5–20% per week (patient's choice). There were also 16 upper- and lower-body resistance band exercises, a plan to complete these exercises each day at a moderate intensity, and the patient was instructed to increase reps, sets, and resistance levels over the weeks of the intervention. Patients were assessed two times: pre-intervention and post-intervention (6 weeks) using two patient-reported outcomes of CIPN severity: 0–10 ratings of numbness/tingling and hot/coldness in hands/feet. **Results.** Adherence to the intervention was moderate, with 77% of patients performing some resistance training, which was done on average every other day (not every day as instructed). The increase in daily steps by exercisers was moderate, increasing by 649 steps/day vs. controls who decreased by 129 steps/day on average. Exercise reduced CIPN symptoms of hot/coldness in hands/feet (-0.46 units, $p = 0.045$) and numbness and tingling (-0.42 units, $p = 0.061$) compared to the control. These were small effect sizes of approximately 0.2 [80]. In addition, exercise reduced CIPN symptoms more for patients who were older ($p = 0.086$), male ($p = 0.028$), or had breast cancer compared to other cancers ($p = 0.076$). **Strengths.** The greatest strengths of this study are its large sample size (355 patients) and its multi-site nature, allowing generalizability to geographically distinct locations in the United States. Due to the large size, the study was also able to explore individual differences that might moderate the effectiveness of exercise on CIPN. **Limitations.** The biggest limitations of this study are that it is an exploratory secondary analysis (the original study was designed to assess fatigue), the limited rigor in assessment of CIPN (only two single-item ratings of CIPN symptoms), and the mild dose of walking exercise delivered (although resistance exercise dose was good, at 3 sessions/week). **Overall impression.** This study suggests that 6 weeks of home-based walking and resistance exercise during neurotoxic chemotherapy partially attenuates the severity of CIPN. Because this is an exploratory secondary analysis (i.e., the study was not designed to assess CIPN), it sets the stage for a follow-up Phase I or II randomized study using the same intervention, which is ongoing (Kleckner entries in Table 8.3).

Highlight Study 3. Streckmann et al. 2019 [78] Design. In this 4-armed randomized, controlled, assessor-blinded trial ($N = 40$), Streckmann et al. compared sensorimotor training ($N = 10$) and whole-body vibration training ($N = 10$) to an oncological control group ($N = 10$) as well as a healthy age- and gender-matched control group for reference values ($N = 10$). The primary study aim was to analyze the potential of neuromuscular stimulating exercise interventions for the reduction of CIPN signs and symptoms, including peripheral deep sensitivity, Achilles tendon reflex (ASR), patellar tendon reflex (PSR), light-touch perception, sense of position, and lower leg strength. The secondary endpoints were nerve conduction velocity and amplitude, balance control, quality of life, and CIPN-related pain. The intervention groups exercised twice a week for 6 weeks. **Results.** Patients exercising improved sensory and associated motor symptoms. Significant intergroup differences were found for the tendon reflexes (ASR $p = 0.017$ and PSR $p = 0.020$), peripheral deep sensitivity ($p = 0.010$), and pain ($p = 0.043$). Furthermore, tendencies were found regarding the subjective improvement of symptoms ($p = 0.075$) and two subscales of the EORTC-QLQ-C30 questionnaire: pain ($p = 0.054$) and dyspnea ($p = 0.054$). Interestingly, the results were symptom-specific: the sensorimotor training group was superior regarding the tendon reflexes with a tendency towards improvements in the subjective report of symptoms, while whole-body vibration was superior regarding the reduction of pain. **Strengths.** The major strengths of this study are that it compares multiple exercise modalities, suggesting that specific exercises can target specific CIPN signs and symptoms, and that it used a wide array of CIPN outcome measures. Next, the study achieved high exercise compliance (97.5%) with no adverse events. Indeed, the exercises are feasible, with low intensity though high impact, ideal for oncological patients in all phases of therapy, and the sensorimotor exercises can furthermore be integrated into daily living at home with little effort and minimal cost. **Limitations.** Due to the small sample size, results should be considered exploratory. Due to the heterogenous sample (taxane, platinum, and vinca-alkaloid chemotherapies), study design was challenging as it had to be feasible for patients exhibiting very different performance levels. **Overall impression.** The results suggest that specific exercises (sensorimotor and whole-body vibration) may reduce CIPN-related symptoms and are likely feasible and safe. It sets the stage for larger and more definitive follow-up studies, which are ongoing (Streckmann entries in Table 8.3).

Summary of all Existing Studies The literature on exercise for CIPN contains 23 studies, including 15 randomized studies (10 comparing exercise vs. non-exercise control) that collectively suggest that exercise is beneficial for the treatment and prevention of CIPN with little to no side effects. However, these studies tend to be either rigorous yet small or large yet simple and exploratory. Future work needs to build up with larger rigorous studies, and ultimately move towards more definitive Phase III studies, as there are currently none published. That said, exercise is among the most promising options for treatment and prevention of CIPN.

8.4.1 Ongoing and Forthcoming Pre-Registered Studies of Exercise and CIPN (Table 8.3)

We searched Clinicaltrials.gov, the European Union (EU) Clinical Trials Register, the Australian New Zealand Clinical Trials Registry, and published protocols[2] plus word-of-mouth to get a sense of forthcoming studies. Although this is not an

---

[2]We searched for: exercise OR "physical therapy" OR "occupational therapy" OR "training" | chemotherapy neuropathy.

exhaustive search of pre-registered studies, Clinicaltrials.gov is one of if not the most popular pre-registration databases for clinical trials. Overall, we found 19 studies registered (Table 8.3), with 9 completed, 6 recruiting, 2 terminated early due to low accrual, 1 in regulatory review, and 1 planned. There are 17 randomized studies and 2 non-randomized, with 10 studies for treatment and 9 studies for prevention of CIPN. Sample sizes range from 6 to 159 with mean $\pm$ SD = 63 $\pm$ 48. These are generally smaller- to medium-sized studies (likely Phase I and Phase II studies) with 13 of 19 having a sample size $\leq$60, whereas 5 studies have at least 96 patients. It is promising to see so many pre-registered studies, as this adds to the rigor of this field of research and helps avoid redundancy for investigators planning new studies. The field is clearly moving forward, with larger studies and most being randomized studies testing existing interventions, helping to delve deeper into that line of work. There are no Phase III RCTs currently registered. Phase III RCTs will likely be planned in the next few years after completion of additional larger Phase II RCTs.

8.5 Mechanisms of How Exercise May Prevent or Treat CIPN

In addition to identifying whether and what dose of exercise might treat or prevent CIPN, we believe it is critical to understand *how* exercise affects CIPN. Mechanistic knowledge can help optimize the use of exercise to best treat CIPN in several ways, such as (1) optimizing exercise dose given a patient's individual characteristics (e.g., particular CIPN symptoms, fitness, preferences), (2) development of biomarkers to diagnose CIPN at earlier timepoints and track the patient's response to exercise, and (3) exploiting psychosocial mechanisms to reduce CIPN and improve exercise adherence. Although a detailed discussion of neurophysiological and psychosocial mechanisms is beyond the scope of this chapter, we briefly review key potential mechanisms based on studies of exercise for CIPN, for other types of neuropathy, and for healthy individuals.

Neurophysiological Mechanisms Through Which Exercise Might Affect CIPN (Fig. 8.1) Exercise produces benefits across multiple types of nerve injury [42] and through a range of mechanisms at different levels of analysis including molecular, subcellular, cellular, and neural circuits, and whole nervous system. Therefore, there may not be one predominant mechanism underlying its efficacy and these distinct effects might work synergistically to improve CIPN symptoms. First, exercise protects against axonal degeneration as shown in several studies in mice with CIPN [59] or other nerve injuries [81–84]. In humans, including patient with diabetes, studies have shown exercise-induced improvements in neuronal health including intraepidermal nerve fiber density [85] or other more subtle measures [86, 87]. Second, exercise increases expression of neurotrophic factors such as GDNF, BDNF, and IGF-1 [88]. However, upregulation of neurotrophic factors is not solely associated with nerve regeneration, and may also be associated with neuropathic pain and its maintenance [89]. Third, exercise has potent anti-inflammatory effects on the body [90] that have been implicated in the treatment of CIPN [91, 92]. Indeed, contracting muscles release pro-inflammatory IL-6 [93], which causes an increase in anti-inflammatory IL-10 and IL-1RA [94], and chronic exercise has been shown to reduce markers of inflammation in patients receiving chemotherapy [95] as well as animal models of nerve injury, involving the BMP-7 transcription pathway [96], increased IL-10, reduced IL-6 [97], and reduced TNFα [83, 97]. Fourth, exercise has beneficial effects on mitochondria—such as BDNF-induced mitochondrial biogenesis and increasing the nervous system's anti-oxidant capacity [98]—and mitochondrial dysfunction and oxidation have been implicated in the pathophysiology of CIPN [99–101] (Fig. 8.1).

Many of these effects of exercise occur not only in the periphery but also in the brain, as exercise can improve mitochondrial function in the brain [98], increase neurogenesis in the brain [60], enhance the brain's descending inhibition of pain [102], and affect regulation of neurotransmitters such as serotonin and norepinephrine/noradrenaline [103], which are pharmacological targets for treating neuropathic pain [104, 105]. At a higher level of analysis, exercise also improves interoception, which is the process by which the nervous system processes and represents the physiological condition of the body [106]. Interoception is supported by large-scale brain networks involving the thalamus, insula, somatosensory cortex, anterior cingulate cortex, amygdala, hippocampus, and periaqueductal gray, among other regions [107]. Interoception is important for CIPN because it is a core process in how mental states—including symptoms that patients report—emerge from bodily sensations [108]; when neurotoxic chemotherapy alters peripheral input to the brain, interoception can be compromised due to unexpected and noisy peripheral signals that inhibit the brain's ability to predict and therefore regulate or control its current neurophysiological and neuromuscular states. Exercise might improve interoception because it recruits interoceptive circuitry per increased functional connectivity between the insula and the amygdala [109], functional connectivity between the insula and the thalamus [110], and blood flow in the insula [111].

Building on the idea of how exercise treats CIPN by way of improving interoception, the total effects of exercise on the entire neuraxis can help explain how patients experience their own bodies and navigate their environment through routine yet complex motor tasks such as walking. Indeed, the function (or dysfunction) of the entire neuraxis will result in some degree of postural stability (or instability), especially with limited visual input or on unpredictable or unstable surfaces where individuals need to rely more on internal bodily cues (i.e., interoception and proprioception). Additionally, one CIPN exercise study found that exercise training helps patients regain balance by emphasizing proprioceptive information rather than vestibular information [74]. Through repeated exercise stimulation, the entire nervous system undergoes changes to better learn how to process peripheral input and move the body in a coordinated, predictable, and desired way [112]; we hypothesize that exercise can facilitate this type of adaptive learning despite noisy peripheral inputs resulting from neurotoxicity.

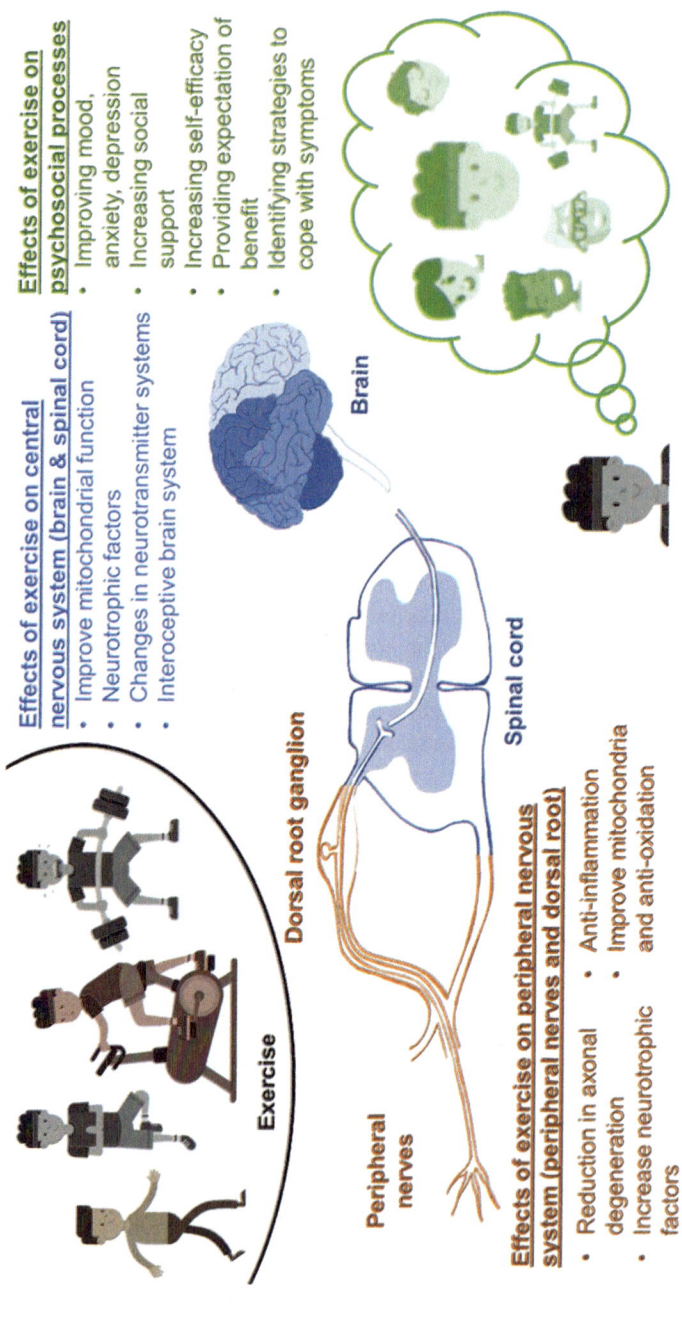

Fig. 8.1 Exercise might treat or prevent CIPN via effects on the peripheral nervous system, the central nervous system, and psychosocial processes

Psychosocial Mechanisms Through Which Exercise Might Affect CIPN Although there is little research on psychosocial aspects of exercise and CIPN, we hypothesize that exercise may exert its effects on CIPN through a wide array of psychosocial processes [113] including but not limited to: (1) improving mood, anxiety, and depression [113–115], as they relate to symptoms of CIPN such as pain [116] and their improvement can alleviate pain [117]; (2) increasing social support if exercise is done with a partner or trainer [118], as social support can reduce depression and inflammation [119], thereby reducing CIPN symptom severity; (3) increasing self-efficacy (i.e., the belief that one can accomplish a specific goal; [120, 121]), thereby helping patients experience less stress in response to challenging situations and discomfort; (4) providing an expectation of benefit [122], including a placebo response, which is a valid and potentially powerful psychosocial-level mechanism for treating symptoms (while also being a key factor that should be accounted for in research) [122]; and (5) identifying strategies to cope with existing symptoms (e.g., finding more stable and comfortable shoes).

Summary of Mechanisms Altogether, these neurophysiological and psychosocial effects of exercise have bidirectional causal relationships that explain how exercise can influence CIPN signs and symptoms. It is likely that not just one or two mechanisms are important, but rather that to understand CIPN we must understand the many simultaneous and interacting mechanisms at different levels of analysis: from molecular through the whole neuraxis to psychosocial and perhaps the healthcare system.

8.6 Conclusions and Future Work

Taken together, there is a very rapidly growing preliminary body of research suggesting that exercise can help to both treat and prevent CIPN. However, these studies are Phase I or Phase II (small to moderate sample size, not definitive in nature; [123]), with much heterogeneity across studies in terms of exercise dose, CIPN outcome measures, and patient populations. Therefore, exercise is not part of current evidenced-based guideline recommendations for CIPN [25].

However, exercise and physical therapy may help patients with CIPN if they are referred to a qualified exercise professional. We envision several avenues for future research, including more definitive answers to *whether* exercise is beneficial, *what dose* of exercise is needed, and *how* it exerts its effects mechanistically. To rigorously advance the body of research studying the effects of exercise on CIPN, we believe these questions should be investigated using preclinical models, tightly controlled clinical trials (e.g., at academic medical centers), and pragmatic real-world trials (e.g., at a variety of community sites).

First, we need more definitive knowledge on *whether* exercise can treat CIPN by way of larger and more definitive studies, including the first Phase III study of exercise for the treatment of CIPN and for the prevention of CIPN, and eventually Phase IV and V studies [123]. We also need additional Phase II and III studies with different populations spanning different neurotoxic chemotherapy agents (taxanes, platinums, vinca alkaloids, etc.) as they have distinct pathophysiology and might respond differently to exercise. Moreover, to best compare results across future studies and support future meta-analyses, it would help if researchers agreed to a standard minimal set of CIPN outcome measures—which patient-reported outcomes, which clinical reported outcomes, and how to conduct those measures [62].

Second, and simultaneously, we need to understand *what dose* of exercise is needed in terms of frequency, intensity, type, duration, and timing. Our review reveals a wide range of exercise modalities including aerobic (walking, cycling), resistance (bands, machines, free weights), sensorimotor, balance, stretching, yoga, dance, etc., and those modalities have distinct effects on the signs and symptoms of CIPN likely due to distinct mechanisms of action. As this research direction continues, we need to systematically consider these factors to identify how to best use exercise to help treat or prevent CIPN. This will require Phase II and III RCTs that randomize patients to one of multiple exercise doses.

Third, we need to understand *how* exercise exerts its effects from a biopsychosocial perspective in terms of neurotrophic growth factors, inflammation, oxidation, effects on the peripheral vs. central nervous system, and both exploiting and accounting for psychosocial mechanisms such as improving mood, social support, self-efficacy, and expectation of benefit. This knowledge, combined with detailed mechanistic insight of an individual patient's CIPN phenotype, can help match patients to the right type and dose of exercise to best treat or prevent CIPN and to maximize adherence to exercise. This is analogous to proposals for mechanism-based classification of a patient's pain for prescribing physical therapy [124–126] or mechanistic-based classification of a patient's cancer-related cardiovascular toxicity [48].

Fourth, it will be essential to translate this knowledge into pragmatic and real-world studies of exercise for the treatment or prevention of CIPN (i.e., Phase IV and V studies; [123]). Indeed, the majority of exercise CIPN studies are conducted at academic medical centers even though the majority of patients with cancer are treated in community oncology clinics [127]. When translating exercise into broader populations, it will be important to help make exercise sustainable (to maintain long-term adherence), accessible (perhaps combinations of supervised and unsupervised training using videos, an app, or a website), easy to disseminate (for clinicians making referrals), cost-effective, and appealing for patients of various ages, cultures, socioeconomic levels, physical abilities, etc. Indeed, exercise must be adapted and validated in more heterogeneous populations and less-well-controlled populations. If exercise is proven useful for CIPN, its role may be simultaneous with duloxetine and other potential treatments for CIPN, and in patients who have complex medical histories including comorbidities and health behaviors that contribute to neuropathy such as diabetes, vascular disease, alcohol consumption, smoking, etc. Therefore, future work should also explore multimodal therapies for CIPN, such as exercise plus duloxetine, exercise plus electrical stimulation, or exercise plus nutrition interventions to hasten adaptations to exercise. In conclusion, research on exercise for the treatment and prevention of CIPN is a rapidly growing area of work, with an immense potential benefit for patients. We are optimistic for the trajectory of this work including seeking definitive answers to whether exercise is beneficial, what dose of exercise is needed, how it exerts its effects mechanistically, and how to best disseminate exercise to patients in the real world.

Acknowledgements The authors acknowledge the following funding mechanisms to support this work, including the National Cancer Institute K07CA221931 to IRK, a National Health and Medical Research Council of Australia Career Development Fellowship (#1148595) to SBP. The authors also thank the Transdisciplinary Research in Energetics and Cancer (TREC) Training Workshop funded by the National Cancer Institute (R25CA203650) and the Multinational Association for Supportive Care in Cancer (MASCC) for facilitating collaboration among the authors. The authors also thank Dr. Amber Kleckner for editorial input and Ms. Kaitlin Chung for help with the literature review.

References

1. Chan A, Hertz DL, Morales M, Adams EJ, Gordon S, Tan CJ et al (2019) Biological predictors of chemotherapy-induced peripheral neuropathy (CIPN): MASCC neurological complications working group overview. Support Care Cancer 27(10):3729–3737. https://pubmed.ncbi.nlm.nih.gov/31363906/
2. Staff NP, Grisold A, Grisold W, Windebank AJ (2017) Chemotherapy-induced peripheral neuropathy: a current review. Ann Neurol 81(6):772–781. https://doi.org/10.1002/ana.24951
3. Gu Y, Menzies AM, Long GV, Fernando S, Herkes G (2017) Immune mediated neuropathy following checkpoint immunotherapy. J Clin Neurosci 45:14–17
4. Reeves BN, Dakhil SR, Sloan JA, Wolf SL, Burger KN, Kamal A et al (2012) Further data supporting that paclitaxel-associated acute pain syndrome is associated with development of peripheral neuropathy: North Central Cancer Treatment Group trial N08C1. Cancer 118 (20):5171–5178
5. Argyriou AA, Cavaletti G, Briani C, Velasco R, Bruna J, Campagnolo M et al (2013) Clinical pattern and associations of oxaliplatin acute neurotoxicity: a prospective study in 170 patients with colorectal cancer. Cancer 119(2):438–444
6. Pachman DR, Qin R, Seisler DK, Smith EM, Beutler AS, Ta LE et al (2015) Clinical course of oxaliplatin-induced neuropathy: results from the randomized phase III trial N08CB (Alliance). J Clin Oncol 33(30):3416
7. Seretny M, Currie GL, Sena ES, Ramnarine S, Grant R, MacLeod MR et al (2014) Incidence, prevalence, and predictors of chemotherapy-induced peripheral neuropathy: a systematic review and meta-analysis. Pain 155(12):2461–2470. https://doi.org/10.1016/j.pain.2014.09.020
8. Hershman DL, Unger JM, Crew KD, Till C, Greenlee H, Minasian LM et al (2018) Two-year trends of Taxane-induced neuropathy in women enrolled in a randomized trial of acetyl-l-carnitine (SWOG S0715). J Natl Cancer Inst 110(6):669–676. https://doi.org/10.1093/jnci/djx259
9. Bandos H, Melnikow J, Rivera DR, Swain SM, Sturtz K, Fehrenbacher L et al (2018) Long-term peripheral neuropathy in breast cancer patients treated with adjuvant chemotherapy: NRG oncology/NSABP B-30. J Natl Cancer Inst 110(2):djx162. https://doi.org/10.1093/jnci/djx162
10. daCosta DBM, Copher R, Basurto E, Faria C, Lorenzo R (2014) Patient preferences and treatment adherence among women diagnosed with metastatic breast cancer. Am Health Drug Benefits 7(7):386–396

11. Lyman GH (2009) Impact of chemotherapy dose intensity on cancer patient outcomes. Journal of the National Comprehensive Cancer Network : JNCCN 7(1):99–108

12. Gewandter JS, Kleckner AS, Marshall JH, Brown JS, Curtis LH, Bautista J et al (2019) Chemotherapy-induced peripheral neuropathy (CIPN) and its treatment: an NIH Collaboratory study of claims data. Support Care Cancer 28(6):2553–2562. https://pubmed.ncbi.nlm.nih.gov/31494735/

13. Pike CT, Birnbaum HG, Muehlenbein CE, Pohl GM, Natale RB (2012) Healthcare costs and workloss burden of patients with chemotherapy-associated peripheral neuropathy in breast, ovarian, head and neck, and nonsmall cell lung cancer. Chemother Res Pract 2012:913848. https://doi.org/10.1155/2012/913848

14. Chan CW, Cheng H, Au SK, Leung KT, Li YC, Wong KH et al (2018) Living with chemotherapy- induced peripheral neuropathy: uncovering the symptom experience and self-management of neuropathic symptoms among cancer survivors. Eur J Oncol Nurs 36:135–141

15. Monfort SM, Pan X, Patrick R, Ramaswamy B, Wesolowski R, Naughton MJ et al (2017) Gait, balance, and patient-reported outcomes during taxane-based chemotherapy in early-stage breast cancer patients. Breast Cancer Res Treat 164(1):69–77. https://doi.org/10.1007/s10549-017-4230-8

16. Wasilewski A, Mohile N (2020) Meet the expert: how I treat chemotherapy-induced peripheral neuropathy. J Geriatr Oncol. https://doi.org/10.1016/j.jgo.2020.06.008

17. Kolb NA, Smith AG, Singleton JR, Beck SL, Stoddard GJ, Brown S et al (2016) The association of chemotherapy-induced peripheral neuropathy symptoms and the risk of falling. JAMA Neurol 73(7):860–866. https://doi.org/10.1001/jamaneurol.2016.0383

18. Gewandter JS, Fan L, Magnuson A, Mustian K, Peppone L, Heckler C et al (2013) Falls and functional impairments in cancer survivors with chemotherapy-induced peripheral neuropathy (CIPN): a University of Rochester CCOP study. Support Care Cancer: official journal of the Multinational Association of Supportive Care in Cancer 21(7):2059–2066. https://doi.org/10.1007/s00520-013-1766-y

19. Zajaczkowska R, Kocot-Kepska M, Leppert W, Wrzosek A, Mika J, Wordliczek J (2019) Mechanisms of chemotherapy-induced peripheral neuropathy. Int J Mol Sci 20(6):1451. Https://doi.org/10.3390/ijms20061451

20. Flatters SJL, Dougherty PM, Colvin LA (2017) Clinical and preclinical perspectives on Chemotherapy-Induced Peripheral Neuropathy (CIPN): a narrative review. Br J Anaesth 119 (4):737–749. https://doi.org/10.1093/bja/aex229

21. Alberti P, Rossi E, Cornblath DR, Merkies IS, Postma TJ, Frigeni B et al (2014) Physician-assessed and patient-reported outcome measures in chemotherapy-induced sensory peripheral neurotoxicity: two sides of the same coin. Ann Oncol: official journal of the European Society for Medical Oncology/ESMO 25(1):257–264. https://doi.org/10.1093/annonc/mdt409

22. Park SB, Alberti P, Kolb NA, Gewandter JS, Schenone A, Argyriou AA (2019) Overview and critical revision of clinical assessment tools in chemotherapy-induced peripheral neurotoxic-ity. J Peripher Nerv Syst 24:S13–S25

23. Postma TJ, Aaronson NK, Heimans JJ, Muller MJ, Hildebrand JG, Delattre JY et al (2005) The development of an EORTC quality of life questionnaire to assess chemotherapy-induced peripheral neuropathy: the QLQ-CIPN20. Eur J Cancer 41(8):1135–1139. https://doi.org/10.1016/j.ejca.2005.02.012

24. Cavaletti G, Frigeni B, Lanzani F, Piatti M, Rota S, Briani C et al (2007) The Total Neuropathy Score as an assessment tool for grading the course of chemotherapy-induced peripheral neurotoxicity: comparison with the National Cancer Institute-Common Toxicity Scale. J Peripher Nerv Syst 12(3):210–215

25. Loprinzi CL, Lacchetti C, Bleeker J, Cavaletti G, Chauhan C, Hertz DL et al (2020) Prevention and management of chemotherapy-induced peripheral neuropathy in survivors of adult cancers: ASCO guideline update. J Clin Oncol 38(28):3325–3348. https://doi.org/10.1200/JCO.20.01399

26. Hou S, Huh B, Kim HK, Kim KH, Abdi S (2018) Treatment of chemotherapy-induced peripheral neuropathy: systematic review and recommendations. Pain Physician 21 (6):571–592

27. Hershman DL, Lacchetti C, Dworkin RH, Lavoie Smith EM, Bleeker J, Cavaletti G et al (2014) Prevention and management of chemotherapy-induced peripheral neuropathy in survivors of adult cancers: American Society of Clinical Oncology clinical practice guideline. J Clin Oncol 32(18):1941–1967. https://doi.org/10.1200/JCO.2013.54.0914

28. Smith EML, Pang H, Cirrincione C, Fleishman S, Paskett ED, Ahles T et al (2013) Effect of duloxetine on pain, function, and quality of life among patients with chemotherapy-induced painful peripheral neuropathy: a randomized clinical trial. JAMA 309(13):1359–1367

29. Able SL, Cui Z, Shen W (2014) Duloxetine treatment adherence across mental health and chronic pain conditions. Clin Econ Outcomes Res CEOR 6:75–81. https://doi.org/10.2147/CEOR.S52950

30. Tham A, Jonsson U, Andersson G, Söderlund A, Allard P, Bertilsson G (2016) Efficacy and tolerability of antidepressants in people aged 65 years or older with major depressive disorder– a systematic review and a meta-analysis. J Affect Disord 205:1–12

31. Brami C, Bao T, Deng G (2016) Natural products and complementary therapies for chemo-therapy- induced peripheral neuropathy: a systematic review. Crit Rev Oncol Hematol 98:325–334. https://doi.org/10.1016/j.critrevonc.2015.11.014

32. Dorsey SG, Kleckner IR, Barton D, Mustian K, O'Mara A, St Germain D et al (2019) NCI Clinical Trials Planning Meeting for prevention and treatment of chemotherapy-induced peripheral neuropathy. J Natl Cancer Inst. https://doi.org/10.1093/jnci/djz011

33. Duregon F, Vendramin B, Bullo V, Gobbo S, Cugusi L, Di Blasio A et al (2018) Effects of exercise on cancer patients suffering chemotherapy-induced peripheral neuropathy undergoing treatment: a systematic review. Crit Rev Oncol Hematol 121:90–100. https://doi.org/10.1016/j.critrevonc.2017.11.002

34. Kanzawa-Lee GA, Larson JL, Resnicow K, Smith EML (2020) Exercise effects on chemo-therapy- induced peripheral neuropathy: a comprehensive integrative review. Cancer Nurs 43 (3):E172–EE85

35. Caspersen CJ, Powell KE, Christenson GM (1985) Physical activity, exercise, and physical fitness: definitions and distinctions for health-related research. Public Health Rep 100 (2):126–131

36. Neil-Sztramko SE, Medysky ME, Campbell KL, Bland KA, Winters-Stone KM (2019) Attention to the principles of exercise training in exercise studies on prostate cancer survivors: a systematic review. BMC Cancer 19(1):321

37. Pedersen BK, Saltin B (2015) Exercise as medicine - evidence for prescribing exercise as therapy in 26 different chronic diseases. Scand J Med Sci Sports 25(Suppl 3):1–72. https://doi.org/10.1111/sms.12581

38. Fiuza-Luces C, Santos-Lozano A, Joyner M, Carrera-Bastos P, Picazo O, Zugaza JL et al (2018) Exercise benefits in cardiovascular disease: beyond attenuation of traditional risk factors. Nat Rev Cardiol 15(12).731–743. https://doi.org/10.1038/s41569-018-0065-1

39. Kvam S, Kleppe CL, Nordhus IH, Hovland A (2016) Exercise as a treatment for depression: a meta-analysis. J Affect Disord 202:67–86. https://doi.org/10.1016/j.jad.2016.03.063

40. Balducci S, Iacobellis G, Parisi L, Di Biase N, Calandriello E, Leonetti F et al (2006) Exercise training can modify the natural history of diabetic peripheral neuropathy. J Diabetes Complicat 20(4):216–223. https://doi.org/10.1016/j.jdiacomp.2005.07.005

41. Dixit S, Maiya AG, Shastry BA (2014) Effect of aerobic exercise on peripheral nerve functions of population with diabetic peripheral neuropathy in type 2 diabetes: a single blind, parallel group randomized controlled trial. J Diabetes Complicat 28(3):332–339. https://doi.org/10.1016/j.jdiacomp.2013.12.006

42. Cooper MA, Kluding PM, Wright DE (2016) Emerging relationships between exercise, sensory nerves, and neuropathic pain. Front Neurosci 10:372. https://doi.org/10.3389/fnins.2016.00372

43. Kleckner IR, Dunne RF, Asare M, Cole C, Fleming F, Fung C et al (2018) Exercise for toxicity management in cancer—a narrative review. Oncol Hematol Rev 14(1):28

44. Lin P-J, Peppone LJ, Janelsins MC, Mohile SG, Kamen CS, Kleckner IR et al (2018) Yoga for the management of cancer treatment-related toxicities. Curr Oncol Rep 20(1):5

45. Campbell KL, Winters-Stone KM, Wiskemann J, May AM, Schwartz AL, Courneya KS et al (2019) Exercise guidelines for cancer survivors: consensus statement from international multidisciplinary roundtable. Med Sci Sports Exerc 51(11):2375–2390. https://doi.org/10. 1249/MSS.0000000000002116

46. Buffart LM, Kalter J, Sweegers MG, Courneya KS, Newton RU, Aaronson NK et al (2017) Effects and moderators of exercise on quality of life and physical function in patients with cancer: an individual patient data meta-analysis of 34 RCTs. Cancer Treat Rev 52:91–104. https://doi.org/10.1016/j.ctrv.2016.11.010

47. Mustian KM, Alfano CM, Heckler C, Kleckner AS, Kleckner IR, Leach CR et al (2017) Comparison of pharmaceutical, psychological, and exercise treatments for cancer-related fatigue: a meta-analysis. JAMA Oncol 3(7):961–968

48. Scott JM, Nilsen TS, Gupta D, Jones LW (2018) Exercise therapy and cardiovascular toxicity in cancer. Circulation 137(11):1176–1191. https://doi.org/10.1161/CIRCULATIONAHA. 117.024671

49. Hile ES, Fitzgerald GK, Studenski SA (2010) Persistent mobility disability after neurotoxic chemotherapy. Phys Ther 90(11):1649–1657. https://doi.org/10.2522/ptj.20090405

50. van Brussel M, Takken T, van der Net J, Engelbert RH, Bierings M, Schoenmakers MA et al (2006) Physical function and fitness in long-term survivors of childhood leukaemia. Pediatr Rehabil 9(3):267–274. https://doi.org/10.1080/13638490500523150

51. Wilkes G (2007) Peripheral neuropathy related to chemotherapy. Seminars in oncology nursing. Elsevier, Amsterdam, pp 162–173

52. Mols F, Beijers AJ, Vreugdenhil G, Verhulst A, Schep G, Husson O (2015) Chemotherapy-induced peripheral neuropathy, physical activity and health-related quality of life among colorectal cancer survivors from the PROFILES registry. J Cancer Surviv: Research and Practice 9(3):512–522. https://doi.org/10.1007/s11764-015-0427-1

53. Greenlee H, Hershman DL, Shi Z, Kwan ML, Ergas IJ, Roh JM et al (2017) BMI, lifestyle factors and taxane-induced neuropathy in breast cancer patients: the pathways study. J Natl Cancer Inst 109(2):djw206. https://doi.org/10.1093/jnci/djw206

54. Kerns SL, Fung C, Monahan PO, Ardeshir-Rouhani-Fard S, Abu Zaid MI, Williams AM et al (2018) Cumulative burden of morbidity among testicular cancer survivors after standard cisplatin-based chemotherapy: a multi-institutional study. J Clin Oncol: Official Journal of the American Society of Clinical Oncology 36(15):1505–1512. https://doi.org/10.1200/JCO. 2017.77.0735

55. Packel LB, Prehn AW, Anderson CL, Fisher PL (2015) Factors influencing physical activity behaviors in colorectal cancer survivors. Am J Health Promot 30(2):85–92. https://doi.org/10. 4278/ajhp.140103-QUAN-7

56. Stevinson C, Steed H, Faught W, Tonkin K, Vallance JK, Ladha AB et al (2009) Physical activity in ovarian cancer survivors: associations with fatigue, sleep, and psychosocial functioning. Int J Gynecol Cancer: Official Journal of the International Gynecological Cancer Society 19(1):73–78. https://doi.org/10.1111/IGC.0b013e31819902ec

57. Thomaier L, Jewett P, Brown K, Gotlieb R, Teoh D, Blaes AH et al (2020) The associations between physical activity, neuropathy symptoms and health-related quality of life among gynecologic cancer survivors. Gynecol Oncol 158(2):361–365. https://doi.org/10.1016/j. ygyno.2020.05.026

58. Kleckner IR, Park SB, Streckmann F, Wiskemann J, Hardy S, Mohile NA (2021) Clinical and practical recommendations in the use of exercise, physical therapy, and occupational therapy for chemotherapy-induced peripheral neuropathy. In: Lustberg MB, Loprinzi CL (eds) Diagnosis, management and emerging strategies for chemotherapy induced neuropathy. Springer, Berlin

59. Park JS, Kim S, Hoke A (2015) An exercise regimen prevents development paclitaxel induced peripheral neuropathy in a mouse model. J Peripheral Nervous System: JPNS 20(1):7–14. https://doi.org/10.1111/jns.12109
60. Slivicki RA, Mali SS, Hohmann AG (2019) Voluntary exercise reduces both chemotherapy-induced neuropathic nociception and deficits in hippocampal cellular proliferation in a mouse model of paclitaxel-induced peripheral neuropathy. Neurobiol Pain 6:100035. https://doi.org/10.1016/j.ynpai.2019.100035
61. Bland KA, Kirkham AA, Bovard J, Shenkier T, Zucker D, McKenzie DC et al (2019) Effect of Exercise on Taxane Chemotherapy-Induced Peripheral Neuropathy in Women With Breast Cancer: A Randomized Controlled Trial. Clin Breast Cancer 19(6):411–422. https://doi.org/10.1016/j.clbc.2019.05.013
62. Gewandter JS, Brell J, Cavaletti G, Dougherty PM, Evans S, Howie L et al (2018) Trial designs for chemotherapy-induced peripheral neuropathy prevention: ACTTION recommendations. Neurology 91(9):403–413. https://doi.org/10.1212/WNL.0000000000006083
63. Andersen Hammond E, Pitz M, Steinfeld K, Lambert P, Shay B (2020) An Exploratory Randomized Trial of Physical Therapy for the Treatment of Chemotherapy-Induced Peripheral Neuropathy. Neurorehabil Neural Repair 34(3):235–246. https://doi.org/10.1177/1545968319899918
64. Clark PG, Cortese-Jimenez G, Cohen E (2012) Effects of Reiki, yoga, or meditation on the physical and psychological symptoms of chemotherapy-induced peripheral neuropathy: a randomized pilot study. J Evid Based Complement Alternat Med 17(3):161–171
65. Dhawan S, Andrews R, Kumar L, Wadhwa S, Shukla G (2020) A randomized controlled trial to assess the effectiveness of muscle strengthening and balancing exercises on chemotherapy-induced peripheral neuropathic pain and quality of life among cancer patients. Cancer Nurs 43(4):269–280. https://doi.org/10.1097/NCC.0000000000000693
66. Kleckner IR, Kamen C, Gewandter JS, Mohile NA, Heckler CE, Culakova E et al (2018) Effects of exercise during chemotherapy on chemotherapy-induced peripheral neuropathy: a multicenter, randomized controlled trial. Support Care Cancer: Official Journal of the Multinational Association of Supportive Care in Cancer. 26(4):1019–1028. https://doi.org/10.1007/s00520-017-4013-0
67. Schwenk M, Grewal GS, Holloway D, Muchna A, Garland L, Najafi B (2016) Interactive sensor- based balance training in older cancer patients with chemotherapy-induced peripheral neuropathy: a randomized controlled trial. Gerontology 62(5):553–563. https://doi.org/10.1159/000442253
68. Streckmann F, Kneis S, Leifert JA, Baumann FT, Kleber M, Ihorst G et al (2014) Exercise program improves therapy-related side-effects and quality of life in lymphoma patients undergoing therapy. Ann Oncol: Official Journal of the European Society for Medical Oncology/ESMO 25(2):493–499. https://doi.org/10.1093/annonc/mdt568
69. Stuecher K, Bolling C, Vogt L, Niederer D, Schmidt K, Dignass A et al (2019) Exercise improves functional capacity and lean body mass in patients with gastrointestinal cancer during chemotherapy: a single-blind RCT. Support Care Cancer: Official Journal of the Multinational Association of Supportive Care in Cancer 27(6):2159–2169. https://doi.org/10.1007/s00520-018-4478-5
70. Vollmers PL, Mundhenke C, Maass N, Bauerschlag D, Kratzenstein S, Rocken C et al (2018) Evaluation of the effects of sensorimotor exercise on physical and psychological parameters in breast cancer patients undergoing neurotoxic chemotherapy. J Cancer Res Clin Oncol 144(9):1785–1792. https://doi.org/10.1007/s00432-018-2686-5
71. Zimmer P, Trebing S, Timmers-Trebing U, Schenk A, Paust R, Bloch W et al (2018) Eight-week, multimodal exercise counteracts a progress of chemotherapy-induced peripheral neuropathy and improves balance and strength in metastasized colorectal cancer patients: a randomized controlled trial. Support Care Cancer: official journal of the Multinational

Association of Supportive Care in Cancer 26(2):615–624. https://doi.org/10.1007/s00520-017-3875-5

72. Thompson BT, Schoenfeld D (2007) Usual care as the control group in clinical trials of nonpharmacologic interventions. Proc Am Thorac Soc 4(7):577–582

73. Andrade C (2015) The primary outcome measure and its importance in clinical trials. J Clin Psychiatry 76(10):1320–1323

74. Courneya KS, McKenzie DC, Mackey JR, Gelmon K, Friedenreich CM, Yasui Y et al (2014) Subgroup effects in a randomised trial of different types and doses of exercise during breast cancer chemotherapy. Br J Cancer 111(9):1718–1725. https://doi.org/10.1038/bjc.2014.466

75. Henke CC, Cabri J, Fricke L, Pankow W, Kandilakis G, Feyer PC et al (2014) Strength and endurance training in the treatment of lung cancer patients in stages IIIA/IIIB/IV. Support Care Cancer: Official Journal of the Multinational Association of Supportive Care in Cancer. 22 (1):95–101. https://doi.org/10.1007/s00520-013-1925-1

76. Kneis S, Wehrle A, Muller J, Maurer C, Ihorst G, Gollhofer A et al (2019) It's never too late - balance and endurance training improves functional performance, quality of life, and alleviates neuropathic symptoms in cancer survivors suffering from chemotherapy-induced peripheral neuropathy: results of a randomized controlled trial. BMC Cancer 19(1):414. https://doi.org/10.1186/s12885-019-5522-7

77. Schonsteiner SS, Bauder Missbach H, Benner A, Mack S, Hamel T, Orth M et al (2017) A randomized exploratory phase 2 study in patients with chemotherapy-related peripheral neuropathy evaluating whole-body vibration training as adjunct to an integrated program including massage, passive mobilization and physical exercises. Exp Hematol Oncol 6:5. https://doi.org/10.1186/s40164-017-0065-6

78. Streckmann F, Lehmann HC, Balke M, Schenk A, Oberste M, Heller A et al (2019) Sensori-motor training and whole-body vibration training have the potential to reduce motor and sensory symptoms of chemotherapy-induced peripheral neuropathy-a randomized controlled pilot trial. Support Care Cancer: Official Journal of the Multinational Association of Supportive Care in Cancer 27(7):2471–2478. https://doi.org/10.1007/s00520-018-4531-4

79. McCrary JM, Goldstein D, Sandler CX, Barry BK, Marthick M, Timmins HC et al (2019) Exercise-based rehabilitation for cancer survivors with chemotherapy-induced peripheral neuropathy. Support Care Cancer: Official Journal of the Multinational Association of Supportive Care in Cancer 27(10):3849–3857. https://doi.org/10.1007/s00520-019-04680-w

80. Kleckner IR, Kamen C, Gewandter JS, Mohile NA, Heckler CE, Culakova E et al (2019) Response to Crevenna and Ashbury, Vallance and Bolam, and Crevenna and Keilani regarding the effects of exercise on chemotherapy-induced peripheral neuropathy. Support Care Cancer: Official Journal of the Multinational Association of Supportive Care in Cancer 27(1):7–8. https://doi.org/10.1007/s00520-018-4528-z

81. Armada-da-Silva PA, Pereira C, Amado S, Veloso AP (2013) Role of physical exercise for improving posttraumatic nerve regeneration. Int Rev Neurobiol 109:125–149. https://doi.org/10.1016/b978-0-12-420045-6.00006-7

82. Sabatier MJ, Redmon N, Schwartz G, English AW (2008) Treadmill training promotes axon regeneration in injured peripheral nerves. Exp Neurol 211(2):489–493. https://doi.org/10.1016/j.expneurol.2008.02.013

83. Bobinski F, Martins DF, Bratti T, Mazzardo-Martins L, Winkelmann-Duarte EC, Guglielmo LG et al (2011) Neuroprotective and neuroregenerative effects of low-intensity aerobic exercise on sciatic nerve crush injury in mice. Neuroscience 194:337–348. https://doi.org/10.1016/j.neuroscience.2011.07.075

84. Cobianchi S, Marinelli S, Florenzano F, Pavone F, Luvisetto S (2010) Short- but not long-lasting treadmill running reduces allodynia and improves functional recovery after peripheral nerve injury. Neuroscience 168(1):273–287. https://doi.org/10.1016/j.neuroscience.2010.03.035

85. Singleton JR, Marcus RL, Jackson JE, K Lessard M, Graham TE, Smith AG. (2014) Exercise increases cutaneous nerve density in diabetic patients without neuropathy. Ann Clin Transl Neurol 1(10):844–849. https://doi.org/10.1002/acn3.125

86. Kluding PM, Pasnoor M, Singh R, Jernigan S, Farmer K, Rucker J et al (2012) The effect of exercise on neuropathic symptoms, nerve function, and cutaneous innervation in people with diabetic peripheral neuropathy. J Diabetes Complicat 26(5):424–429. https://doi.org/10.1016/j.jdiacomp.2012.05.007

87. Singleton JR, Marcus RL, Lessard MK, Jackson JE, Smith AG (2015) Supervised exercise improves cutaneous reinnervation capacity in metabolic syndrome patients. Ann Neurol 77 (1):146–153. https://doi.org/10.1002/ana.24310

88. Park JS, Höke A (2014) Treadmill exercise induced functional recovery after peripheral nerve repair is associated with increased levels of neurotrophic factors. PLoS One 9(3):e90245. https://doi.org/10.1371/journal.pone.0090245

89. Cobianchi S, Arbat-Plana A, Lopez-Alvarez VM, Navarro X (2017) Neuroprotective effects of exercise treatments after injury: the dual role of neurotrophic factors. Curr Neuropharmacol 15 (4):495–518. https://doi.org/10.2174/1570159x14666160330105132

90. Gleeson M, Bishop NC, Stensel DJ, Lindley MR, Mastana SS, Nimmo MA (2011) The anti-inflammatory effects of exercise: mechanisms and implications for the prevention and treatment of disease. Nat Rev Immunol 11(9):607–615. https://doi.org/10.1038/nri3041

91. Wang XM, Lehky TJ, Brell JM, Dorsey SG (2012) Discovering cytokines as targets for chemotherapy-induced painful peripheral neuropathy. Cytokine 59(1):3–9. https://doi.org/10.1016/j.cyto.2012.03.027

92. Brandolini L, d'Angelo M, Antonosante A, Allegretti M, Cimini A (2019) Chemokine signaling in chemotherapy-induced neuropathic pain. Int J Mol Sci 20(12):2904. https://doi.org/10.3390/ijms20122904

93. Fischer CP (2006) Interleukin-6 in acute exercise and training: what is the biological relevance? Exerc Immunol Rev 12:6–33

94. Steensberg A, Fischer CP, Keller C, Moller K, Pedersen BK (2003) IL-6 enhances plasma IL-1ra, IL-10, and cortisol in humans. Am J Physiol Endocrinol Metab 285(2):E433–E437. https://doi.org/10.1152/ajpendo.00074.2003

95. Kleckner IR, Kamen C, Cole C, Fung C, Heckler CE, Guido JJ et al (2019) Effects of exercise on inflammation in patients receiving chemotherapy: a nationwide NCORP randomized clinical trial. Support Care Cancer 27(12):4615–4625

96. de la Puerta R, Carcelén M, Francés R, de la Fuente R, Hurlé MA, Tramullas M (2019) BMP-7 protects male and female rodents against neuropathic pain induced by nerve injury through a mechanism mediated by endogenous opioids. Pharmacol Res 150:104470. https://doi.org/10.1016/j.phrs.2019.104470

97. Chen YW, Chiu CC, Hsieh PL, Hung CH, Wang JJ (2015) Treadmill training combined with insulin suppresses diabetic nerve pain and cytokines in rat sciatic nerve. Anesth Analg 121 (1):239–246. https://doi.org/10.1213/ane.0000000000000799

98. Raefsky SM, Mattson MP (2017) Adaptive responses of neuronal mitochondria to bioenergetic challenges: Roles in neuroplasticity and disease resistance. Free Radic Biol Med 102:203–216

99. Bennett GJ, Doyle T, Salvemini D (2014) Mitotoxicity in distal symmetrical sensory peripheral neuropathies. Nat Rev Neurol 10(6):326–336. https://doi.org/10.1038/nrneurol.2014.77

100. Ma J, Kavelaars A, Dougherty PM, Heijnen CJ (2018) Beyond symptomatic relief for chemotherapy-induced peripheral neuropathy: targeting the source. Cancer 124 (11):2289–2298. https://doi.org/10.1002/cncr.31248

101. Waseem M, Kaushik P, Tabassum H, Parvez S (2018) Role of mitochondrial mechanism in chemotherapy-induced peripheral neuropathy. Curr Drug Metab 19(1):47–54. https://doi.org/10.2174/1389200219666171207121313

102. Ellingson LD, Stegner AJ, Schwabacher IJ, Koltyn KF, Cook DB (2016) Exercise strengthens central nervous system modulation of pain in fibromyalgia. Brain Sci 6(1):8

103. Heijnen S, Hommel B, Kibele A, Colzato LS (2016) Neuromodulation of aerobic exercise—a review. Front Psychol 6:1890
104. Marks DM, Shah MJ, Patkar AA, Masand PS, Park G-Y, Pae C-U (2009) Serotonin-norepinephrine reuptake inhibitors for pain control: premise and promise. Curr Neuropharmacol 7(4):331–336
105. Obata H (2017) Analgesic mechanisms of antidepressants for neuropathic pain. Int J Mol Sci 18(11):2483
106. Khalsa SS, Adolphs R, Cameron OG, Critchley HD, Davenport PW, Feinstein JS et al (2018) Interoception and Mental Health: A Roadmap. Biol Psychiatry Cogn Neurosci Neuroimag 3 (6):501–513. https://doi.org/10.1016/j.bpsc.2017.12.004
107. Kleckner IR, Zhang J, Touroutoglou A, Chanes L, Xia C, Simmons WK et al (2017) Evidence for a large-scale brain system supporting allostasis and interoception in humans. Nat Hum Behav 1(5):1–14
108. Kleckner IR, Quigley KS (2014) An approach to mapping the neurophysiological state of the body to affective experience. In: Barrett LF, Russell J (eds) The psychological construction of emotion. Guilford, New York
109. Weng TB, Pierce GL, Darling WG, Falk D, Magnotta VA, Voss MW (2017) The acute effects of aerobic exercise on the functional connectivity of human brain networks. Brain Plast 2 (2):171–190. https://doi.org/10.3233/BPL-160039
110. Rajab AS, Crane DE, Middleton LE, Robertson AD, Hampson M, MacIntosh BJ (2014) A single session of exercise increases connectivity in sensorimotor-related brain networks: a resting-state fMRI study in young healthy adults. Front Hum Neurosci 8:625. https://doi.org/10.3389/fnhum.2014.00625
111. Williamson JW, Nobrega AC, McColl R, Mathews D, Winchester P, Friberg L et al (1997) Activation of the insular cortex during dynamic exercise in humans. J Physiol 503:277–283
112. Taube W, Gruber M, Gollhofer A (2008) Spinal and supraspinal adaptations associated with balance training and their functional relevance. Acta Physiol 193(2):101–116
113. Biddle SJH, Mutrie N, Gorely T (2015) Psychology of physical activity: determinants, well-being and interventions. Routledge/Taylor & Francis, London
114. Kandola A, Ashdown-Franks G, Hendrikse J, Sabiston CM, Stubbs B (2019) Physical activity and depression: towards understanding the antidepressant mechanisms of physical activity. Neurosci Biobehav Rev 107:525–539
115. Loh KP, Kleckner IR, Lin PJ, Mohile SG, Canin BE, Flannery MA et al (2019) Effects of a home-based exercise program on anxiety and mood disturbances in older adults with cancer receiving chemotherapy. J Am Geriatr Soc 67(5):1005–1011
116. Bonhof CS, van de Poll-Franse LV, Vissers PA, Wasowicz DK, Wegdam JA, Révész D et al (2019) Anxiety and depression mediate the association between chemotherapy-induced peripheral neuropathy and fatigue: Results from the population-based PROFILES registry. Psychooncology 28(9):1926–1933
117. IsHak WW, Wen RY, Naghdechi L, Vanle B, Dang J, Knosp M et al (2018) Pain and depression: a systematic review. Harv Rev Psychiatry 26(6):352–363
118. Kamen C, Heckler C, Janelsins MC, Peppone LJ, McMahon JM, Morrow GR et al (2016) A dyadic exercise intervention to reduce psychological distress among lesbian, gay, and heterosexual cancer survivors. LGBT Health 3(1):57–64
119. Hughes S, Jaremka LM, Alfano CM, Glaser R, Povoski SP, Lipari AM et al (2014) Social support predicts inflammation, pain, and depressive symptoms: longitudinal relationships among breast cancer survivors. Psychoneuroendocrinology 42:38–44
120. Higgins TJ, Middleton KR, Winner L, Janelle CM (2014) Physical activity interventions differentially affect exercise task and barrier self-efficacy: a meta-analysis. Health Psychol 33(8):891
121. Hughes D, Baum G, Jovanovic J, Carmack C, Greisinger A, Basen-Engquist K (2010) An acute exercise session increases self-efficacy in sedentary endometrial cancer survivors and controls. J Phys Act Health 7(6):784–793

122. Lindheimer JB, O'Connor PJ, Dishman RK (2015) Quantifying the placebo effect in psychological outcomes of exercise training: a meta-analysis of randomized trials. Sports Med 45 (5):693–711
123. Onken LS, Carroll KM, Shoham V, Cuthbert BN, Riddle M (2014) Reenvisioning clinical science: unifying the discipline to improve the public health. Clin Psychol Sci 2(1):22–34
124. Kumar SP (2011) Cancer pain: a critical review of mechanism-based classification and physical therapy management in palliative care. Indian J Palliat Care 17(2):116
125. Kumar SP, Saha S (2011) Mechanism-based classification of pain for physical therapy management in palliative care: a clinical commentary. Indian J Palliat Care 17(1):80
126. Kumar SP, Prasad K, Kumar VK, Shenoy K, Sisodia V (2013) Mechanism-based classification and physical therapy management of persons with cancer pain: a prospective case series. Indian J Palliat Care 19(1):27
127. Community Oncology Alliance (COA). What is community oncology. https://www.communityoncology.org/wp-content/uploads/2017/08/What-is-Comm-Onc.pdf2017
128. Cammisuli S, Cavazzi E, Baldissarro E, Leandri M (2016) Rehabilitation of balance disturbances due to chemotherapy-induced peripheral neuropathy: a pilot study. Eur J Phys Rehabil Med 52(4):479–488
129. Fernandes J, Kumar S (2016) Effect of lower limb closed kinematic chain exercises on balance in patients with chemotherapy-induced peripheral neuropathy: a pilot study. Int J Rehabil Res 39(4):368–371. https://doi.org/10.1097/MRR.0000000000000196
130. Galantino ML, Tiger R, Brooks J, Jang S, Wilson K (2019) Impact of somatic yoga and meditation on fall risk, function, and quality of life for chemotherapy-induced peripheral neuropathy syndrome in cancer survivors. Integr Cancer Ther 18:1534735419850627. https://doi.org/10.1177/1534735419850627
131. Galantino ML, Brooks J, Tiger R, Jang S, Wilson K (2019) Effectiveness of somatic yoga and meditation: a pilot study in a multicultural cancer survivor population with chemotherapy-induced peripheral neuropathy. Int J Yoga Therap. https://doi.org/10.17761/2020-D-18-00030
132. Kneis S, Wehrle A, Dalin D, Wiesmeier IK, Lambeck J, Gollhofer A et al (2020) A new approach to characterize postural deficits in chemotherapy-induced peripheral neuropathy and to analyze postural adaptions after an exercise intervention. BMC Neurol 20(1):23. https://doi.org/10.1186/s12883-019-1589-7
133. Moonsammy SH, Guglietti CL, Santa Mina D, Ferguson S, Kuk JL, Urowitz S et al (2013) A pilot study of an exercise & cognitive behavioral therapy intervention for epithelial ovarian cancer patients. J Ovarian Res 6(1):21. https://doi.org/10.1186/1757-2215-6-21
134. Wonders KY, Whisler G, Loy H, Holt B, Bohachek K, Wise R (2013) Ten weeks of home-based exercise attenuates symptoms of chemotherapy-induced peripheral neuropathy in breast cancer patients. Health Psychol Res 1(3):e28. https://doi.org/10.4081/hpr.2013.e28
135. Worthen-Chaudhari L, Lamantia MT, Monfort SM, Mysiw W, Chaudhari AMW, Lustberg MB (2019) Partnered, adapted argentine tango dance for cancer survivors: a feasibility study and pilot study of efficacy. Clin Biomech (Bristol, Avon) 70:257–264. https://doi.org/10.1016/j.clinbiomech.2019.08.010
136. Streckmann F, Hess V, Bloch W, Decard BF, Ritzmann R, Lehmann HC et al (2019) Individually tailored whole-body vibration training to reduce symptoms of chemotherapy-induced peripheral neuropathy: study protocol of a randomised controlled trial-VANISH. BMJ Open 9(4):e024467. https://doi.org/10.1136/bmjopen-2018-024467
137. Streckmann F, Balke M, Lehmann HC, Rustler V, Koliamitra C, Elter T et al (2018) The preventive effect of sensorimotor- and vibration exercises on the onset of Oxaliplatin- or vinca-alkaloid induced peripheral neuropathies – STOP. BMC Cancer 18(1):62. https://doi.org/10.1186/s12885-017-3866-4

Clinical and Practical Recommendations in the Use of Exercise, Physical Therapy, and Occupational Therapy for Chemotherapy-Induced Peripheral Neuropathy

9

Ian R. Kleckner, Susanna B. Park, Fiona Streckmann, Joachim Wiskemann, Sara Hardy, and Nimish Mohile

Abstract

Chemotherapy-induced peripheral neuropathy (CIPN) is a highly prevalent, severe, and dose-limiting toxicity of several chemotherapy regimens for the treatment of multiple cancers including lung, breast, prostate, gastrointestinal, blood, and others. Patients with CIPN may experience numbness, tingling, pain, and cramping in the hands and feet, as well as problems with balance and gait that increase the risk of falls, reduce physical function, and hinder activities of daily living. At this point there are extremely limited treatment options for CIPN. Fortunately, a growing body of preliminary evidence suggests that exercise, physical therapy, and occupational therapy may help prevent, treat, and manage

I. R. Kleckner (✉) · S. Hardy · N. Mohile
University of Rochester Medical Center, Rochester, NY, USA
e-mail: Ian_Kleckner@URMC.Rochester.edu; Sara_Hardy@URMC.Rochester.edu;
Nimish_Mohile@URMC.Rochester.edu

S. B. Park
Brain and Mind Centre, Faculty of Medicine and Health, University of Sydney, Sydney, Australia
e-mail: Susanna.Park@Sydney.edu.au

F. Streckmann
Department of Sport, Exercise and Health, University of Basel, Basel, Switzerland

Department of Oncology, University Hospital Basel, Basel, Switzerland

German Sport University Cologne, Köln, Germany
e-mail: fiona.streckmann@unibas.ch

J. Wiskemann
Heidelberg University Clinic and National Center for Tumor Diseases, Heidelberg, Germany
e-mail: Joachim.Wiskemann@nct-heidelberg.de

M. Lustberg, C. Loprinzi (eds.), *Diagnosis, Management and Emerging Strategies for Chemotherapy-Induced Neuropathy*,
https://doi.org/10.1007/978-3-030-78663-2_9

CIPN. Although there is not definitive evidence for the benefits of exercise on CIPN due to lack of Phase III randomized controlled trials, exercise is generally helpful in the cancer treatment continuum and poses low risk for patients with the help of a qualified professional. Therefore, we present clinical suggestions in the use of exercise for CIPN, including assessments of patient risk factors and other considerations. We conclude with an example exercise prescription that a qualified exercise professional can adapt for the specific needs, risks, and abilities of each individual patient.

Keywords

Exercise · Physical therapy · Occupational therapy · Chemotherapy · Neuropathy · CIPN · Clinical

9.1   Introduction

Chemotherapy-induced peripheral neuropathy (CIPN) is a severe toxicity that occurs in 58–78% of patients on average [1] receiving platinum-based agents (oxaliplatin, cisplatin, carboplatin), taxanes (paclitaxel, docetaxel), vinca alkaloids, bortezomib, and thalidomide analogs [2, 3], as well as certain immunotherapies [4]. These drugs are used to treat lung, breast, prostate, gastrointestinal, cervical, ovarian, testicular, blood, bone marrow, and other cancers. CIPN can involve acute signs and symptoms that present in the hours and days after an infusion [5–7] and can last for years after completion of chemotherapy [8, 9]. CIPN is a dose-limiting toxicity, meaning that it gives cause to lower chemotherapy doses or terminate it altogether [10], which may compromise anti-cancer treatment [11]. CIPN is also stressful on the healthcare system—medical claims data suggest that CIPN is under-diagnosed [12] and that patients with CIPN typically require 12 more outpatient visits, 3 more hospital days, and $17,000 USD more in medical expenses than matched patients without CIPN [13]. The symptoms of CIPN occur in a stocking-glove distribution (hands and feet) with some combination of numbness, tingling, shooting or stabbing pains, burning pain, cramping, and hypersensitivity to cold [3, 14]. The consequences of CIPN include loss of tactile or vibration sensitivity, walking gait and balance problems (i.e., postural instability; especially with eyes closed) [15, 16], increased risk of falls [17, 18], compromised participation in activities of daily living (e.g., walking, texting, writing, buttoning clothes), and, in rare cases, damage to the autonomic nervous system leading to impaired organ function (e.g., orthostatic hypotension) [3].

The 2020 American Society for Clinical Oncology (ASCO) Guidelines for CIPN concluded no recommended methods to prevent CIPN, and only one established method to treat CIPN: a moderate recommendation of the drug duloxetine to treat CIPN-related pain [19]. There are no recommended supplements, integrative therapies [20], devices, or behavioral interventions for CIPN [19] due to lack of multiple definitive Phase III randomized controlled trials (RCTs). Therefore, research on promising treatments for CIPN is a high-priority area of inquiry to ultimately identify and optimize additional treatments for CIPN [21]. One of the most promising treatments for CIPN is exercise, as shown by a growing body of studies [22–24]. At this point, no recommendations can be made on the use of exercise for the treatment or prevention of CIPN due to lack of larger and more definitive studies that would confirm efficacy and clarify risks [19].

However, exercise is generally helpful in the cancer treatment continuum and poses low risk for patients with the help of a qualified professional [25–28]. For CIPN, many patients already take on self-management using exercise and other strategies [29]. Although some clinicians consider exercise and physical therapy as part of CIPN symptom management [16], prior research has shown that few patients are referred to physical therapy during or after treatment [30]. The lack of referrals misses a critical window of opportunity given the potential preventive effect of exercise on CIPN as well as other toxicities (e.g., fatigue, distress) [28]. Indeed, exercise should be recommended at diagnosis rather than waiting until CIPN symptoms appear [28].

Given the inconsistent use of exercise for CIPN management, and the lack of Phase III randomized controlled trials indicating definitive benefits of specific exercise programs, it may be challenging for patients, clinicians, and exercise professionals (including physical therapists and occupational therapists) to have a starting point for an exercise program that may treat or prevent CIPN.

In this chapter, we provide general recommendations for the use of exercise for the prevention and treatment of CIPN. We begin with basic definitions of key exercise-related terms and principles. We present clinical considerations for assessing patient risks before starting an exercise program. Then we provide a suggested exercise program that should be tailored to each individual patient by a qualified exercise professional. We conclude with the likely future trajectory of the use of exercise for CIPN.

9.2 Features of an Exercise Program and Definitions of Key Terms

This chapter considers the roles of physical activity, exercise, physical therapy, and occupational therapy, which are related but distinct terms and approaches (see definitions in Table 9.1). Here we focus on the term exercise as this reflects the types of interventions used in most studies on treating CIPN using exercise and related interventions (physical therapy, occupational therapy, etc.) [24]. In studies or prescriptions of exercise, it is important to consider the dose, which comprises the frequency, intensity, type, and duration (definitions in Table 9.1).

The appropriate dose of exercise for treating CIPN can be determined by considering principles of exercise training including (1) specificity, (2) progression, (3) overload, (4) initial values, (5) diminishing returns, and (6) reversibility [33]. Typically, dose is progressively increased over several weeks and may be periodized into cycles of higher doses (a larger stimulus to ultimately drive physiological adaptation) alternated with lower doses to overcome training adaptation barriers (i.e., the principle of diminishing returns).

Table 9.1 Key terminology in the use of exercise and related interventions

Term	Definition
Physical activity	Any bodily movement produced by skeletal muscles that results in energy expenditure. Physical activity can be categorized into occupational, sports, conditioning, household, or other activities [31].
Exercise	Subset of physical activity that is planned, structured, and repetitive and has as a final or an intermediate objective to improve or maintain physical fitness [31].
Physical therapy	A branch of passive rehabilitative measures (e.g., manual therapy, massage, traction, ultrasound, electrical stimulation) to help patients regain or improve their physical abilities
Occupational therapy	Treatments that help people with injuries do what they want and need to do via therapeutic use of daily activities, thus enabling patients to live life to its fullest, including activities of daily living in the occupational, recreational, and household setting (American Occupational Therapy Association).
Exercise dose	Dose is comprised of four key features, sometimes referred to by the acronym FITT [26, 32] • Frequency—How often an exercise session is performed (e.g., 3 sessions per week). • Intensity—Based on percent maximum heart rate, percent maximum force production, or perceived exertion. • Type—The broad class of movements, methods, and energy systems utilized (e.g., aerobic, resistance, mixed). • Time or duration—How long each exercise session lasts (e.g., in minutes).

9.3 Clinical and Practical Suggestions for Using Exercise for CIPN

After having identified signs and symptoms of CIPN via routine screening or by asking the patient, yet prior to referral for an exercise program, patients should be screened for risk factors that impact what type of exercise program might be most appropriate. Routine cardiac screening is not necessary in the absence of a high risk history as defined by the American College of Sports Medicine (ACSM) [34, 35]. However, patients with exposure to hormone therapy should be screened for osteopenia/osteoporosis and fracture risk should be evaluated in patients with bone metastases [25]. Clinicians can then make a referral to a qualified exercise instructor or physical therapist.

A physical therapist can prescribe a safe set of exercises to improve strength, mobility, and reduce risk of falls. An evaluation for assistive devices in patients with significant mobility issues can also help ensure patient safety as CIPN may increase risk of falls or dropping objects. Interventions to engage patients in a regimen of physical activity are particularly important in patients who may also have weight loss, cachexia, fatigue, osteoporosis/osteopenia, and chronic hospitalizations leading to deconditioning [28, 36]. Patients with more severe symptoms may require exercises that do not rely on balance such as use of a stationary bike [37].

Local community support groups sometimes can direct patients to exercise facilities and trainers that may have experience in working with patients with cancer or neurologic disabilities. In addition, qualified exercise professionals and their programs can be found worldwide through the Exercise Program Registry (https://www.exerciseismedicine.org/support_page.php/moving-through-cancer/) of the Moving through Cancer Initiative of the ACSM. However, in some locations this is not available, so patients and clinicians are left without the ability to make connections with exercise professionals.

9.4 Example Exercise Program

Table 9.2 provides a suggestion of an exercise program but it must be adapted to the individual patient by a qualified professional (i.e., exercise physiologist, trainer, physical therapist, occupational therapist) depending on the patient's abilities, goals, symptoms, and risks. Indeed, based on the current status of knowledge, it is not possible to provide detailed training recommendations regarding frequency, intensity, type, or duration. Our suggestions begin with the ACSM guidelines, which indicate that patients with cancer start with a small amount of exercise and slowly build to up to 150 min/week of moderate-intensity aerobic exercise, or 75 min/week of vigorous intensity exercise, combined with 2–3 sessions/week of strength training across all major muscle groups, plus regular stretching [27]. We also considered published recommendations on the use of whole-body vibration training [38]. We emphasize the idea of an inverted-U association between exercise dose and exercise response [39] (i.e., a moderate dose and slow progression of exercise is best to avoid over-training). From there, we drew from the published literature and our experiences working with patients. An exercise intervention designed for patients with CIPN can vary depending on the patient's specific signs and symptoms. In other words, different exercise modalities might influence different CIPN signs and symptoms. For example, sensory symptoms in the feet seem to be prevented or reduced by sensorimotor exercise training, resistance training, as well as by multimodal approaches (combination of sensorimotor, resistance, and endurance). Regarding CIPN-associated functional limitations, sensorimotor training seems to improve postural control (including balance), while resistance training is effective for muscular strength. Therefore, to treat or prevent as many symptoms of CIPN as possible it is advisable to recommend a multimodal training approach consisting of at least sensorimotor and resistance training plus specific exercises for the hands. Training sessions requiring high levels of coordination and risk (e.g., elevated balance tasks) should be performed in a supervised setting initially and can then slowly be transferred into a non-supervised setting. Exercise with a lower demand can be done with patients unsupervised after a training session by a qualified instructor. The stability of the foot is also crucial for patients to feel secure enough to do other exercises; therefore, patients with balance loss, foot drop, or absent reflexes should be recommended to work on stability in the feet and lower extremities first, to give them security and confidence to be more active again in general.

Table 9.2 Suggestion for an example exercise program for patients with cancer in the treatment or prevention of CIPN to be adapted by a qualified professional for the specific needs, risks, and abilities of each individual patient

Type of exercise	Frequency	Duration and volume	Intensity	Example exercises	Progression
Sensorimotor training	2–3 sessions/week	15–35 min/session Balance or coordination tasks 3 Sets of 20–30 seconds per repetition.	The patient should be trained at their personal limit Ensure that training does not place the patient at excessive risk of fall	Balance or coordination tasks standing on one foot, tandem walk, etc. Hand exercises (e.g., vibration dumbbells, therapeutic plasticine, therapy balls that can be compressed)	Task difficulty should increase throughout the training period (weeks and months), e.g. by modifying the support surface and/or visual control (e.g., eyes open vs. closed), and/or include dual-task aspects (e.g., concurrent cognitive challenge).
Resistance training	2–3 sessions/week	1–3 sets of 10–15 repetitions of 5–15 different exercises	50–70% 1-repetition maximum rating of perceived exertion 14–16 Out of 20	Upper and lower body exercises using pushing and pulling movement patterns on machines or using resistance bands	Increasing repetitions, sets, and resistance levels slowly over the training period (weeks and months) while staying within the desired range of perceived exertion
Endurance training	2–3 sessions/week Home-based walking can be performed daily	10–40 min/session	55–75% heart rate reserve or 70–80% HR_{max}	Treadmill, walking, elliptical trainer, or cycle ergometer (especially if balance is problematic)	Increasing duration (minutes per week) slowly over the training period (weeks and months) to achieve ACSM recommendation (150 min/week)
Whole-body vibration training	2–3 sessions/week	4 sets of stands (forefoot) 30–60 sec	Amplitude of 2–4 mm; frequency of 25–35 Hz	Forefoot stance, slightly bent knees, head held high, vary the width of the feet or go into semi-tandem, tandem stance	Progression should increase, starting by adjusting the frequency parameter

• ACSM guidelines suggest that patients start with a small amount of exercise and slowly build to up to 150 min/week of moderate-intensity aerobic exercise, or 75 min/week of vigorous intensity exercise, combined with 2–3 days of strength training across all major muscle groups, plus regular stretching [27]

• Training sessions requiring high levels of coordination and risk (e.g., elevated balance tasks) should be performed in a supervised setting initially and can then slowly be transferred into a non-supervised setting. Exercise with a lower demand can be done with patients unsupervised after a training session by a qualified instructor

The length of this intervention should correspond to the total length of the individual's chemotherapy, or perhaps longer. For patients starting an exercise program with the intent to reduce CIPN symptoms after completing chemotherapy, this rehabilitative measure should take place as long as patients need the program to reduce symptom severity and/or to develop compensation techniques. When starting a training program, clinicians, health care professionals, and trainers should be aware of proposed adaptation phases for a sedentary individual when starting an exercise program [40]: (1) adoption phase, more emphasis should be placed on changing psychological mechanisms and not overwhelming physiological systems due to lack of physiological conditioning, (2) maintenance phase, both psychological and physiological mechanisms at play, and (3) habituation phase on physiological mechanisms and influence of behavioral conditioning.

It is critically important to consider patient preferences and intrinsic motivation. Patients who enjoy exercise and are more motivated will be more likely to integrate exercise into their daily life, and therefore continue to exercise regularly over a long period of time. Our companion review paper [24] revealed distinct exercise modalities using dance [41], a computer game [42], and other options such as sports to elicit aerobic or balance training effects. Patients may also find it very motivating to emphasize the relevance of exercise during and after chemotherapy to maintain physical function.

9.5 Conclusions and Future Work

In summary, it is clear that exercise prescriptions are mostly in line with the current general exercise recommendations for cancer patients and may therefore not only be effective in the prevention and rehabilitation of CIPN, but also have a positive effect on other treatment-related side effects such as fatigue and distress [27, 28]. Regardless of which exercise modalities, durations, and intensities appear to be effective now, it is important to take patient preference into account. Patient preference is important because it is less demanding to achieve an effective training stimulus if patients appreciate the program and therefore exercise regularly over a long period of time. To that end, ongoing and future clinical trials of exercise and related interventions for CIPN will continue to reveal whether and how much exercise (frequency, intensity, type, duration) can best target CIPN for a particular set of symptoms and patient characteristics (age, sex, comorbidities, fitness level, physical abilities) and patient preferences. Overall, we are optimistic for the rapidly growing body of research on the use of exercise for CIPN. Identifying successful treatments for CIPN will ultimately help patients, caregivers, families, providers, and the entire healthcare system.

Acknowledgements The authors acknowledge the following funding mechanisms to support this work, including the National Cancer Institute K07CA221931 to IRK, a National Health and Medical Research Council of Australia Career Development Fellowship (#1148595) to SBP. The authors also thank the Transdisciplinary Research in Energetics and Cancer (TREC) Training

Workshop funded by the National Cancer Institute (R25CA203650) and the Multinational Association for Supportive Care in Cancer (MASCC) for facilitating collaboration among the authors. The authors also thank Dr. Amber Kleckner for editorial input.

References

1. Seretny M, Currie GL, Sena ES, Ramnarine S, Grant R, MacLeod MR et al (2014) Incidence, prevalence, and predictors of chemotherapy-induced peripheral neuropathy: a systematic review and meta-analysis. Pain 155(12):2461–2470. https://doi.org/10.1016/j.pain.2014.09.020
2. Chan A, Hertz DL, Morales M, Adams EJ, Gordon S, Tan CJ et al (2019) Biological predictors of chemotherapy-induced peripheral neuropathy (CIPN): MASCC neurological complications working group overview. Support Care Cancer:1–9. https://pubmed.ncbi.nlm.nih.gov/31363906/
3. Staff NP, Grisold A, Grisold W, Windebank AJ (2017) Chemotherapy-induced peripheral neuropathy: a current review. Ann Neurol 81(6):772–781. https://doi.org/10.1002/ana.24951
4. Gu Y, Menzies AM, Long GV, Fernando S, Herkes G (2017) Immune mediated neuropathy following checkpoint immunotherapy. J Clin Neurosci 45:14–17
5. Reeves BN, Dakhil SR, Sloan JA, Wolf SL, Burger KN, Kamal A et al (2012) Further data supporting that paclitaxel-associated acute pain syndrome is associated with development of peripheral neuropathy: north central Cancer treatment group trial N08C1. Cancer 118(20):5171–5178
6. Argyriou AA, Cavaletti G, Briani C, Velasco R, Bruna J, Campagnolo M et al (2013) Clinical pattern and associations of oxaliplatin acute neurotoxicity: a prospective study in 170 patients with colorectal cancer. Cancer 119(2):438–444
7. Pachman DR, Qin R, Seisler DK, Smith EM, Beutler AS, Ta LE et al (2015) Clinical course of oxaliplatin-induced neuropathy: results from the randomized phase III trial N08CB (alliance). J Clin Oncol 33(30):3416
8. Hershman DL, Unger JM, Crew KD, Till C, Greenlee H, Minasian LM et al (2018) Two-year trends of Taxane-induced neuropathy in women enrolled in a randomized trial of acetyl-L-carnitine (SWOG S0715). J Natl Cancer Inst 110(6):669–676. https://doi.org/10.1093/jnci/djx259
9. Bandos H, Melnikow J, Rivera DR, Swain SM, Sturtz K, Fehrenbacher L et al (2018) Long-term peripheral neuropathy in breast cancer patients treated with adjuvant chemotherapy: NRG oncology/NSABP B-30. J Natl Canc Inst 110(2):djx162. https://doi.org/10.1093/jnci/djx162
10. daCosta DBM, Copher R, Basurto E, Faria C, Lorenzo R (2014) Patient preferences and treatment adherence among women diagnosed with metastatic breast cancer. Am Health Drug Benefits 7(7):386–396
11. Lyman GH (2009) Impact of chemotherapy dose intensity on cancer patient outcomes. Journal of the National Comprehensive Cancer Network : JNCCN 7(1):99–108
12. Gewandter JS, Kleckner AS, Marshall JH, Brown JS, Curtis LH, Bautista J et al (2019) Chemotherapy-induced peripheral neuropathy (CIPN) and its treatment: an NIH Collaboratory study of claims data. Support Care Cancer:1–10. https://pubmed.ncbi.nlm.nih.gov/31494735/
13. Pike CT, Birnbaum HG, Muehlenbein CE, Pohl GM, Natale RB (2012) Healthcare costs and workloss burden of patients with chemotherapy-associated peripheral neuropathy in breast, ovarian, head and neck, and nonsmall cell lung cancer. Chemother Res Pract 2012:913848. https://doi.org/10.1155/2012/913848
14. Chan CW, Cheng H, Au SK, Leung KT, Li YC, Wong KH et al (2018) Living with chemotherapy- induced peripheral neuropathy: uncovering the symptom experience and self-management of neuropathic symptoms among cancer survivors. Eur J Oncol Nurs 36:135–141
15. Monfort SM, Pan X, Patrick R, Ramaswamy B, Wesolowski R, Naughton MJ et al (2017) Gait, balance, and patient-reported outcomes during taxane-based chemotherapy in early-stage breast

cancer patients. Breast Cancer Res Treat 164(1):69–77. https://doi.org/10.1007/s10549-017-4230-8

16. Wasilewski A, Mohile N (2020) Meet the expert: how I treat chemotherapy-induced peripheral neuropathy. J Geriatr Oncol. https://doi.org/10.1016/j.jgo.2020.06.008

17. Kolb NA, Smith AG, Singleton JR, Beck SL, Stoddard GJ, Brown S et al (2016) The Association of Chemotherapy-Induced Peripheral Neuropathy Symptoms and the risk of falling. JAMA Neurol 73(7):860–866. https://doi.org/10.1001/jamaneurol.2016.0383

18. Gewandter JS, Fan L, Magnuson A, Mustian K, Peppone L, Heckler C et al (2013) Falls and functional impairments in cancer survivors with chemotherapy-induced peripheral neuropathy (CIPN): a University of Rochester CCOP study. Supportive Care Cancer: Official Journal of the Multinational Association of Supportive Care in Cancer 21(7):2059–2066. https://doi.org/10.1007/s00520-013-1766-y

19. Loprinzi CL, Lacchetti C, Bleeker J, Cavaletti G, Chauhan C, Hertz DL et al (2020) Prevention and management of chemotherapy-induced peripheral neuropathy in survivors of adult cancers: ASCO guideline update. J Clin Oncol 38(28):3325–3348. https://doi.org/10.1200/JCO.20.01399

20. Brami C, Bao T, Deng G (2016) Natural products and complementary therapies for chemotherapy-induced peripheral neuropathy: a systematic review. Crit Rev Oncol Hematol 98:325–334. https://doi.org/10.1016/j.critrevonc.2015.11.014

21. Dorsey SG, Kleckner IR, Barton D, Mustian K, O'Mara A, St Germain D et al (2019) NCI clinical trials planning meeting for prevention and treatment of chemotherapy-induced peripheral neuropathy. J Natl Cancer Inst. https://doi.org/10.1093/jnci/djz011

22. Duregon F, Vendramin B, Bullo V, Gobbo S, Cugusi L, Di Blasio A et al (2018) Effects of exercise on cancer patients suffering chemotherapy-induced peripheral neuropathy undergoing treatment: a systematic review. Crit Rev Oncol Hematol 121:90–100. https://doi.org/10.1016/j.critrevonc.2017.11.002

23. Kanzawa-Lee GA, Larson JL, Resnicow K, Smith EML (2020) Exercise effects on chemotherapy-induced peripheral neuropathy: a comprehensive integrative review. Cancer Nurs 43(3):E172–EE85

24. Kleckner IR, Park SB, Streckmann F, Wiskemann J, Hardy S, Mohile NA (2021) Systematic review of exercise for prevention and management of chemotherapy-induced peripheral neuropathy. In: Lustberg MB, Loprinzi CL (eds) Diagnosis, management and emerging strategies for chemotherapy induced neuropathy. Springer, Berlin

25. Campbell KL, Weller S, Cormie P, Lane KN, Rauw JM, Goulart J (2020) Enhancing safety of exercise for individuals with bone metastases: screening recommendations developed through Delphi consensus process. Am Soc Clin Oncol. https://ascopubs.org/doi/abs/10.1200/JCO.2020.38.15_suppl.e24042

26. Campbell KL, Neil SE, Winters-Stone KM (2012) Review of exercise studies in breast cancer survivors: attention to principles of exercise training. Br J Sports Med 46(13):909–916

27. Campbell KL, Winters-Stone KM, Wiskemann J, May AM, Schwartz AL, Courneya KS et al (2019) Exercise guidelines for Cancer survivors: consensus statement from international multidisciplinary roundtable. Med Sci Sports Exerc 51(11):2375–2390. https://doi.org/10.1249/MSS.0000000000002116

28. Kleckner IR, Dunne RF, Asare M, Cole C, Fleming F, Fung C et al (2018) Exercise for toxicity management in cancer—a narrative review. Oncology & hematology review 14(1):28

29. Speck RM, DeMichele A, Farrar JT, Hennessy S, Mao JJ, Stineman MG et al (2012) Scope of symptoms and self-management strategies for chemotherapy-induced peripheral neuropathy in breast cancer patients. Support Care Cancer 20(10):2433–2439

30. Binkley JM, Harris SR, Levangie PK, Pearl M, Guglielmino J, Kraus V et al (2012) Patient perspectives on breast cancer treatment side effects and the prospective surveillance model for physical rehabilitation for women with breast cancer. Cancer 118(S8):2207–2216

31. Caspersen CJ, Powell KE, Christenson GM (1985) Physical activity, exercise, and physical fitness: definitions and distinctions for health-related research. Public Health Rep 100 (2):126–131
32. Wasfy MM, Baggish AL (2016) Exercise dose in clinical practice. Circulation 133 (23):2297–2313
33. Neil-Sztramko SE, Medysky ME, Campbell KL, Bland KA, Winters-Stone KM (2019) Attention to the principles of exercise training in exercise studies on prostate cancer survivors: a systematic review. BMC Cancer 19(1):321
34. Kenjale AA, Hornsby WE, Crowgey T, Thomas S, Herndon JE (2014) Pre-exercise participation cardiovascular screening in a heterogeneous cohort of adult cancer patients. Oncologist 19 (9):999
35. American College of Sports Medicine (2017) ACSM's guidelines for exercise testing and prescription. Lippincott, Williams, & Wilkins, Baltimore, MD
36. Cole CL, Kleckner IR, Jatoi A, Schwarz EM, Dunne RF (2018) The role of systemic inflammation in cancer-associated muscle wasting and rationale for exercise as a therapeutic intervention. JCSM Clin Rep 3(2):1–19
37. Wolin KY, Schwartz AL, Matthews CE, Courneya KS, Schmitz KH (2012) Implementing the exercise guidelines for cancer survivors. J Support Oncol 10(5):171
38. Streckmann F, Balke M (2018) Bewegungstherapie bei Polyneuropathie. DGNeurologie 1 (1):47–57
39. Carayol M, Bernard P, Boiche J, Riou F, Mercier B, Cousson-Gélie F et al (2013) Psychological effect of exercise in women with breast cancer receiving adjuvant therapy: what is the optimal dose needed? Ann Oncol 24(2):291–300
40. Boutcher SH (1993) Emotion and aerobic exercise. In: Singer RN, Murphey M, Tennant LK (eds) Handbook of research on sport psychology. Macmillan, New York, pp 799–814
41. Worthen-Chaudhari L, Lamantia MT, Monfort SM, Mysiw W, Chaudhari AMW, Lustberg MB (2019) Partnered, adapted argentine tango dance for cancer survivors: a feasibility study and pilot study of efficacy. Clin Biomech (Bristol, Avon) 70:257–264. https://doi.org/10.1016/j.clinbiomech.2019.08.010
42. Schwenk M, Grewal GS, Holloway D, Muchna A, Garland L, Najafi B (2016) Interactive sensor- based balance training in older Cancer patients with chemotherapy-induced peripheral neuropathy: a randomized controlled trial. Gerontology 62(5):553–563. https://doi.org/10.1159/000442253

Natural Course of Neurotoxicity after Immune Checkpoint Inhibitor (ICI) Exposure

10

Andreas A. Argyriou

Abstract

The blockade of immune checkpoint inhibitors (ICIs) with monoclonal antibodies has revolutionized the therapeutic management of several cancer types as these treatments have achieved higher objective response rates and prolonged overall survival. However, targeting of CTLA-4 and PD-1/PD-L1 dysregulates the homeostasis of immune system, thereby increasing the relative risk of systemic immune-related overactivation and immune-related adverse events (irAEs).

Neurological irAEs (NirAEs) are relatively rare but potentially severe and life-threatening. The clinical phenotype of NirAEs greatly varies to involve a wide spectrum of neurological manifestations, although neuromuscular involvement, in the form of myositis, myasthenia gravis, and demyelinating polyradiculoneuropathy, is more frequently disclosed than central nervous system involvement clinically encountered as meningoencephalitis, encephalitis, vasculitis, myelitis, CNS demyelination, neuro-opthalmological events, and cranial neuropathies. Early NirAEs diagnosis, prompt ICIs discontinuation, and induction treatment with immune-modulating therapies, e.g., corticosteroids, IVIG, plasma exchange, and immune suppressants, are factors of paramount importance to optimize clinical outcomes.

Keywords

Immune checkpoint inhibitors · Immune-related adverse events · Neurological immune-related adverse events · Neuromuscular involvement · Natural course · Management · Prognosis

A. A. Argyriou (✉)
"Saint Andrew's" State General Hospital of Patras, Patras, Greece

© The Author(s), under exclusive license to Springer Nature Switzerland AG 2021
M. Lustberg, C. Loprinzi (eds.), *Diagnosis, Management and Emerging Strategies for Chemotherapy-Induced Neuropathy*,
https://doi.org/10.1007/978-3-030-78663-2_10

10.1 Main Text

Over the last decade, the blockade of the immune checkpoints by monoclonal antibodies (mAb) has revolutionized the treatment of several cancer types. Physiologically, T-lymphocyte-associated protein 4 (CTLA-4) and programmed cell death protein 1 (PD-1) are receptors that help to maintain immune tolerance. Targeting these receptors with immune checkpoint inhibitors (ICIs), given as monotherapy or combined with other conventional agents enhances T-cell adaptive immunity against the tumor by blocking immune inhibitory signals, thereby leading to strikingly improved clinical outcomes [1, 2]. The main representatives of this modern class of cancer therapy are ipilimumab, an antibody targeting cytotoxic T-lymphocyte-associated antigen 4 (CTLA-4), nivolumab, an antibody affecting programmed cell death 1 (PD-1), pembrolizumab, cemiplimab, atezolizumab, durvalumab, and avelumab, which target the anti-PD-1 ligand (PD-L1).

Nonetheless, given the regulatory roles that CTLA-4 and PD-1/PD-L1 play in the homeostasis of the immune system, it is expectable that blocking these pathways could increase the risk of various immune-related adverse events (irAEs), mainly due to removal of self-tolerance [3]. Up to 65% of ICI-exposed patients experience irAEs involving the skin, gastrointestinal tract, endocrine system, and liver. Neurological immune-related adverse events (NirAEs) are relatively rare, occurring in up to 6% of patients treated with ICIs and are clinically manifested in the form of various disorders affecting both the central and peripheral nervous system (PNS). Although less frequently encountered than the rest of irAEs, NirAEs merit special attention in their prompt diagnosis and management as, in some cases, these toxicities can be severe or even lethal [4].

As mentioned, ICIs can evoke damage to either the peripheral or central nervous system (CNS). In the general context, neuromuscular adverse events are more frequently encountered than toxicities involving the CNS (5.5% and 0.46%, respectively) after exposure to pembrolizumab, nivolumab, and ipilimumab therapy [5]. The most commonly encountered CNS clinical syndromes include meningoencephalitis, encephalitis, vasculitis, myelitis, CNS demyelination, neuroopthalmological events, and cranial neuropathies, occurring as a result of neuroinflammation. Conversely, the most commonly encountered neuromuscular irAEs include peripheral neuropathy, myositis, and myasthenia gravis. Myositis appears to be the most common clinical syndrome in nivolumab-treated patients, while peripheral neuropathies rather than myositis are more frequently seen after ipilimumab exposure [6].

Usually, NirAEs are late effects after ICI exposure, suggesting that CNS events require a greater median number of ICIs cycles received and a more prolonged time period to NirAEs onset, compared to neuromuscular toxicities. The latter view is supported by the results of a recently published study that showed that the time to presentation of PNS, compared to CNS syndromes, was significantly shorter, i.e., median 70 vs 119 days, respectively [5]. Finally, as opposed to irAEs involving other organs, there is no evidence to support that combination schemes comprising of CTLA-4 plus PD-1 inhibitors, compared to ICI monotherapy, increase the incidence of NirAEs [7].

10.2 CNS Neurotoxicity

10.2.1 Encephalitis/Meningoencephalitis

Patients with encephalitis or meningoencephalitis commonly present with fever, headache, emesis, altered mental status, in keeping with an increased intracranial pressure syndrome. Neurological deficits, including seizures, may also occur. Neuroimaging usually reveals non-specific inflammatory changes, while the analysis of cerebrospinal fluid (CSF) typically shows increased opening pressure and evidence of albumino-cytologic dissociation with mild lymphocytic pleocytosis and elevated CSF albumin levels. Slowing of basal rhythm and evidence of non-specific abnormalities are present in electroencephalography [8].Tellingly, in the metastatic setting, it is often challenging to diagnose ICI-related CNS infections. Although CSF paraneoplastic and autoimmune antibody assays are negative, further extensive diagnostic testing is usually needed to exclude autoimmune encephalitis or cerebellitis, especially when taking into account that ICIs exposure could augment or trigger sporadic paraneoplastic or autoimmune disorders [9].

Nonetheless, ICI-related encephalitis might be a serious (grade 3) adverse event with a relatively high mortality rate [10]. Affected patients should have an inpatient vigilant monitoring and be treated with high-dose IV methylprednisolone at a dose of 1g per day during 5 consecutive days, followed by prednisone 1 mg/kg in progressive dose reduction.Infusion of intravenous immunoglobulin (IVIG) might also be given in severe or progressive symptoms to diminish the mortality risk.

10.2.2 Vasculidities

Contrary to ICI-related encephalitis, primary CNS vasculidities, in the form of giant cell arteritis or isolated retinal vasculitis, have a much more benign natural course and bear a minimal mortality risk [11]. The clinical phenotype of ICIs-related vasculitis was recently described in a systematic review, which identified 20 cases that developed large vessel CNS vasculitis, in particular, after commencing 1–15 treatment cycles of anti-PD-1 therapy [12].

10.2.3 CNS Demyelination

Exacerbation of known multiple sclerosis or *de novo* manifestation of CNS demyelination has also been reported after ICIs exposure. Rapid progression of a case with radiographically isolated syndrome into clinically definite multiple sclerosis has been described 4 months after ipilimumab initiation [13]. Blocking the interaction between PD-1/PD-L1 or CTLA-4 in lymphocytes resident or infiltrating the nervous system could increase local inflammation or reveal latent central inflammation. This mechanism might be responsible for the exacerbation of multiple sclerosis, as documented in some experimental models of CNS inflammation [14]. Conversely, *de novo* CNS demyelination, although very rarely encountered, has been associated with enhanced responses of myelin-reactive peripheral CD4+ T cells [15]. A case with presumed anti-aquaporin-4 antibody-positive neuromyelitis optica spectrum disorder has been also recently described in a patient with lung squamous cell carcinoma two months after treatment with nivolumab [16].

Nonetheless, the overall prognosis of ICI-related central demyelination is favorable, with most cases having partial or complete response to steroids after discontinuation of the offending ICI agent [17].

10.2.4 Neuro-opthalmological IRAEs and Other Cranial Neuropathies

Optic neuritis complicating ICI therapy is rarely encountered and occurs in less than 1% of exposed patients. Literature contains single reports of unilateral or bilateral optic neuritis after therapy with either ipilimumab [18, 19], nivolumab monotherapy [20], or combined with a peptide vaccine [21], atezolizumab [22] and durvalumab [23]. Typically, neuro-opthalmological IRAEs present 2–12 weeks after commencing treatment and in the majority of cases are reversible and responsive to steroid treatment [24]. Steroid-responsive cranial nerve palsies, involving nerve III (oculomotor), nerve VI (abducens), VII (facial), and combined cranial nerve VI and VII palsy have also been rarely reported 4-13 months after initiation of ICIs treatment [25, 26].

10.2.5 Cognitive Decline

ICIs might also cause very late neurotoxicity-related neuropsychiatric effects, including cognitive disorders, fatigue, and mood disorders. This is because of their ability to cross the blood–brain barrier, evoking changes in microglial activation and increasing the levels of cytokines and chemokines in the inner temporal structures, such as the hippocampus [27]. Thus far, this issue has not been thoroughly addressed in the clinical setting although it definitely merits attention, as cognition and mood are strong determinants of daily living activities and quality of life.

10.3 PNS Neurotoxicity

10.3.1 Peripheral Neuropathies

ICIs are generally less toxic for the peripheral nerves than conventional cytotoxic chemotherapy. Specifically, available data show that about 1% of patients exposed to therapy with PD-1/PD-L1 inhibitors will manifest any grade of peripheral nerve damage in the form of axonal sensory peripheral neuropathy, compared to 8.6% of patients receiving conventional chemotherapy. Likewise, treatment-emergent grade 3 neurotoxicity is much less likely to occur with PD-1/PDL-1 therapy (0.3%) than with conventional chemotherapy agents (1.1%) [28]. Moreover, adjunctive use of neurotoxic chemotherapy with ICIs does not seem to increase the risk of either more frequent or more intense treatment-emergent grade 3 or 4 peripheral neuropathy [29].

Tellingly, it is difficult to be confident about the true incidence and severity of peripheral neuropathies after ICI exposure because most of the affected patients are usually pretreated with other neurotoxic chemotherapeutic agents before the initiation of ICIs therapy. In any case, events of *de novo* development sensory axonal peripheral neuropathy, with evidence of symmetrical numbness and paresthesia in a stocking and glove distribution and reduced or abolished tendon reflexes, usually appear after commencing 3–7 cycles of ICIs and after a median time of 70 days from immunotherapy initiation to the neurological adverse event's onset. Patients usually recover soon after ICIs discontinuation even without any intervention [5].

Apart from axonal sensory neuropathies, cases of immune-related demyelinating polyradiculoneuropathy (irDP) can occur in up to 7.6% of patients exposed to 3–4 courses of PD-1/PDL-1 therapy [30] and at a median time of 59 days from the initiation of immunotherapy [5]. Patients usually develop acute or subacute sensory-motor symptoms and cranial nerve involvement with bulbar symptoms and dyspnea. CSF results in these patients is in keeping with albuminocytological dissociation. Nerve conduction studies show a demyelination pattern with marked motor conduction slowing and F waves prolongation, while antiganglioside antibodies are generally absent [31]. irDP is usually responsive to corticosteroid treatment, which should be considered as a first-line treatment.

Second- or third-line treatment with IVIG or plasma exchange should be administered in patients who remain unresponsive to corticosteroids [32].

10.3.2 Myositis

Immune-related myositis (irMyositis) is the most common neuromuscular toxicity of anti-PD-1/anti-PDL1 and anti-CTLA-4 therapy. Elderly male patients are more liable to develop irMyositis within the first two months after the initiation of ICIs treatment [33], although a more prolonged median time of 97 days has also been reported [5].

Patients usually develop diffuse myalgias in the back and proximal limbs, reduced tendon reflexes, and proximal muscle weakness, mainly in the pelvic girdle. These symptoms peak to maximal severity in a median of 10 days [34]. Ocular involvement in the form of ptosis or ophthalmoparesis can occur in up to 70% of patients with irMyositis, while facial weakness and involvement of bulbar muscles is less frequently reported (40–50%) [35].

Increased CK levels up to fivefold over normal, muscle sampling with needle electroneuromyography, showing myopathic motor unit potentials (defined as the presence of polyphasic, short-duration, or low amplitude motor unit action potential with normal or early recruitment) and muscle biopsy with evidence of necrotic myofibers associated with inflammatory infiltrates consistent with necrotizing myopathic changes, are strongly supportive of irMyositis [35], although not all patients present abnormal findings in these tests.

The majority of patients with irMyositis experience a favorable clinical outcome after ICIs discontinuation and administration of immunomodulatory treatment with corticosteroids as first-line treatment and plasma exchange or IVIG as induction therapeutic options. Nonetheless, up to 20% of patients ultimately require non-invasive or mechanical ventilation, due to evidence of treatment-resistant progressive generalized muscle weakness and respiratory or cardiac muscle involvement [36].

10.3.3 Myasthenia Gravis

Immune-related myasthenia gravis (irMG), either developed *de novo* or as a relapsing pre-existing myasthenia gravis (MG), is the most emerging and life-threatening neuromuscular toxicity expected within 2–12 weeks (average 6 weeks) after single or combined ICIs treatment [35, 37]. The diagnosis in these cases is challenging because there is evidence of irMG and irMyositis co-occurrence in the majority of cases and this is significantly associated with triggering a myasthenic crisis requiring ventilator support [38].

Typically, patients present with fluctuating muscle weakness involving ocular, bulbar, and/or respiratory muscles [6]. Response to cholinesterase inhibitors (pyridostigmine or edrophonium test) and positive assays to antibodies against acetylcholine receptor (AChR) have a relatively low (60–80%) diagnostic sensitivity [39].Early recognition of irMG and discontinuation of the offending ICI agent is of paramount importance in order to promote a favorable neurological outcome. In any case, mortality rates remain significant (30%) despite the adequate inpatient administration of induction therapy with corticosteroids, IVIG, plasma exchange, and immune suppressants [37].

Patients with irMG and concurrent irMyositis with irMyocarditis yield an even higher mortality risk [40].

References

1. Emens LA, Ascierto PA, Darcy PK et al (2017) Cancer immunotherapy: opportunities and challenges in the rapidly evolving clinical landscape. Eur J Cancer 81:116–129
2. Hargadon KM, Johnson CE, Williams CJ (2018) Immune checkpoint blockade therapy for cancer: an overview of FDA-approved immune checkpoint inhibitors. Int Immunopharmacol 62:29–39
3. Johnson DB, Balko JM (2019) Biomarkers for immunotherapy toxicity: are cytokines the answer? Clin Cancer Res 25(5):1452–1454
4. Fellner A, Makranz C, Lotem M et al (2018) Neurologic complications of immune checkpoint inhibitors. J Neuro-Oncol 137(3):601–609
5. Bruna J, Argyriou AA, Anastopoulou GG et al (2020) Incidence and characteristics of neuro-toxicity in immune checkpoint inhibitors with focus on neuromuscular events: experience beyond the clinical trials. J PeripherNerv Syst 25(2):171–177
6. Psimaras D, Velasco R, Birzu C et al (2019) Immune checkpoint inhibitors-induced neuromus-cular toxicity: from pathogenesis to treatment. J PeripherNerv Syst 24(Suppl 2):S74–S85
7. Xu M, Nie Y, Yang Y, Lu YT, Su Q (2019) Risk of neurological toxicities following the use of different immune checkpoint inhibitor regimens in solid tumors: a systematic review and Meta-analysis. Neurologist 24(3):75–83
8. Larkin J, Chmielowski B, Lao CD et al (2017) Neurologic serious adverse events associated with Nivolumab plus Ipilimumab or Nivolumab alone in advanced melanoma, including a case series of encephalitis. Oncologist 22(6):709–718
9. Yshii LM, Hohlfeld R, Liblau RS (2017) Inflammatory CNS disease caused by immune checkpoint inhibitors: status and perspectives. Nat Rev Neurol 13(12):755–763
10. Haugh AM, Probasco JC, Johnson DB (2020) Neurologic complications of immune checkpoint inhibitors. Expert Opin Drug Saf 19(4):479–488
11. Choi J, Lee SY (2020) Clinical characteristics and treatment of immune-related adverse events of immune checkpoint inhibitors. Immune Netw 20(1):e9
12. Daxini A, Cronin K, Sreih AG (2018) Vasculitis associated with immune checkpoint inhibitors-a systematic review. Clin Rheumatol 37(9):2579–2584
13. Gerdes LA, Held K, Beltrán E et al (2016) CTLA4 as immunological checkpoint in the development of multiple sclerosis. Ann Neurol 80(2):294–300
14. Perrin PJ, Maldonado JH, Davis TA, June CH, Racke MK (1996) CTLA-4 blockade enhances clinical disease and cytokine production during experimental allergic encephalomyelitis. J Immunol 157(4):1333–1336
15. Cao Y, Nylander A, Ramanan S et al (2016) CNS demyelination and enhanced myelin-reactive responses after ipilimumab treatment. Neurology 86(16):1553–1556
16. Narumi Y, Yoshida R, Minami Y et al (2018) Neuromyelitis optica spectrum disorder second-ary to treatment with anti-PD-1 antibody nivolumab: the first report. BMC Cancer 18(1):95
17. Kumar N, Abboud H (2019) Iatrogenic CNS demyclination in the era of modern biologics. MultScler 25(8):1079–1085
18. Wilson MA, Guld K, Galetta S, Walsh RD, Kharlip J, Tamhankar M et al (2016) Acute visual loss after ipilimumab treatment for metastatic melanoma. J Immunother Cancer 4:66
19. Yeh OL, Francis CE (2015) Ipilimumab-associated bilateral optic neuropathy. J Neuroophthalmol 35:144–147
20. Kartal Ö, Ataş E (2018) Bilateral optic neuritis secondary to nivolumab therapy: a case report. Medicina (Kaunas) 54(5):82
21. Weber JS, Kudchadkar RR, Yu B, Gallenstein D, Horak CE, Inzunza HD et al (2013) Safety, efficacy, and biomarkers of nivolumab with vaccine in ipilimumab-refractory or -naive mela-noma. J Clin Oncol 31:4311–4318
22. Mori S, Kurimoto T, Ueda K, Enomoto H, Sakamoto M, Keshi Y, Yamada Y, Nakamura M (2018) Optic neuritis possibly induced by Anti-PD-L1 antibody treatment in a patient with non-small cell lung carcinoma. Case Rep Ophthalmol 9:348–356

23. Noble CW, Gangaputra SS, Thompson IA et al (2020) Ocular adverse events following use of immune checkpoint inhibitors for metastatic malignancies. Ocul Immunol Inflamm 28(6):854–859
24. Teufel A, Zhan T, Härtel N, Bornschein J, Ebert MP, Schulte N (2019) Management of immune related adverse events induced by immune checkpoint inhibition. Cancer Lett 456:80–87
25. Dalvin LA, Shields CL, Orloff M, Sato T, Shields JA (2018) Checkpoint inhibitor immune therapy: systemic indications and ophthalmic side effects. Retina 38(6):1063–1078
26. Jaben KA, Francis JH, Shoushtari AN, Abramson DH (2019) Isolated Abducens nerve palsy following Pembrolizumab. Neuroophthalmology 44(3):182–185
27. McGinnis GJ, Raber J (2017) CNS side effects of immune checkpoint inhibitors: preclinical models, genetics and multimodality therapy. Immunotherapy 9(11):929–941
28. Nishijima TF, Shachar SS, Nyrop KA, Muss HB (2017) Safety and tolerability of PD-1/PD-L1 inhibitors compared with chemotherapy in patients with advanced Cancer: a Meta-analysis. Oncologist 22(4):470–479
29. Schmid P, Adams S, Rugo HS, Schneeweiss A, Barrios CH, Iwata H et al (2018 Nov 29) Atezolizumab and Nab-paclitaxel in advanced triple-negative breast cancer. N Engl J Med 379 (22):2108–2121
30. Man J, Ritchie G, Links M, Lord S, Lee CK (2018 Jun) Treatment-related toxicities of immune checkpoint inhibitors in advanced cancers: a meta-analysis. Asia Pac J Clin Oncol 14 (3):141–152
31. Touat M, Talmasov D, Ricard D, Psimaras D (2017 Dec) Neurological toxicities associated with immune-checkpoint inhibitors. Curr Opin Neurol 30(6):659–668
32. Psimaras D (2018) Neuromuscular complications of immune checkpoint inhibitors. Presse Med 47(11–12):e253–e259
33. Shah M, Tayar JH, Abdel-Wahab N, Suarez-Almazor ME (2019 Feb) Myositis as an adverse event of immune checkpoint blockade for cancer therapy. Semin Arthritis Rheum 48 (4):736–740
34. Liewluck T, Kao JC, Mauermann ML (2018) PD-1 inhibitor-associated myopathies: emerging immune-mediated myopathies. J Immunother 41(4):208–211
35. Touat M, Maisonobe T, Knauss S, Ben Hadj Salem O, Hervier B, Auré K et al (2018) Immune checkpoint inhibitor-related myositis and myocarditis in patients with cancer. Neurology 91 (10):e985–e994
36. Moreira A, Loquai C, Pföhler C, Kähler KC, Knauss S, Heppt MV et al (2019) Myositis and neuromuscular side-effects induced by immune checkpoint inhibitors. Eur J Cancer 106:12–23
37. Makarious D, Horwood K, Coward J (2017) Myasthenia gravis: an emerging toxicity of immune checkpoint inhibitors. Eur J Cancer 82:128–136
38. Kao JC, Brickshawana A, Liewluck T (2018) Neuromuscular complications of programmed cell Death-1 (PD-1) inhibitors. Curr Neurol Neurosci Rep 18(10):63
39. Suzuki S, Ishikawa N, Konoeda F, Seki N, Fukushima S, Takahashi K et al (2017 Sep 12) Nivolumab- related myasthenia gravis with myositis and myocarditis in Japan. Neurology 89 (11):1127–1134
40. Konstantina T, Konstantinos R, Anastasios K et al (2019) Fatal adverse events in two thymoma patients treated with anti-PD-1 immune check point inhibitor and literature review. Lung Cancer 135:29–32

Ants and Needles and Pins: Living with Neuropathy

Cynthia Chauhan and Mary Lou Smith

Keywords

CIPN · Chemotherapy · Neuropathy · Patient experience

Life is a journey of choices, known to the research community as risk/benefit analyses and to the lay community as choosing the best ways to live as successfully as possible with success being individually defined. For those of us with chemo-induced neuropathy, the neuropathy is a sequalae of one of those hard choices. A question we need to consider is if the choice was an informed one. Did we know both the negative and positive consequences of the choice to have chemotherapy and how do we live with the consequences of our choice? We dealt with the short-term negative effects such as profound nausea and hair loss. The question now becomes dealing with the long-term effects. Some of us were unaware that all side effects do not end when treatment ends.

Neuropathy is a constant presence, sometimes simply annoying us and sometimes overwhelming us. We often experience a compelling numbness that is best described for people who do not have neuropathy as that awful feeling that you experience when your limb has "fallen asleep" and is in the process of awakening. For some of us that feeling is omnipresent. So, that is the base upon which the other symptoms build, including a feeling that ants have set up an anthill in one's leg and are busy building their nest, not just ordinary sugar ants but fire ants. Or, there is that painful awareness of one's feet when one is trying to go to sleep and, in an attempt to settle down, enmeshed in perceived pinpricks, one's feet refuse to be still. We finally get to

C. Chauhan (✉) · M. L. Smith
Wichita, Kansas, USA

Research Advocacy Network, Plano, TX, USA
e-mail: mlsmith@researchadvocacy.org

sleep only to awaken in the morning to the literally painful decision of which clothes and shoes to wear. Weakened by neuropathy and subject to muscle spasms if we move just wrong, we manage to get ourselves out of bed, careful to check the placement of our feet and thoughtful about keeping our balance. Climbing stairs or looking up while standing can be a life-threatening exercise as we may lose balance and possibly fall. Now, what can we manage to put on by ourselves when our hands are not cooperating with us and independent balance is precarious? Losing the sensation of warmth or cold, how do we make sure we do not inadvertently pick up a too hot pot? How can we dress appropriately for the weather? Does the fire or cold in our limbs reflect external reality? What shoes will be least painful as we accommodate ourselves to the loss of proprioception?

Parts of life that were always automatic in the past now are daily conscious decisions. What household chores can we now do with effort that we once sped through easily? Which are simply beyond our ability now or even endangering as we bend over or reach up and lose balance? How many glasses and dishes do we break before we realize that we may need special kitchen and dining utensils? How do we prepare our meals now that peeling and cutting foods are perilous activities? What foods that we always ate with utensils are now more likely to get eaten if we treat them as finger foods? How much easier is it to drink soup from a cup rather than spooning it from a bowl? What social graces can we hold on to and which must we reluctantly forego?

Through it all, the pain and the combination of lost and intensified physical sensations and sensitivity, we need to remember to remain active, social beings. What about the things we do for pleasure and creative engagement—If one is a painter, how to control the brushes and the flow of paints? If we enjoy playing the piano will Chopin become chopsticks? How difficult does reading become when one can no longer hold a book or turn the pages automatically? For those of us who write, how do we control our fingers on the computer keyboard? How do we manage the neuropathy without foisting our issues on others?

As we deal with the physical and social complications of neuropathy, we learn that unremitting pain not only can affect our quality of life but also can lead to depression and social reactivity. Social interactions that were once automatic and pleasant sometimes become tiresome tasks. Social life may become attenuated by the omnipresent pain and/or discomfort and loss of function. One may or may not recognize the developing depression or, recognizing it, deny its importance and influence.

If you are getting bored with this recitation of symptoms and life adjustments, let yourself consider how taxing it is to us to deal with this never-ending cascade of symptoms and conscious choices of things that used to be automatic. When you woke up this morning, did you have to make a conscious decision to get out of bed, carefully planning each move or did you just groan and roll out?

So, now that you have endured the litany of pain and discomfort and loss of automatic decision-making on life's basic functions, let us think together about what can be done.

First, it behooves treating physicians, to make sure the patient is aware of the importance of early symptom reporting as one goes through treatment. Clinicians should genuinely inform their patients that there is no symptom too small to report. That gives them and the patient a head start on handling neuropathy when it first rears its ugly head.

Second, because some patients are reluctant to ask too much or are simply overwhelmed by the disease and treatment, clinicians should initiate regular discussions of possible negative effects of the treatment including being observant of behaviors such as how patients walk or if they are having difficulty buttoning their shirt or coat. Along those lines, clinicians should give patients the time and interest they would want if they were the patient.

Affected patients can and do live with chemo-induced neuropathy and appreciate that, although difficult, it is a consequence of attempts to halt or slow the progression of the cancer. However, it is a consequence that needs to be understood and carefully addressed. Ignoring it does not make it go away and will make it more dangerous.

Printed by Books on Demand, Germany